AF660766

Minor Forest Produce used in Pharmaceutical and other Industries

Minor Forest Produce used in Pharmaceutical and other Industries

T. Krishna Murthy
IFS (Retd.)

PharmaMed Press
An imprint of Pharma Book Syndicate
A unit of BSP Books Pvt. Ltd.
4-4-309/316, Giriraj Lane,
Sultan Bazar, Hyderabad - 500 095.

© 2015, *by Publisher*

All rights reserved. No part of this book or parts thereof may be reproduced, stored in a retrieval system or transmitted in any language or by any means, electronic, mechanical, photocopying, recording or otherwise without the prior written permission of the publishers.

T. Krishna Murthy
IFS (Retd)
4-7-143, Isamia Bazar
Hyderabad 500 027, A.P. (India)

Published by

PharmaMed Press
An imprint of Pharma Book Syndicate
A unit of BSP Books Pvt. Ltd.
4-4-309/316, Giriraj Lane, Sultan Bazar, Hyderabad - 500 095.
Phone: 040-23445605, 23445688; Fax: 91+40-23445611
E-mail: info@pharmamedpress.com

ISBN: 978-93-5230-058-7 (HB)

Narayana Datt Tiwari

GOVERNOR
ANDHRA PRADESH

RAJ BHAVAN
HYDERABAD - 500 041

15th July, 2008

Message

I congratulate Mr. T. Krishna Murthy, IFS, (Retd), on his excellent compendium titled 'Minor Forest Products of India'. This book also features certain bio-diesel plants which are of current interest and research value. The author has given a vivid picture of their availability, exploitation, utilisation and development. Coloured images in the book will help in easy identification of plants and their products. This effort will also help to uplift the economic conditions of the people living in and around the forest, who are dependent on the collection of Minor Forest Products.

The book is very much relevant to all the developing countries where forest produce is abundant and needs to be utilised in a scientific way for the welfare of the humanity.

I am sure the practising Foresters of the Indian Forest Service, Rangers, Forest Guards, Eco-Biologists, Students, Farmers and Traders will greatly benefit from this book. I once again congratulate the Author on this excellent work.

(Narayana Datt Tiwari)

Preface

The Forests in India yield a large number of products, which play an important role in the economy of the country. Among these, Timber and Fire wood are known as Major Forest Products and all other products are called as Minor Forest Products, recently known as Non-Timber Forest Products. The Minor Forest Products individually may not be of great economic importance, but collectively their value is immense. Large quantities of the products are consumed locally. They support a large number of indigenous industries besides finding their way into external trade. There are a number of forest products, which, although not come into the markets so far, are likely to gain economic importance, if properly exploited. In part I of this book, the economic importance of Minor Forest Product, modes of collection, trade, method of estimation of yield and suggestion for development have been dealt with, in general. In Part II, the Minor Forest Products have been individually described in 12 chapters, after grouping them mostly according to utility, i.e., Medicinal Plants; Essential Oils; Spices and Condiments and Savoury Plants; Edible wild Plants; Gums, Resins and Oleo-Resins; Fatty oils; Tanning materials; Natural organic colouring materials; Katha and Cutch. In each chapter, a short note is included dealing with general importance, occurrence, extraction, storage and uses of the products of the group along with a list in alphabetical order of the more important product – yielding plants with their rough distribution in the country and utilisation. Details on the cultivation and processing, exploitation of some of the more important species in the group have also been given in each chapter. The book though published in 1993 required changes as only a few black and white images of the products were given therein. This doesn't really help much to realise the importance of the product and to identify the plants in the field. Earlier coloured images were not available and even a few, if available couldn't be procured. By seeing the coloured images, one can get an idea of the product, which can help in realising its economic and commercial importance. As such, the author was in constant pressure to incorporate as many coloured images as possible to facilitate the students / trainees, producer, collector, supplier, trader and consumer in their easy identification. The main aim of the book is to maximise production and utilisation of the products.

–Author

Acknowledgments

I am extremely thankful to Sri P.S. Rao, IFS, Chief Conservator of Forests, R & D, Govt. of A.P. for liberally permitting me to use his images on Forest species in my book. He would have helped me in bringing out this book but for the tragedy in his family. I am also thankful to Sri A.V. Joseph, IFS, Chief Conservator of Forests, Wild Life Management for going through the chapters on Animal Products and making useful suggestions. My thanks are due to Sri B. Sunder, IFS, C.F. Wild Life Management for supplying information on musk. I must also acknowledge the help of Sri Subbarayudu, Asst. Director, Horticulture, A.P. for supplying information on vanilla. I am grateful to Dr. Suhasini, Asst. Director, Horticulture, A.P. and Dr. Hemalata, Horticultural Officer, A.P. for supplementing information on mushroom and honey-bee respectively. I must also place on record my gratitude to the Forest Research Institute and Colleges, Dehradun – my Almamater, where I served and gained much of the knowledge on the subject, which helped me in writing up this book. The information supplied by Chemistry of Forest Products Division, FRI on modern methods of extraction of katha and cutch on cottage scale basis has been usefully incorporated in the book. I must also acknowledge my gratitude to M/s Google as I have used some of their images in my book. In the end, I must appreciate and acknowledge the tenacity and untiring effort of my grand daughter Miss Sharmila Siddhanti for computerising all the information from scanning, editing, paraphrasing and formatting and taking out the print-outs and giving a final shape to the book.

I am grateful to Mr. Anil Shah director of PharmaMed Press for publication of this Edition. All the staff of PMP and in Particular Mr. Laxminarayana Jakkani and also M/s Sandhya and Mr. Kamesh who worked diligently. PMP has put in its best to bring out this excellent technical publication.

–Author

Contents

Chapter 5

Chapter 6

Chapter 7

Chapter 8

Abbreviations

Name of the language or place abbreviated.

Arab.	Arabic
Assam.	Assamese
Beng.	Bengali
Bhuta.	Bhutan
Bomb.	Bombay
Bush.	Bushahr
Cacha	Cachar
Eng.	English
Garh.	Garhwal
Guj.	Gujarathi
Haz.	Hazara
Hind.	Hindi
Jaun.	Jaunsar
Kan.	Kanerese
Kashm.	Kashmir
Khas.	Khasi
Konk.	Konkani
Kum.	Kumaon
Lada.	Ladakh
Lep.	Lepcha.
Mal.	Malayalam
Mar.	Marathi
Nep.	Nepali
Ori.	Oriya

Pers.	Persian
Punj.	Punjab
Rajas.	Rajasthan
Sans.	Sanskrit
Sim.	Simla
Tam.	Tamil
Tel.	Telugu

List of Illustrations

Part - I

1 Introduction

Man's dependence on plants for his existence dates back to the beginning of the human race. In the early days he had only limited needs like, food, shelter, and clothing. But with the advancement of civilization, his requirements also grew. The present day man depends heavily on a very large number of plants and animal products to meet his daily needs. The forest products obtained are converted into useful articles, contributing to his comfortable life. He needs wood as a structural material, as a source of energy, for manufacturing of paper, rayon and many other fashioned and finished products. The drugs obtained from plants and plant-parts can cure many of the human sufferings starting from the simple headache to the serious heart ailments. India is fortunately endowed with a wide variety of climatic and edaphic conditions resulting in a rich and varied flora and fauna. Most of the plants are wild in nature and only a few are cultivated. Although effort has been made to include quite a large number of economically important plants in this book, the lists are by no means exhaustive. Forests in India also produce a wide variety of animal products. For a precise understanding of the term minor forest product it is necessary to define timber and fuel-wood. Timber in general, is defined as wood whose value is mainly on account of its mechanical properties and fuel-wood whose value is only on account of its calorific contents. Minor forest products comprise all forest products other than timber and fuel-wood and include medicinal plants, essential oils, spices, edible wild plants, gums, resins and oleoresins, fatty oils, tanning materials, natural organic colouring materials, katha and cutch, oxalic acid, fibres and flosses, beverages and narcotics, fodder and forage plants, saponins, fish poisons, insecticides and raticides, green manure, beads, rubber plants, plants useful for paper, basket and wicker work including canes, beedi leaf, miscellaneous materials including thatching and broom materials. Besides these plant products, animal products such as lac, honey, horns, hooves, ivory and hides (of forest origin) are included amongst minor forest produce. These products seem to be minor in comparison to timber and wood-fuel but are important when their utility, money value and importance in local economy are taken into consideration. Minor forest products worth millions of rupees are extracted annually and revenue realised in addition to many millions worth of products (including grass and grazing) which are collected/utilized free or at concessional rates by local forest dwellers. Cork is the most important product of the Algerian forest. In Malaysian plantation programmes, production of rubber occupies a more prominent place than timber. In Sudan the most valuable forest product is gum arabic obtained mainly from *Acacia Senegal* (L) Willd. In Karnataka State in India, oil from sandal (*Santalum album* L.) stands out as a very important product and this tree at one time was declared as a royal tree. With the advancement of science and technology more and more species may come into prominence and even the present products may acquire greater importance in future. As an example the case of *Rauvolfia serpentina* Benth. a plant found wild in India may be mentioned. Till recently there was no demand for this plant. Schlittler and Muller isolated an alkaloid reserpine from this plant in 1950 which was found efficacious in the treatment of hypertension. Now the drug based on its alkaloid is used all over the world and the plant has a world wide demand. Therefore to treat all such products as minor is unreasonable. However, in the recent past there has been an attempt to change the nomenclature of such products to Non-wood forest products. It is not understood as to how the important products such as sandal wood oil and agar wood oil etc. which are exclusively obtained from the wood of sandal wood tree and agar wood tree can fit in under the nomenclature of non-wood forest products. As such the changed nomenclature does not give proper definition for such products.

Therefore by conventional usage this expression Minor Forest Products has acquired a standing, the same is used in this context incorporating the changed name in the brackets. There are certain plants which apparently look to be agricultural or horticultural but because of their economic importance they are also being grown on forest lands & as such this book also includes some of the more important plants of this type.

The diversity of minor forest products in India with a bewildering range of uses is perhaps unrivalled in any other country. The products individually may not appear to be of great value but collectively they are of immense value to the country as a whole. India is predominantly an agricultural country with an enormous Livestock population of about 470 millions (1992 census). The agricultural residues and fodder crops supply only a fraction of all fodder needed by this live-stock. The forests provide the rest in the form of grasses and forage plants. Few plants are more useful to the forest tribes in the country than bamboo. It is often referred to as

"Green-Gold" and is considered as the poor man's timber meeting a variety of his needs. Besides other uses it is used for making huts, mats, baskets, cots, scaffolding, spears, lance-shafts, etc. In times of food scarcity its seed is collected and eaten by the poorer classes. Canes and rattans are also equally important being used as a substitute for ropes, walking sticks, fishing rods, spears, lance-shafts and for making baskets, furniture and sports goods. There are hundreds of wild plants whose fruits, seeds, leaves, roots etc., form an important adjunct to cultivated crops, and in a heavily populated country like India they make an important contribution to the diet of millions. Fibres are used for making ropes, cordages, twines, brushes, brooms, baskets, mats, upholstery and for paper making. Some of the grasses besides being used for making paper are useful for the manufacture of ropes, cordages, brooms, mats, thatches, brushes, baskets, etc. A common man is often dependent for relief in cases of petty sickness on medicinal plants available in the forests. Many kinds of alcoholic drinks, beverages and narcotics are derived from forest plants. Oils derived from some of the forest plants are used for cooking and illumination besides other uses. Tanning materials which are solely obtained from the forests are used for processing and curing leather. A number of industries based on minor forest products have developed. Among these, mention may be made of resin and turpentine, beedi (leaf cigarettes), tannin, lac, katha (catechu) and cutch, canes, sports goods, soap, cosmetics, perfumes, disinfectants and drugs. The industries play an important role in the economy of the country besides providing employment to tens of thousands of people both in the forest and in the factories. A large export trade of both raw materials and finished products has been built up and a variety of products are also exported earning foreign exchange. A correct assessment of the total minor forest produce wealth of our country has not been made so far and its potential value remains almost unknown.

There are many exotic plants like *Acacia mearnsii* de Willd. (Wattle), *Agave sisalana* Perr. (Sisal), *Cymbopogan winterianus* Jowitt. (Java Citronella), *Hevea brasiliensis* H.B.K. Muell. Arg. (Rubber), *Eucalyptus* spp. (medicinal and aromatic oils), which yield products of great economic importance and as such they require to be propagated in our country in a big way. Already some species like *Anacardium occidentale* L. (Cashew), *Coffea arabica* L. (Coffee) are being cultivated on a large-scale in the country. These products besides meeting the internal demand are being exported earning valuable foreign exchange. But much is still to be done to meet the increasing demand of the products. Although we have a large variety of canes in our country, still Malaccan canes like *Calamus caesius* Blume and *C.ornatus* Blume ex Schult. f. are regarded as the best for rattaning purpose.

Therefore such exotic species are required to be grown to produce products of good quality to meet the market trends. *Agave sisalana* Perr. a Brazilian species yields commercial sisal fibre. It is a very hardy species coming up even under adverse climate, edaphic and biotic conditions. It is an ideal species for introduction in places where soil conservation measures are required to be taken and incidentally produce Sisal fibre. It is also seen that many of the economically important exotic species are not being grown by farmers since their gestation period is long and as such they require to be encouraged under forestry practices. Our country offers ideal conditions for propagation of such exotic species and therefore it is necessary to grow them, to increase the supplies not only to meet the internal demand but have surplus stocks for export trade. Considering the above factors some of such important species have been dealt under the appropriate chapters.

2

Collection Methods and Marketing of Minor Forest Produce

Collection Methods and Marketing of Minor Forest Produce: Minor Forest Products have certain inherent disadvantages with regard to exploitation. They are bulky and most of them also do not occur in compact areas but widely scattered making economic exploitation difficult. They are often found in difficult and inaccessible terrain where mode of transport is either non-existent or primitive. All these factors are bound to increase the cost of collection and transport. *As* most of these are easily perishable commodities, they present serious problems in harvesting and storage. Lack of precise information on the best season and proper methods of collection, their processing, drying, grading and storage and such other factors have stood in the way of proper exploitation and utilization of a number of products.

The right of collection of minor forest products is sold by a system of auction or lease and it is only in a few cases like sandal, beedi leaf, where systematised and controlled collection of material is attempted through departmental agency. The products are sold singly or in groups. Usually those which are of greater commercial importance are sold singly. The areas are put up to auction by the forest department. The lease period is usually for only year but in rare cases depending on the nature and value of the produce it may be longer or shorter. The lessee having taken the area on lease engages agents who in turn approach the villagers to collect and deliver the produce to them. Wages are paid usually on a quantity basis. Collection of the produce is a secondary occupation for the villagers. The products are then sold to the whole-sale dealers. These products pass through several hands before they actually reach the consumer.

Very often minor forest products are collected haphazardly without caring for the quality of the produce. For example gum karaya which is collected and marketed, usually has a good proportion of bark and other impurities. Sometimes such sub-standard materials bring disrepute to the Indian trade in the export market. Therefore efforts will have to be made to explain to the collector the need for collecting material of good quality. It has also been observed that many of the collectors do not know where and whom to sell their produce, partly for want of this information and partly without realising the need for good quality material they make a haphazard collection of the produce and sell locally at whatever price is offered to them with the consequent low revenue to the State. Therefore in the best interest of the State and the collector, the latter will have to be made aware of the marketing trend and prospective buyers.

Motive of larger gains has often led to adulteration of the products. In the initial stages it may look quite lucrative but when once the real worth of the material is known, it will fall into disrepute with the consequent set-back to the trade. This practice of adulteration has caused great concern in drug industry. It will not only reduce the potency of the drug but some times will have harmful effects on the health of the people. Therefore stringent measures to stop this practice in whatever form are necessary. The minor forest products which are important in the external trade today are myrobalans (crushed, whole and extract), gumkaraya, gum-arabic and other gums, olibanum, sandalwood oil, kokum oil, sal oil, cutch (catechu extracts) shellac and other forms of lac, beedies and beedi wrapper leaves, nux-vomica (seeds, alkaloids and tincture) and serpentina roots.

Among the important minor forest products which hold promise in external trade are essential oils like sandalwood oil and palmarosa oil. Indian sandalwood oil has a certain inherent superior quality in it. This quality coupled with the scientific management of sandal forest has enabled the country to retain a virtual monopoly in the foreign trade even though it is meeting competition with the cheaper Australian oil obtained from *Eucaria spicata* (Syn. *Fusanus spicatus*). Palmarosa oil (Rosha grass oil) from *Cymbopogon martini* (Roxb.) Wats var. *motia* and ginger grass oil from *C.martini* (Roxb.) Wats var. *Sofia,* are also of great commercial importance. Of the two, the former is in greater demand. East Indies and Seychelles have however begun to produce these oils and unless the yield and quality of the Indian oil is improved the trade is bound to suffer by the competition. At one time, India met the world's requirement of vetiver oil from *Vetiveria Zizanioides* (L.) Nash. but ever since the cultivation of the grass in Java, the trade in this commodity is slipping from our hands. The only way to retain the trade is to increase the yield of oil and reduce the cost so that it could be sold at a cheaper rate.

Important medicinal plants amongst the minor forest products are *Rauvolfia serpentina* Benth. and nux-vomica from *Strychnos-nuxvomica* (L.). *Rauvolfia serpentina* Benth. which is used in the treatment of hypertension and certain forms of insanity was in great demand till recently. India met all this demand from natural resources without ever thinking of supplementing it by artificial regeneration. The natural resources gradually got depleted and an acute shortage for the drug was felt subsequently. As a result, the two African plants *R.vomitoria* Afzel. and *R.caffra,* which have similar medicinal properties, came into the market and the demand for Indian rauvolfia started decreasing. Indian rauvolfia has certain advantages over the African plants, with a higher percentage of total alkaloids and higher reserpine content. Although reserpine has been synthesized, owing to its complicated structure it has not yet been found feasible to produce the drug synthetically on a commercial scale. Therefore even now the trade in this drug can be protected if the supplies are made possible to the world market at a cheaper price.

The trade in nux-vomica seed, strychnine and brucine, is practically the monopoly of India although Sri Lanka and the Philippine Islands are the competitors. Sri Lanka's supply of nux-vomica seeds is limited. The Philippines (Islands) produce *S.ignatii* Bergius seeds which has a reasonable amount of strychnine and brucine, the latter also being used as a rodent poison and as an insecticide. Hitherto India used to export all the seed produced, but now strychnine and brucine are being manufactured in the country and both nux-vomica seed as well as its alkaloids, from India are finding their way into the world market.

Of the gums, karaya or hog gum derived from *Sterculia urens* Roxb. and *S.villosa* Roxb. has recently acquired a great commercial importance and during the year 2002-2003, 1119.838 tonnes of gum valued at Rs. 131845718.00 was exported besides other gums. Among fibres, sisal and rhemi are of great commercial value. Efforts are being made to increase the production of sisal by raising plantations but rhemi is still not receiving the required attention. As India's production of cane is not sufficient to meet the internal requirements, java cane was at one time being imported. With import restrictions, an acute shortage of cane is being felt and people have started using synthetic cane for rattening chairs. Natural cane has certain superior inherent properties over synthetic cane. It does not sag and has a longer life. As such the demand for natural cane will remain steady and there are bright prospects for increasing its yield.

At one time natural dye-stuffs, of which forests were the main source, were being used in textiles, now chemical dyes have completely replaced them in this industry. However, annatoo a natural dye obtained from *Bixa orellana* still holds out good prospects in the trade. It is used for colouring butter, cheese, margarine and other food stuffs. It is non-toxic as against the synthetics and as such it has been able to retain the market. Since supplies of this dye are limited, efforts made to increase the production would be financially viable. If production of the dye is not increased, sooner or later substitutes will come in or other countries will score over us in the supplies, and our trade in this product will dwindle.

These are only a few among the many important minor forest products with export potential and there is still considerable scope to increase the trade. Sometimes a motive of greater gains has led to adulteration and if this is practised many of the products will lose foreign markets. In order to have a fair name in the trade, this practice will have to be stopped and materials of only the standard qualities brought into the market. Grading, freedom from impurities and adopting proper marketing strategy with an attractive packing are some of the ways to enhance the value of raw materials. Although we are exporting a large number of minor forest products, a variety of other products are being imported by us to meet our internal demands. Clove oil, citronella oil, menthol, peppermint oil and even spices like cinnamon and cassia bark are being imported. It is a matter for regret that a country on which the whole world depended once for the supplies of cinnamon, should now have to depend on imports, it is because of large-scale exploitation made in the past without even thinking of increasing its production by cultivation.

Import and export figures of minor forest products for the year 2002-2003 are given below. These figures are only indicative in nature. Since some of these are imported for processing purposes and are re-exported. The record of the exact quantities of imports and re-exports are not maintained. Among these, some of the export products like coffee, cashew, kuth, lemongrass oil, rubber, walnut are derived from the joint venture i.e. non-forest as well as forest. It is difficult to say the exact contribution of each of the sectors in this trade. However, it is felt that non-forest sector has a major share in the production and export trade. However, some of these products like *gum karaya*, *myrobalans* and their derivatives are purely forest origin.

Minor forest produce in trade and commerce. (Source Monthly Statistics of the foreign trade of India Vol. I & II Exports and re-exports, and imports for year ending March 2003.)

S.No.	Item	2002- 2003 March			
		Imports		Exports	
		Quantity in Kgs	Amount in Rs.	Quantity in Kgs	Amount in Rs.
1	Natural honey	338649	177603757	6646589	529047844
2	Mushroom spawn			33831	1542063
3	Asparagus fresh or chilled	1025	739342	48242	1279889
4	Mushroom fresh or chilled	100	60768	14315113	517696278
5	Green pepper	3940	1385201	342680	21765595
6	Green pepper in brine			729166	41563964
7	Mushrooms & dried Morrels	15392	10163411	97082	151063403
8	Truffles	7840	175978		
9	Asparagus dried	45313	20487859	7537	384153
10	Sago pith			303355	4247435
11	Cashewnuts raw	394726564	12089354976	6943871	1167074585
12	Others	8255480	273074348		
13	Cashew kernel broken	516600	22678802	20242478	2977615891
14	Cashew kernel whole	257989	10431374	101821274	16316626518
15	Walnut fresh or dried in shells			655123	90986341
16	Shelled walnuts fresh or dried	34090	548583	6976119	1121295592
17	Tamarind fresh			3138990	43880606
18	Sitaphal			57520	1046160
19	Custard apple (ata)			194376	3421821
20	Tamarind dried			8318958	135450557
21	Coffee Arabica Plantation-A	2997386	65159835	45651574	2281613375
22	Coffee Arabica Plantation-B	36000	1884690	7093070	344909862
23	Coffee Arabica Plantation-C			4536939	203048035
24	Coffee Arabica Plantation-other grades	2048	451084	5041644	255529770
25	Coffee Arabica - Cherry AB	5133634	44771288	5234098	223817211
26	Coffee Arabica - Cherry PB			1288076	47738702
27	Coffee Arabica - Cherry C			1567800	53598267
28	Coffee Arabica - B/B/B/			468978	13088897
29	Coffee Arabica - Other grade			5490997	264833812
30	Coffee Robusta Parchment - AB			8890239	355697066
32	Coffee Robusta Parchment - C			765710	22074329
33	Coffee Robusta Parchment - Other grade			11792876	484521793
34	Coffee Rob. Cherry 'AB'			26084478	877654648
35	Coffee Rob. Cherry 'PB'			2423661	85976059
36	Coffee Rob. Cherry 'C'			1062111	38246307
37	Coffee Rob. Cherry 'B/B/B'			250100	12683202
38	Coffee Rob. Cherry 'Bulk'			16322197	527020723
39	Coffee Rob. Cherry other grade			164571927	6928060248

Contd...

S.No.	Item	2002- 2003 March			
		Imports		Exports	
		Quantity in Kgs	Amount in Rs.	Quantity in Kgs	Amount in Rs.
40	Not roasted but decaffeinated coffee			187470	5542109
42	Roasted decaffeinated coffee			146809	20092790
43	Coffee husks & skin			385263	1917691
44	Pepper long			416755	66317372
45	Pepper pinheads			616976	42664412
46	Freez dried green pepper			65897	43112473
47	White pepper	226605	25094954	219874	27360791
48	Other pepper	15480155	1238101198		
49	Crushed or ground pepper	154910	7170307	3599788	247534093
50	Pepper Jamaica	849121	61010015	200632	11520966
51	Vanilla bean and powder	2569	499361	68404	19664400
52	Cinnamon bark	74638	3074088	4345	323618
53	Cinnamon Tree flower	720	103897		
54	Cinnamon & tree flowers ground or crushed	132000	4462973	36100	3389513
55	Cardamom large	4261527	424369293	1548038	266810951
56	Cardamom small	302526	79307725	317798	182208783
57	Cardamom small coorg green			1733	821867
58	Cardamom bleached, half bleached, bleachable			10217	4288499
59	Cardamom small seeds			22842	17933837
60	Cardamom small mixed			69000	52233098
62	Tejpat beans			476289	10858699
63	Cassia tora			1863440	16465711
64	Tejpat powder			63091	2708416
65	Cardamom husk			21428	1946029
66	Flour meal of sago			204049	6272210
67	Flour meal of tamarind			597354	9829247
68	Powder of tamarind			2973723	66302314
69	Kokam dried pericarp of Garcinia indica			45956	4303722
70	Tamarind seeds			1740845	29539646
71	Hops cones, powdered dried or fresh	96779	36571742		
72	Liquorice roots fresh, dried, powdered	783170	10269680	332896	20652705
73	Belladona roots			460597	24752748
74	Galangal roots	123500	4809505	342168	9032706
75	Serpentina roots	36160	544199		
76	Kuthroots	176130	2423893		
77	Zedovary roots			24000	571595
78	Other gensing roots	129538	19339629	1456198	122087895
79	Agar wood	124544	30881223	202311	22851805
80	Belladona leaves			10260	515580

Contd...

S.No.	Item	2002- 2003 March			
		Imports		Exports	
		Quantity in Kgs	Amount in Rs.	Quantity in Kgs	Amount in Rs.
82	Chirata	2500	49500	59225	13098965
83	Nux-vomica dried, ripe seed			3900	329614
84	Pyrethrum	148500	17058660		
85	Sandalwood chips & dust			726780	317930280
86	Sarasaparilla	30501	3248386	25683	674896
87	Tukmaria			278963	10344104
88	Ayurvedic & Unani herbs	2170032	81670609	7865163	242560985
89	Sandalwood chips			40100	35630336
90	Vinca rosea			682548	29245297
91	Gamboge	1296317	76313317	57775	34457063
92	Neem leaves & powder			22225	606717
94	Cinchona bark	1793193	73159077		
95	Kokam flowers			58526	12138778
96	Shellac hand made	40958	7693978	948139	168341579
97	Shellac machine made	83350	4392183	3927770	538147170
98	Shellac			415429	43854379
99	Seed lac	461081	28194482		
100	Stick lac			4025	552063
101	Other lac	27250	2684358		
102	Dewaxed and decolourised lac			91800	18979286
103	Bleached lac			74645	12957359
104	Garnet lac			3202	609754
105	Button lac			5762	691298
106	Garnet lac			1960	216787
107	Other lacs			154352	14436513
108	Gum arabic	7341249	166074541	56380	18005924
109	Asian gum	75010	1535401	285333	12024273
110	African gum	22000	540100	129319	7965252
111	Asafoetida	514230	138884615	473914	43999424
112	Banjameen ras	118508	4873533	1200	242897
113	Benjamin gowrie			1845	85298
114	Karaya gum Indian tragacanth			1119838	131845718
115	Tragacanth	4898	359037		
116	Other natural gums	777732	37213876	645535	36222984
117	Copal	1593932	30294879	49664	2814460
118	Dammarbatu	4206353	83947270	222	12472
119	Other resins	147048	9286789	69522	5301381
120	Myrrh	12200	570998		
121	Olibanum	16250	435987	6500	909985
122	Mastic gum	4725	417305	640	37010

Contd...

S.No.	Item	2002- 2003 March			
		Imports		Exports	
		Quantity in Kgs	Amount in Rs.	Quantity in Kgs	Amount in Rs.
123	Xanthium gum	52888	15759941	7000	427370
124	Other gum resins	148670	19384009	968705	45864217
125	Balsam of tolu	68569	10914831	10	10496
126	Pine oleoresins			150265	76813435
127	Galangal oleoresins			45	406994
128	Other balsam & oleoresins			349080	255963435
129	Extracts of hops	66439	40597813		
130	Sap / extracts of pyrethrum or of the roots of plants containing rotenene			100	448131
131	Extract of belladonna	27894	36663190		
132	Extract of cascara sagrada	1099	2079949	2208	1289121
133	Ginsing extract	13144	46059667		
134	Extract of Nuxvomica			6505	4824310
135	Extracts of neem			15737	31402125
136	Gymnema extract			43059	24481832
137	Gambose extract	129720	15840680	867848	542239261
138	Bamboos	98086	2677479	245852	4345445
139	Cashew shell liquid			5929067	78968947
140	Cashewnut shell liquid purified & distilled (Cardanol)			209751	6641466
141	Rattans	104713	2554928		
142	Canes used as Plaiting	132235	10067657	31500	1965244
143	Kapok for stuffing	29688	1868993	34847	2262922
144	Broom corn			56975	450986
145	Palm fibre for brushes	131117	3316699	4863375	140898460
146	Red sandal wood powder used in dying			18098	7135194
147	Henna leaves			149388	9563053
148	Henna powder			2383099	206076295
149	Gallnuts	78364	2484600		
150	Myrobalan amla in tanning			254417	6849055
151	Myrobalans other whole/cut used in tanning			246745	9790611
152	Other tanning substances			115123	2370988
153	Beedi wrapper leaves	19473	687144	3505164	218817184
154	Soap nuts			10166	865412
155	Soap nut powder			3051	89675
157	Rudraksha seed	124142	4180567	300	14250
158	Quebracho extract	784332	37858515	194865	23336421
159	Wattle extract	12384314	430854204	1000	37337
160	Gambier extract	11771001	811157955		

Contd...

S.No.	Item	2002- 2003 March			
		Imports		Exports	
		Quantity in Kgs	Amount in Rs.	Quantity in Kgs	Amount in Rs.
162	Myrobalan fruit extract	10590	756250	590100	28684991
163	Gallotannic acid	14000	819802		
164	Tannin derivatives	1628935	143739653		
165	Cutch extract	246408	40535251	3187331	72958079
166	Lac-dye			6850	1718029
167	Indigo natural	24000	5700112	26131	2248390
168	Citronella oil java	17659	9415910	13316	5262462
169	Citronella oil ceylon	500	134000	71621	25427469
170	Essential oil or pepperment; mentha arvensis			1280800	52027866
171	Mentha piperita oil			2643114	915097021
172	Dementhalised Japanese mint oil			505230	262434547
173	Vetever oil	6810	10947161	2090	688833
174	Cajuput oil	1515	1352700		
175	Cassia oil			812	734660
176	Cedarwood oil	8540	3573669	1425	483164
177	Cinnamon bark oil	3656	1331776	1095	714248
178	Cinnamon leaf oil	8476	2852401	16263	6950939
179	Clove leaf oil			10905	5910792
180	Eucalyptus oil	108713	18850341	10640	4067940
181	Gingergrass oil	300	365345	215	592511
182	Palmarosa oil	500	1195899	30661	20723364
183	Patchouli oil	114028	68894355	5155	2813998
184	Sandalwood oil	500	150080	6135	113951783
185	Rosa oil	13010	6570089	753	5460342
186	Camphor oil			1951	2686304
187	Lemongrass oil	755	215043	45387	28288918
189	Greater galangal oil	1008	980504	1545	386924
190	Juniper oil			5337	1008974
191	Agar oil	14381	7391995	8792	550851
192	Ambrett seed oil			14136	18552464
193	Pine oil			22838	1450368
194	Sandalwood	11937	16170811	2127	31458977
195	Natural corks & stopper	12036	4840824	750	185270
196	Other articles of natural corks	12524	4756253	59301	3731172
197	Raw furskins of fox with or without whole or without head, tail or paws	200	74457		

3

Prospects of Development

Prospects for Development: With the increasing demand of minor forest products in industries and trade there is a need for ascertaining their potentialities. Although information on some of the important species like sandal and resin yielding pine, *Pinus roxburghii* Sar. may be available, in the case of a large number of other important species, data is either *not* available or incomplete. Very often inquiries pour in both from industries and traders about the distribution and yield estimates of certain minor *forest* products available in the regions and forest officers for lack of precise information find themselves helpless. Thus many of the potential products remain unexploited. In order to exploit these resources fully it is necessary for forest officers to know, what, when, where and how much he can supply from his region. It is therefore essential to conduct surveys for finding out the distribution and exploitable yield of at least the more important species. Along with this data, information about the method of collection, season of collection, processing details, grading, storage, transportation facilities and marketing aspects of these products is essential, for proper utilization of the produce.

Sampling methods for estimating the yield depend on various factors such as the kind of produce, geography and purpose for which data are required. In some cases fruits, seeds, flowers or leaves are valuable while in some others bark, stem or roots. Therefore these products cannot be estimated with the same yard-stick. The method and intensity of survey also varies according to the importance of the species. However, for the general purpose of getting a fair idea of the potentialities of the produce, the method should be such that it does not involve high cost and time and *at* the same time covers extensive areas. In order to avoid duplication of work and to reduce cost of labour and time, surveys of the important minor forest products can be taken up *at* the time of the preparation of the working plan. Separate stock maps for very important minor forest produce may be attempted. Depending on the density of stocking of each Minor Forest Produce, the area may be divided into different strata / zones. In each strata random sample survey with intensity varying from 0.01 % in case of less important to 5% in respect of more important species may be done. Higher intensity of sampling may be required only in the case of species of great economic value. Complete count of the plants of each species in each sample unit noting their size, shape and quality which have a bearing on the yield of the produce may be done. Yield particulars of flower, fruit, leaf, seed, bark, root, wood etc., depending on the nature of the produce required, should be collected from the sample trees of each species separately.

The other variables which affect the yield like season, good, moderate and poor yield years should also be kept in view while assessing the yield. When the counting of the plants is not possible as in the case of grasses, the yield of material in the sample is estimated by weight. The yield data so collected for sample trees/units etc., may be compiled for the whole area.

The increasing population results in increasing pressure through Government and politicians to clear large areas of forests and convert them into agricultural land to relieve the food shortage. Whether a particular land and soil is suitable for agriculture or forestry, they tend to think, as merely a scientific refinement and hold that with intensive management any land could be converted into an agricultural land. If the forester does not look beyond the customary role of a timber grower to keep forestry active and in harmony with socioeconomic needs, sooner or later all the fertile forest land will be snatched away from his hand and converted into agricultural land. He will be left with only unproductive and rocky hill tops. To meet the socio-economic requirements of the country the forester should try to modify the age-old management practices. To relieve food shortages the concept of multiple use of land, combining agri-silvicultural practices i.e. growing agricultural crops in conjunction with forest crops wherever possible should be thought of. The practice of agri-silviculture (Taungya) is nothing new to Indian forestry. Taungya plantation was started in the year 1866 in Myanmar (Burma). The object of this was to provide more food and also to provide employment to the toiling masses. This method of Taungya or agri-silviculture can also be extended for cultivation of medicinal and other economic plants. There are a large number of wild plants whose seeds, fruits, roots etc., make an important contribution to the diet of poor people particularly those living near the forests. During the fruiting season *Annona squamosa* L. (Seethaphal, custard apple) for example which is a delicacy besides giving considerable revenue to the exchequer, this fruit forms an important part of the diet of the rural poor in the regions where it is found. This species is gregarious in occurrence and can conveniently be cultivated in other areas where climatic and edaphic conditions are similar. There is a large number of other wild plants for e.g., *Pinus gerardiana* Wall., *Limonia acidissima* L. *Juglans regia* L., (Walnut); *Buchanania lanzan.,* L., Spring.; *Mimosops elengi* L.; *Manilkara hexandra* (Roxb.) Dub. etc., which produce edible fruits and seeds. But many of these are found scattered in nature and are difficult to economically exploit. If efforts are made to grow such species which have commercial potentiality, we may not only be in a position to meet local requirements but also to export. There is a large number of other plants in the forests which have great commercial and industrial value. Many of them are found scattered in nature making economic exploitation difficult. As a result they are either left uncared for or are exploited only to a limited extent. Therefore efforts will have to be made to increase the yield and supply of such raw-materials at competitive rates. This is possible only when concentrated crops are available. With the increasing tempo of industrialization and other socio-economic developments the demand for a large number of other products is increasing and industrialists and traders are looking to forest departments for their supplies. Although we obtain tanning materials from indigenous plants like *Acacia nilotica* (L.) Del. (Syn. *A.arabica* Willd.) *Cassia auriculata* L., C. *fistula* L. and *Terminalia chebula* Retz. and exotics like *Acacia mearnsii* de willd. Syn. *A.mollisima* Willd. (Wattle) which are grown in the Nilgiris, the supplies are insufficient to meet the industrial and commercial requirements. If wattles and other tanning materials are grown on a commercial-scale, we will not only be able to meet our own requirement of raw materials and their extracts but also will be able to build up surpluses for export. Though steps have been taken to grow the fibre-yielding *Agave sisalana* Perr. (sisal fibre) in arid areas, not much attention is being paid to growing an equally important *species.Boehmeria nivea* (L.) Gaud. (rhemi fibre). It has not only a great commercial potentiality but also is required in the textile industry within the country. Many other commercially important essential oils can also be produced in India. Citronella oil which is obtained from *Cymbopogon nardus* Rendle (Ceylon citronella oil) *C.winterianus* Jowitt (Java citronella oil), *Eucalyptus citriodora* Hook., is being imported. The former two are perennial grasses and can be grown in the country. They require high temperature and well-distributed rainfall of at least 1500-1750 mm per annum. They do not require much care and tend to propagate naturally. *Eucalyptus citriodora* Hook. is again quite a suitable species for citronella oil. It can also be grown extensively in tropical and sub-tropical regions in the country. Juniper berries which are used as a flavouring agent in brewery are in great demand but no efforts have been made to raise the species at least over a few hundred hectares. *Bursera penicillata* Sesse and Moc. ex. DC. Eng. (Syn. *B.delpechiana* Poiss) Linaloe yields an essential oil which has a great future. There is considerable demand for the oil in perfumery in the country and also for export trade. Karnataka and Andhra Pradesh have made a humble beginning to grow it in forest areas. It appears from the experience that it can successfully be raised in low rainfall and even in denuded and eroded areas. Thus the so called less potential areas can also be put to some productive use. However, economic considerations favour its cultivation on good sites. *Sterculia urens Roxb.* the source of karaya gum is very much scattered in many places of its distribution, thereby resulting in a high cost of collection of gum and rendering systematic and scientific tapping difficult. If the species is planted on a larger scale, we may, in course of time, develop a greater trade. These are only a few among the many other important minor forest products which need the attention of foresters. Many of them, like timber species, need long term investments with distant rewards. Since farmers in India are always anxious to have quicker and greater returns from their crops, these minor forest products are not receiving attention under agricultural practices. It is only the forest department which can afford to make such an investment and not be impatient for immediate cash returns. Therefore the burden of increasing the yield of such products lies on the forester. In afforestation areas where the object is to conserve soil and moisture, and where production of timber is not aimed at, advantage should be taken of the multiple use of the species and choose only those which are economically important. As far as possible they may be raised in pure stands. This will render easy and economic collection of the produce. There is a large number of species which have varying light requirements. Being shade tolerant or shade bearing they can be grown in association with the timber species without hampering the growth of the latter. In Tamilnadu camphor yielding ocimum, *Ocimum kilimandscharicum* Guerke, was introduced in teak plantations as an under-growth. It is growing very luxuriently tapping only the surface layer of the soil leaving the rest to teak. Similarly, *Strychnos nuxvomica* L., (nux-vomica) which yields medicinal seed and *Schleichera oleosa* (Laur.). Oken (Kusum) whose seed yields fatty oil are shade tolerant and can be tried as a middle storey or as an associate crop in teak and other timber plantations. These species not only yield economically important products but also, being semi ever-green to ever-green in nature, protect the soil from desication during the hot season when teak is leafless. The only objection which often occurs to the mind of a forester is that in such multiple storey plantations there is root competition for the main timber species. This can be avoided if the lower storey species are introduced only when the main timber species have become fully established and sent out their roots to deeper layers of soil. And further, selection of species for the under storey in addition to various other considerations, should be such that at no stage of their growth they should create root competition with the main timber species. In this way the same land can be used for production of timber and for the increase in the yield of some minor forest products. Only when such efforts are made to increase the intensity of existing important species which are scattered in nature and to introduce the economically important exotic species for which our country offers suitable soil and climatic conditions, we will be in a position to meet our own requirements and also be in a position to increase our export trade.

Although some work has been done on the utilisation aspects of some of the species, still much remains to be done. There is still a large number of species found abundantly but their utilization is not known. As such these species have remained unexploited. Therefore planned investigations on chemical and commercial aspects of these species are necessary to find out their uses.

There is a large number of other species in the country which owing either to lack of processing knowledge or to financial limitations have not assumed economic importance. For example, Cedar wood oil is being imported into the country for perfumery industry. This valuable essential oil can be substituted by our Himalayan cedar wood, or *Cedrus deodara* Loud. This species is found abundantly in the Western and central Himalayas in the temperate region. The heart-wood on steam or water-cum-steam distillation yields the oil which has almost the same properties as those of the imported oil. Canada balsam which is used for mounting microscopic slides is being imported into the country. It has been found that if pure and good quality gum-oleoresin of *Boswellia serrata* Roxb. ex. Colebr. (salai) is mixed with 50 per cent xylene it produces almost the same binding and other properties of Canada balsam. Although results of these investigations have been known long back they have not received the attention of manufacturers, traders and consumers. Similarly we are importing clove oil in large quantities. It has been found that the oil obtained from the leaf of *Cinnamomum zeylanicum* Bl. is rich in eugenol and can be used as a substitute for the imported clove oil. This species is found in Kerala and Karnataka, but no steps have so far been taken to extract oil on a commercial scale. Many of such products have very simple manufacturing processes. And in order to exploit the forest wealth to the best advantage possible it is necessary that correct and useful information is disseminated by popular articles and even by demonstration. Only when people are convinced about the economic potentialities and financial benefits of the products, greater utilization of the wealth can be expected.

Part - II

4

Medicinal Plants

4.1 Growth of Knowledge of Medicinal Plants

Forests in India provide hundreds of medicinal plants some of which are pharmocopoeial while others are used in indigenous medicine. From earliest times mankind has used plants in an attempt to cure diseses and relieve physical sufferings.They derived this knowledge as a result of trial and error. The history of medicine in India can be traced to very ancient times. Medicinal properties of plants have been mentioned in Rig Veda which is one of the oldest repositories of human knowledge which is supposed to have been written between 4500 and 1600 B.C. Mention has been made of Soma plant perhaps Ephedra gerardiana Wall. and its effects on man. In Ayurveda which is the basis and foundation of the ancient medical science of India, definite properties of drugs and their uses have been given. According to western scholars the age of Ayurveda is fixed at about 2500 B.C. to 600 B.C. Charaka, a branch of Ayurveda and which is supposed to have been written about 1000 B.C. deals with medicine. From this period down to the Mohammaden invasion of India, Ayurvedic medicine flourished. Many of the civilized countries of that time eagerly sought for information and drug plants. During the period of Alexander the Great, Indian medicine was at its zenith. The work of the great physician Dioscorides mentions a variety of Indian plants used in medicine. Gradually various branches of medicine passed into the hands of priests, and drugs and herbs partly gave way to charms and amulets. After the invasion of Mohammadens, Ayurvedic medicine started its decline and gave place to Yunani or Mohammaden medicine. Thereafter there was an intimate exchange of knowledge between the Ayurvedic and Mohammaden medicine and a rich store of combined materia medica came up. When the British rule was established in India the western system of Allopathic medicine was introduced. Gradually its demand extended all over the country.

4.2 Chemical Constituents

The active principles found in medicinal plants may be alkaloids, glucosides or other similar compounds of a very complex nature. Some of them when taken in large doses have very deliterious effects on health and some times even cause death but when they are taken in small doses they are very useful in the treatment of various ailments and diseases. These active principles may be found in any one part of the plant or in more than one in varying proportions. They may be found in root, bark, stem, leaf, fruit or seed. The physiological function of these active principles in the body of the plant is not fully understood. Some however consider that they are waste products of the metabolism of plants. They are extracted with appropriate solvents.

Abelmoschus moschatus Medik

4.3 Collection

Collection of medicinal plants requires scientific approach. It is necessary to understand in which part of the plant or at which stage of its growth medicinal constituents are high. If the collection is made prematurely or is delayed these medicinal constituents decrease affecting the potency of the drug. When this drug is an essential oil, the raw material so collected should be processed at the appropriate time as otherwise the essential oil gets evaporated out resulting in low yields. Improper storage of the drug plant results in deterioration of its quality and potency. Most of these drug plants are collected by petty contractors or the local people. This collection is seldom systematized and controlled with the result that many of the drug plants have come into disrepute on account of the haphazard collection, non-grading and improper care in drying and storage. Sometimes adulteration with spurious plants is practiced with a motive of larger gains. This amounts to tampering with the quality or strength of the medicines. Unless and until this practice is stopped, the trade in Indian medicinal plants will not improve and the use of drugs in the treatment of diseases will not be efficacious.

4.4 Medicinal Plants

This chapter deals briefly with individual species giving their distribution and medicinal uses. In case of more important plants, methods of their propagation, processing etc., are given. Some of these are believed to contain un-known or little known active principles. There are others whose chemical and therapeutic values have been studied and are of great pharmacological and commercial importance.

Acalypha indica L.

1. *Abelmoschus moschatus* Medik Syn. *Hibiscus abelmoschus* L.; Malvaceae;

Na : Arabic: Habulmusk; Assamese: Gorukhia Korai; Bengali, Gujarathi & Hindi: Mushkdana; English: Ambretteseed; Kanerese: Kasturi bendi; Malayalam: Kattukasturi; Marathi: Kasturibhenda; Tamil: Varttilai Kasturi, Katukasturi; Telugu: Kasturi benda.

Hab : It is an erect hispid herb with polymorphic leaves having 5 to 7 lobes. It is found in drier parts of the country.

The seeds are small and scented and are used as tonic, carminative, stomachic stimulant and anti-spasmadic.They are also used in nervous disorders and a paste of them is used in skin diseases.

2. *Acalypha indica L.;* Euphorbiaceae;

Na : Beng.: Muktajhuri; Guj.: Dadano; Hindi & Marathi: Khokte, Kuppi; Kan.: Kuppigida; Sans.: Haritamanjari; Tam. & Mal.: Kuppaimeni; Telugu: Kuppichettu.

Hab : It is a small annual herb common in wasteland throughout India. The decoction of the plant is used as a laxative. The juice of the leaf is used in skin diseases and is said to be very useful in certain forms of jaundice.

Achyranthus aspera L.

3. *Achyranthus aspera* L.; Amaranthaceae

Na : Beng.: Apang; Gujarathi: Aghedo; Hindi: Latjira; Kanerese: Uttarane; Malayalam: Katalati; Marathi: Agadha; Sanskrit: Apamarga; Tamil: Nayrivi; Telugu: Uttarane; Persian: Khare vazhun;

Hab : It is a stiff erect weed common in waste lands during rainy period. The fruits are thorny. The seeds are stated to reduce appetite. They are also used in stomach troubles by the local people. The decoction of the plant is said to be diuretic.

Aconitum chasmanthum **Stapf ex Homes**

4. *Aconitum chasmanthum* Stapf ex Homes; Ranunculaceae.

Na : Hazara: Mohri; Kashmir: Ban-bal-nag.

Hab : It is an herb found in the western temperate to alpine zones in the Himalayas from Hazara to Kashmir.

The tuberous roots are generally collected late in September. Mother tubers are deeply grooved and wrinkled black outside and brown right through. Roots are very bitter and contain 4.3 per cent total alkaloids and are the source of a drug "aconite" which is used to relieve pain and fever.

5. *A.heterophyllum* Wall.; Ranunculaceae.

Na : Hindi: Atis; Sanskrit: Ativisha.

Hab : It is an herb found in the western temperate to alpine regions in the Himalayas.

The roots are bitter and are used medicinally as astringent, stomachic, diarrhea, dysentery, aphrodisiac and as an excellent tonic. The active principle is atisine (0.4%) which is an amorphous alkaloid. This plant is officinal in Indian pharmacopoeia.

A.heterophyllum **Wall.**

6. *A. ferox* Wall.; Ranunculaceae.

Hab : It is a rare and a poisonous species found in the northern Himalayas in Kashmir and also in Nepal.

It has the same uses as those of other species of Aconitum. A preparation of the roots is also used in the hilly district in India as an arrow-poison.

7. *Acorus calamus* L.; Acoraceae

Na : Eng.: Sweet flag; Hindi & Bengali: Safed bach; Kanerese: Bajegida; Malayalam: Vavambu; Marathi & Guj.: Vekhand; Tam.: Vasambu; Telugu: Vasa.

Hab : It is an aromatic herb found wild or cultivated upto 1800 m altitude in the Himalayas particularly in marshy and moist places. For its propagation cuttings of rhizome are planted at an espacementof 30 cm and profuse watering done when dry conditions are prevailing.

Harvesting: After about 10 months of planting, the crop is ready for harvesting. When the lower leaves start yellowing and drying, watering has to be stopped. The land can be deeply ploughed or dug out for rhizomes. Adhering soil to the rhizome should be removed and cleaned and cut into small pieces and thoroughly dried in shade, bringing moisture content to a maximum of 15%. The rhizome to be used for planting should be kept in a

Acorus calamus **L.**

Adhatoda vasica Nees

pit covered up with dry soil. Such material can be used for planting. A hectare of plantation may yield about 2.5 to 3 tonnes of roots.

Uses: Dry rhizomes contain 1.5 to 3.5% of a yellow aromatic volatile oil. Roots also contain a glucoside named acorin. The rhizome is used medicinally as carminative, stimulant and tonic. It produces beneficial results in cases of dyspepsia and chronic diarrhoea. The powdered rhizome is also useful against bed-bugs, moths, lice etc.

8. *Adhatoda vasica* Nees (Syn. Justicea adhatoda L.); Acanthaceae.

Na : Beng.: Bakas; Eng.: Malabar nut; Gujrathi: Alduso; Hindi: Arusha, Vasaka; Kanerese: Adusoge; Mal.: Atalotakam; Marathi: Adulsa; Pers.: Rus. Sanskrit: Vasaka; Tamil: Adododai; Telugu: Adasaramu;

Hab : It is a semi-evergreen to evergreen shrub commonly found in North India and also in other regions in the country. The leaves are the source of an indigenous drug known as 'vasaka' which is used as an expectorant and to relieve coughs.

Aegle marmelos L. – fruit

9. *Aegle marmelos* L. Corr.; Rutaceae.

Na : Hindi: Bel; Sanskrit: Bilva; Telugu: Bilwa, Maredu.

Hab : It is a moderately tall thorny tree found in dry tropical forests in Andhra Pradesh, Maharashtra, Madhya Pradesh and other parts up to the foot hills of Himalayas. Roots are said to improve appetite. They are one of the ingredients of Dasamula drug of Ayurveda.

Aegle marmelos L. Leaf

10. *Aerva lanata* L. Juss.; Amaranthaceae

Na : Sanskrit: Pashana bedha; Telugu: Kondapindi, Pindikura.

Hab : It is a small herb of 60 to 70 cm with alternate leaves found in warmer parts in the country. The plant is said to remove bladder stone and is also used by local people in other urinary ailments.

11. *Alangium salvifolium* L. f. (Wang.); Alangiaceae.

Na : Hindi & Sanskrit: Ankola; Telugu: Ankolamu, Udugu.

Hab : It is a common plant in the dry tropical forests. Leaves and seeds are used in skin diseases. Seed oil and germinating seedlings are stated to be used for witch-work.

Alangium salvifolium L. f. (Wang.)

12. *Aloe barbadensis* Mill. (Syn. A.vera Tourn-ex L.) Agavaceae

Na : Bengali & Sanskrit: Ghritakumari; Eng.: Barbados aloe; Hindi: Ghikanvar; Kanerese: Lolisara; Malayalam: Kumari; Marathi: Korphad; Tamil: Chirukuttali; Telugu: Chinna kalabanda.

Hab : The plant is originally native of West Indies, but is now growing wild in warm and dry regions all over India. It comes up in waste lands, sandy and rocky areas. The plant grows to a height of about 60 cm. Its leaves are about 45 cm in length and about 7-8 cm in width tapering towards the tip. The margin and tip of the leaf bear spines. When the leaves are slit, fleshy gel comes out. Hence it is called "Ghruta Kumari" in Sanskrit and "Ghee Ghuwar" in Hindi. In autumn the plant produces slender flowering stalk in the centre. The leaves are used medicinally.

Aloe barbadensis **Mill**

Soil and Climate : The plant is drought resistant and comes up in a variety of soils-red soils, sandy and BC soils. Water logged areas are not suitable for its growth. Hot and warm areas with good sub-soil moisture are suitable. It grows in areas having a rainfall range of 350-500 mm.

Propagation: It should be planted with the commencement of monsoon after a good shower. A well grown plant produces a large number of root suckers all around, every year. They should be pricked out leaving 1 or 2 suckers and planted at site and an espacement of 45 × 45 cm or 60 × 60 cm depending upon the fertility of the soil. Closer spacing is adopted in poorer soils. Although the plant is drought resistant, irrigation or watering the plants during the dry periods will greatly enhance the yield of leaves. 1 or 2 weedings may be done depending on the intensity of the weeds. The root suckers developed around the plants in the plantation should be pricked out and planted in the nursery for planting at site at a later stage.

Prophylactic treatment : There is not much damage by insects and pests. However, if any damage is noticed spraying neem oil or decoction of neem leaves may reduce the damage.

Harvest : The crop is ready for harvesting after about 8 months. The leaves should be cut and harvested at a height of 5 cm from ground level. Leaves can be harvested even earlier after about 3-4 months, depending on the demand. The leaves can be marketed as such or slit and the gel is extracted and bottled. The gel on drying becomes powdery and is called "Mushambaram" in Indian medicine.

Yield : The plantation can last for 3-4 years. Thereafter it has to be replaced. Each hectare of plantation can produce about 50-75 tonnes of fresh leaves every year.

Uses : Aloe gel has a cooling effect if taken orally and keeps the bowels clean. It helps in digestion and also gives energy. It is used in skin ailments, face powders, beauty parlour products, gynaecological ailments. Aloe has anti-bacterial properties.

13. *Anamirta cocculus* (L.) Wight & Arn. (Syn. Menispermum cocculs L.); Menispermaceae.

Anamirta cocculus
(L.) Wight & Arn.

Na : Eng.: Fish berry, Levant nut; Hindi & Sanskrit: Kakmari.

Hab : It is a climbing shrub found wild in the Khasia hills, Orissa, Karnataka and Kerala states. The fruits are used locally in chronic skin diseases.

Andrographis paniculata; **(Burm.f.) Wall. ex Nees**

14. *Andrographis paniculata;(Burm.f.)* Wall. ex Nees Acanthaceae

Na : Hindi: Kalamegh.

Hab : The plant, an herb is half a meter in height. All parts of the plant i.e., leaves, stem, flowers, fruits and roots are bitter in taste and are used medicinally. Flowers are small whitish in colour. Fruits are thin, elongated dehiscing longitudinally. Seeds are small. The plant is found growing along with other plants. After 1 or 2 years growth, the plant dries up. In the ensuing rains, the fallen seeds germinate producing a bunch of seedlings. Leaves contain more of medicinal properties.

Soil and Climate: The plant is found mainly in tropical to sub-tropical regions in central and south India – temperature ranging from up to 40-45 °C. It does not grow in water logged or heavy rainfall areas. It has a very wide distribution.

Propagation: The plant can be raised either by seed or stem cuttings. It is preferable to raise by seed. Seed mixed with 10 times of sand is broadcast on prepared beds. Beds are watered with a rose-can. After about a month or month and half, the seedlings grown on nursery beds are pricked out and transplanted in the field at a distance of 30-45 cm. Transplanting should be done with a regular good shower of monsoon. Since the seed is small, a hectare of plantation requires about 250-300 grams of seed. Rich soils are preferable to grow this crop. The crop is ready in about 4 months when the plant reaches a height of 1 meter. If it is only a rain fed crop, only 1 harvest can be had. If it is irrigated, another harvest can also be had. Application of fertilizers, farmyard manure at about 10 tonnes per hectare in poor soils will greatly increase the yield of crop. When the fruits are showing signs of drying, they should be collected and seed separated and stored for raising future crop. After harvesting the crop, if the plants are irrigated, second harvest can also be obtained. At each harvest about 2-2.5 tonnes of dried crop per hectare can be obtained.

Argemone mexicana **L.**

Prophylactic treatment: The entire plant is bitter and as such it is not palatable and is also not damaged by pests.

Uses: In household medicine, leaves and stems are powdered and mixed with cardamom and cloves and small quantities of sugar to alleviate bitter taste and given for stomachic, diarrhoea and fever. The drug is also used in the treatment of jaundice and liver trouble. It has good external market.

15. *Argemone mexicana L.;* Papavaraceae.

Na : Sanskrit: Swarnakshira; Telugu: Jerupotu chettu, *Nelarakkisa.*

Hab : It is a common spiny herb with yellow flowers and stems and leaves yield yellow juice. Roots are used in skin diseases by the local people.

Aristolochia bracteolata **Lamk.**

16. Aristolochia bracteolata *Lamk.; Aristolochiaceae.*

Na : Guj. & Hindi: Kidmar; Kanerese: Adumuttadagida; Sanskrit: Dhumra patra; Tamil: Aduthinapalai; Telugu: Gadapaku, Gadida.

Hab : It is a climbing shrub found in Uttar Pradesh, Bengal, Madhya Pradesh and Andhra Pradesh. It is used locally as a purgative and anthelmintic. The roots are used for expelling round-worms. It is used by the local people in skin diseases.

Artemisia maritima L. var. brevifolia Wall.

17. *A.indica* L.; Aristolochiaceae

Na : Arabic & Persian: Zuravande; Hindi and Guj.: Arkmula; Marathi: Spanana; Sanskrit: Iswari; Telugu: Nalla Iswari.

Hab : It is a perennial climber of the dry deciduous forests in Andhra Pradesh and other places. Roots are used in skin diseases in Ayurvedic system of medicine.

18. *Artemisia maritima* L. var. brevifolia Wall.; Asteraceae (Compositae.)

Na : Eng.: Worm-seed, santonica; Hindi: Kirmala; Marathi: Kirmaniowa; Persian: Dirmonab.

Hab : It is found in the low rainfall areas of western Himalayas in Kashmir and Himachal Pradesh at an altitude of 2100 to 2700 m. It is a shrubby aromatic plant about 1 m in height. It is officinal in Indian pharmacopoeia. The flowering tops are used as an anthelmintic and also as the source of santonin-the wormifuge. Artemisia seed is best sown during spring either towards the end of February or early in March in well prepared nursery beds in shallow drills and covered with a thin layer of soil. The seed being minute is mixed with sand before sowing to effect its even distribution. Artemisia seed starts germinating in 8 to 14 days. Transplanting is done with the onset of monsoon at an espacement of 60 to 90 cm each way. Aerial parts are harvested just before the flower buds open since santonin content at that time is high. Content of santonin in stocks is less. Hence they are discarded.

A.vulgaris L.

19. *A.vulgaris* L.; Asteraceae syn. Compositae.

Na : Hindi & Bengali: Nagdona; Marathi: Dhordavana; Sanskrit: Nagadamani; Tamil and Telugu: Machipatri.

Hab : It is a common weed in waste lands. Leaves contain an essential oil. They are used by local people in skin diseases.

20. *Asparagus adscendens* Roxb.; Asparagaceae.

Na : Guj.: Ujilimusli, Dholimusli; Hindi; Safed musli.

Hab : It is a small herb found in the drier tropical forests up to about 1600 m in the Himalayas. The root is used as a demulcent, in diarrhoea and dysentery. It is also used as an aphrodisiac.

Asparagus adscendens Roxb.

21. *A. racemosus*. Willd.; Asparagaceae.

Na : Hindi: Musli; Sanskrit: Satavari; Telugu: Chandamamagadda, Pilli pichara, pillitega; Ramunijatta, Seethamma Jatta.

Hab : It is an ornamental, spiny perennial climber. It is not browsed by cattle. It has fibrous roots. Each plant has as many as 100 or more fibrous roots. As such the plant is called "Satavari" in Sanskrit.

The roots are used medicinally. The leaves are needle like. The flowers are whitish in colour and found in bunches. Fruits are round and green in colour. On ripening they become reddish.

Climate and soil: It grows well in warm and dry climate, ranging from 10-50 °C with moderate rainfall of 1000–2000 mm. It is not fastidious about soil but grows well in well drained red to sandy loam soils.

A.racemosus. **Willd.**

Cultivation: It can be grown either by seeds or tuffts of roots. In tuffts, roots are attached to the bottom portion of the stem. 3-4 seeds are sown in polythene bags filled with fertile soil about 2-3 months before planting. They should be watered in case of dry periods. Germination is about 50-60 per cent and is complete in about a month. In about 1-2 months of germination, seedlings are fit for planting. It is preferable to plant at least 2 month old seedling. Incase of tufft planting, freshly dug plants are used for planting. The tuffts are split vertically. Care should be taken to see that at least 2-3 roots are attached to the bottom of the stem. After a good shower of rain, the seedlings, after removal of polythene covers or tuffts as the case may be are planted at an espacement of 60 × 60 cm in pits of 30 × 30 × 30 cm. Closer spacing can also be adopted. Pits are required to be filled with good soil mixed with well rotted farm-yard manure. After planting, the soil around the seedlings should be well pressed. If it is intended to raise as a plantation crop, the plants may be watered once or twice a week during dry period. In a good density forest, the seedlings find support themselves. In open areas supports are planted nearby.

Prophylactic treatment: Spraying of insecticides, particularly when larvae/ insects are noticed on the plants will greatly help the crop.

Manure: About 10-15 tonnes of farm yard manure and about 30-40 kg of potash per hectare in less fertile areas will boost up the crop.

Harvesting: Since only roots have medicinal value, at least 2 year old crop is harvested. The roots are to be dug out only when the soil is moist. Otherwise the roots may break and remain in the soil. The tuffts can be used for planting. When they are not required for planting, the adhering soil along with root-bark should be removed and the roots are thoroughly dried in the sun. If the roots are not dried well, they may develop mould. Each hectare of plantation may yield 10-15 tonnes of dried roots.

Uses: Roots are used as nerving tonic in debility and as aphrodisiac. Even in raw or cooked form or soup, the roots are eaten. They are said to improve milking capacity. They are in great demand in western countries.

22. *Atropa acuminata* Royle ex Lindley; Solanaceae.

Na : Eng.: Indian belladonna; Hindi: Sagangur. Indian belladonna (*Atropa acuminata* Royle ex Lindiley) differs from the European belladonna (*Atropa belladonna* L.) by its more acuminate leaves and yellowish green flowers. The flowers are yellowish-purple in European, belladonna.

Distribution: Belladonna grows in the Western Himalayas at an altitude of 1800 to 3300 m from Kashmir to Shimla.

Parts used as drug: The leaves, flowering tops and roots are used in medicine in various forms. They have a fair market in India and abroad. The drug is officinal in Indian pharmacopoeia. The leaves, flowering tops and roots are used in the manufacture of belladonna tincture and plasters.

Cultivation: In India the drug has been collected mostly from wild plants, although small-scale cultivation has been attempted from time to time. Recently, however, its cultivation is being tried in Kashmir State. Its cultivation on an experimental basis has also been successfully carried out at Chakrata (UttaraAnchal).

Atropa acuminata
Royle ex Lindley

Climate: The plant thrives in temperate regions having less rainfall. When grown in sunny situations, the plants are said to be richer in alkaloidal content than those in shade and also they produce more leaf.

Soil: It requires loamy soils with adequate subsoil drainage, supplemented by abundant balanced nitrogenous manuring organic or inorganic. In general, well-drained loams suitable for growing vegetables are good for belladonna cultivation. In ill- drained soils a large number of plants die.

Nursery and Transplanting: About 150 grams of seeds for a hectare of plantation may be sown in beds having well-pulverised soil in February-March. When the seedlings are 10 to 12 cm high, they are transplanted in rows 50 cm apart, with a spacing of 50 cm between the plants in a row. With this spacing, about 40,000 seedlings will be required for a hectare of plantation.

Habit of growth: It flowers from July to October and fruits in November after which the sub-aerial portion dries. The rootstocks throw fresh shoots in the ensuing spring.

Harvesting and Preparation: If growth is satisfactory, a small crop of leaves may be cut about 15 cm above ground in the autumn of the first season but the first main crop is collected during June in the year following in which the seed was sown. Usually, a second crop of leaves can be collected in August and sometimes the third late in September. In bad seasons only 2 crops are available. The larger stalks may be rejected as they are poor in alkaloids. The plant being perennial can be cropped in the same manner for a second and third summer. The collected herb must be carefully handled and dried rapidly in the shade in order to get the maximum alkaloidal content in the crude drug and also to retain its colour. On an average 700 kg of leaves per hectare per annum can be obtained under favourable conditions. Belladonna plants are at their best upto third or fourth year after

A.belladonna L.

which the roots are also dug up cleaned and dried as they are also important medicinally. The yield of dry roots may be from 180 to 225 kg per hectare at the end of 3 years growth. The leaves should be stored in the dark and in air-tight containers. Drying must be done quickly to avoid deterioration.

Standard: Dried leaves or leaves and other aerial parts should contain not less than 0.3 per cent and roots not less than 0.4 per cent of the alkaloids calculated as hyoscyamine.

23. *A.belladonna* L.; Solanaceae

Na : Eng.: Deadly night shade, Belladonna; Hindi: Sagangur.

Hab : It has been introduced into India and is being grown in the temperate regions in the Himalayas. The roots, leaves and flowering tops are used as a sedative, stimulant and antispasmodic and also in the manufacture of tinctures and plasters. Cultivation and processing details are similar to *A. acuminate* Royle ex Lindley.

Balanites aegyptiaca **(L) Delie Tree**

24. *Balanites aegyptiaca* (L.) Delile (Syn. B.roxburghii Planch.); Zygophyllaceae.

Na : Guj.: Regorea; Hindi: Hingar, Hingot; Marathi: Hinganbet; Sanskrit: Ingudi; Tamil: Nanjunda; Telugu: Gari.

Hab : It is a small spiny tree found in tropical regions in the country. The bark and oil from the seeds are sometimes locally used to treat cuts and wounds and other skin diseases. Local people use the roots in birth control measures.

Balanites aegyptiaca **(L)** ***Fruits***

25. *Berberis aristata* DC.; Berberidaceae.

Na : Beng.: Darhaldi; Eng.: Indian barberry; Hindi: Dar-haldi, Rasaunt, Kashmal; Mar.: Daruhald; Sans.: Daruharidra.

Hab : It is a spiny shrub found in the North West temperate Himalayas from 1800 to 3000 m and also in the Nilgiris. The roots yield an extract called Rasaunt; Rasavanti, Rasanjin. The active principle is berberine, which can be obtained in the form of its salts. It is used as tonic, alterative and stomachic. It is also used locally in the treatment of ophthalmic diseases.

Berberis aristata **DC**

26. *B.lycium* Royle; Berberidaceae

Na : Garhwal: Kirmora; Hindi: Darhaldi, Rasaunt; Jaunsar: Chatroi; Simla: Kasmal.

Hab : The plant is found in the North Western Himalayas in the temperate regions. Its properties and uses are the same as those of B. aristata DC. It is also used for piles, as a tonic and laxative.

27. *Boerhavia diffusa* L.; Nyctaginaceae.

Na : Bengali: Punurnava; Guj.: Vakhakhaparo; Marathi: Tambadivasu; Tamil: Mukaratee-Kirei; Hindi, Sanskrit and Telugu: Punernava.

Hab : It is a trailing herb with opposite leaves. One of them is larger. Fruits are slimy. Decoction of the plant is used in Kidney trouble.

Boerhavia diffusa L.

28. *Bombax ceiba* L. (Syn. *Bombax malabaricum* DC.; *Salmalia malabarica* (DC.) Schott & End.); Bombacaceae.

Na : Eng.: Silk cotton tree; Hindi: Semul; Telugu: Burgu.

Hab : It is a tall and prickly tree with whorled branching found throughout tropical forests in the country. The roots are said to possess tonic, aphrodisiac properties. They are ground and eaten or taken with milk.

***Bombax ceiba* L.**

29. *Boswellia serrata* Roxb. ex Colebr.; Burseraceae.

Na : Bengali, Hindi & Marathi: Salai; Eng.: Indian frankincense; Kanerese: Madi; Sanskrit: Kunduru, Sallaki; Tamil: Parangisambrani; Telugu: Anduk, Andugu.

Hab : It is a large tree found practically throughout the tropical forests in India. The gum-oleo-resin, obtained on tapping the bark, is used locally in rheumatism.

***Boswellia serrata* Roxb.ex Colebr. Branch**

30. *Caesalpinia bonducella* Flem. Syn. C.crista L. Caesalpiniaceae.

Na : Arabic: Akhit makit; Beng.: Nata; Hindi: Kat, karanj, karanju; Kanerese: Gajagakai; Malayalam: kazanchik-kuru; Marathi: Gajaga, sagargota; Persian: Khayade-ibilis; Tamil: Kazhichikey; Telugu: Gachakaya.

Hab : It is a prickly much branched shrub found wild in the hot and drier parts in the country. The powder of the seeds mixed with black pepper is used locally as a tonic and febrifuge. The bark is also employed as a tonic.

***Caesalpinia bonducella* Flem. Plant**

31. *Canarium strictum* Roxb.: Burseraceae.

Na : Bengali, Gujrathi, Hindi: Kaladamar; Eng.: Black damar; Kanerese: Halemaddu; Malayalam: Karuttakungiliyam; Marathi: Dhup, Raldhup; Tamil: Karapu; Kongilium, Karinkunthirikkum; Telugu: Nallarojanamu; Trade: Black Dhup, Indian white Mahogany;

Hab : It is a tree found in South India particularly in Karnataka. The resin which is obtained by tapping the tree is used as a substitute for Burgundy pitch. It is used for medicinal plasters.

Caesalpinia bonducella* Flem. *Fruit and Seeds

***Cannabis sativa* L**

***Carum carvi* L.**

***Cassia alata* L.**

32. *Cannabis sativa* L., Cannabinaeae.

Na : Bengali, Gujrathi and Hindi: Bhang, Charas, Ganja, Jia, Siddhi; Kanerese: Bangi; Sanskrit: Bhanga, Vijaya; Tamil: Ganja, Bangi; Telugu: Ganjai.

Hab : It is a small herbaceous plant found wild in some places particularly in the foot hills of Himalayas. The extract of leaves is used in dandruff. It is also said to promote growth of hair and prevent hair-fall.

33. *Carum carvi* L.; Umbelliferea.

Na : Eng.: Caraway; Hindi: Shia zira, Zira.

Hab : It is a biennial herb 30 to 90 cm high with small white flower borne in dense umbels. Fruits 3-7 mm long and 1.5-2 mm in dia. The species is found naturally growing in the north Himalayan regions but economic collection of fruits is not feasible due to its scattered occurrence in inaccessible areas. The caraway plant requires a dry temperate climate. It is not fastidious about soil but the yield is higher when grown on medium light soils which are rich in organic matter. The crop can be raised either by direct sowing or by transplanting. However, the performance of direct sown crop is better. The seeds mixed with 10 times of sand, are sown in early spring in hills either by broadcast or in a row 30 cm apart. It is cultivated as a summer crop in the hills and winter crop in the plains. Germination is more than 60 per cent and is complete in about a month. In case of congestion the seedlings are thinned out and the final espacement is kept at 30 cm. About 3.5 kg of seed is required for raising one hectare of plantation. Weeding is carried out as and when required. Towards the end of October when winter sets in the leaves start browning and ultimately all sub-aerial parts die before snow-fall in December. But with the commencement of spring they sprout and flower. In summer months periodic irrigation may be necessary. Fruits ripen in early June. Reaping is done just before the seeds are fully ripe. Dry fruits fall and are lost during reaping. It should be harvested early in the morning when the dew is still on the plants. The plants are dried and the fruits threshed out. Cleaned and stored in bags. The yield of seed is variable and may be about 675 kg per hectare. The seed is officinal in Indian pharmacopoeia. According to the Indian pharmacopoeial limit seed should not have less than 3.5 per cent of oil on zero moisture basis. Caraway sold in Indian markets is mostly imported. Both caraway and its oil are used as mild stomachic, carminative and diuretic. Caraway water is used in the treatment of flatulent colic. The oil is used to correct the nauseating and griping effects of some medicines. It is also used in treatment of scabies in admixture with alcohol and castor oil.

34. *Cassia alata* L.; Caesalpiniaceae;

Na : Eng.: Ring-worm cassia; Hindi: Dad mardan.

Hab : It is a small shrub found in West Bengal and Tamilnadu. The leaves are used locally for ring-worms and skin diseases.

35. *C.fistula* L.; Caesalpiniaceae:

Na : Bengali: Sundali; Amultas; Eng.: Indian labernum, Golden shower; Gujrathi: Garmala; Hindi: Amaltas; Kanerese: Kakke; Marathi: Bhava; Sanskrit: Suvarnaka, Rajataru; Tamil: Konnei; Telugu: Rela.

Hab : It is a beautiful ornamental plant with yellow flowers and is found abundantly in the tropical forests particularly in the drier regions. Pulp of the fruit, root bark seed and leaves are used as a purgative.

C. fistula L.Tree

C. fistula L. *Inflorescense*

36. *Catunaregam spinosa* (Thunb.) Triveng. *Syn. Randia dumetorum*; *Gardenia dumetorum; Roxb.*; Rubiaceae

Na : Assamese: Gurol, Behmona, Mona; Bengali & Hindi: Mainphal; Guj: Mindhal; Kanarese: Kare,

Banegara, Mangari; Khasia: Diengmakasing-khlaw; Kashmir: Kirkla, Kokoa; Malayalam: Kara; Marathi; Ghela, Peralu, Mindhal, Wagatta; Oriya: Palova; Sanskrit; Madana; Tamil: Marukkalankay, Madkarai; Telugu; Manga.

Hab : It is a shrub or a small thorny tree found in the sub-Himalayan tract, Maharastra, Gujarat, Uttar Pradesh, Madhya Pradesh and Andhra Pradesh. The pulp of the fruit is said to be valuable emetic and is used as a substitute for ipecac. It is anthelmintic and is also used for dysentery.

Catunaregam spinosa (Thunb.) Triveng. Branch and fruit

37. *Cedrus deodara* Loud. (Syn. *C.libani* Barrel var. deodar Hook. f. Pinus deodara Roxb.); Pinaceae.

Na : Eng.: Deodar; Hindi: Daru, Deodar; Sanskrit: Devadaru, Deodaru.

Hab : It is a tall tree extensively found in the temperate regions of Central and Western Himalayas. Oil obtained from the wood is used locally in skin diseases and for ulcers. (Pl. see essential oils).

Cedrus deodara Loud.

38. *Celastrus paniculatus* Willd.; Celastraceae.

Na : Sanskrit: Jyotishmati; Hindi and Bengali: Malkanjri; Telugu: Teegapalleru; Malaria teega.

Hab : It is a perennial herb found in dry tropical forests in Andhra Pradesh. Fruits ripen in July/August. Seed oil is used as a mosquito repellent and is also said to be useful in mental disorders.

39. *Centella asiatica* (L.) Urban (Syn. Hydrocotyle asiatica L.) Apiaceae.

Na : Beng.: Tholkhuri; Eng.: Asiatic or Indian pennywort; Hindi: Brahmi, Brahma manduki; Mar.: Karinga, Karivana; Sanskrit: Mandukaparni; Tamil: Vallarei; Tel.: Brahmi, Saraswataku.

Celastrus paniculatus **Willd. - Plant with fruits**

Celastrus paniculatus **Willd. Seeds**

Centella asiatica **(L.) Urban**

Hab : It is a creeping faintly aromatic herb found wild throughout India particularly in moister places. It is said to be useful as an antidote against cholera, as diuretic alterative, tonic, leprosy and also in insanity. It is also said to improve memory. The whole plant is used either in a powder form or in the form of infusion 5 to 10 grams of powder is taken thrice daily. In larger doses it is said to have some side effects. An essential oil can be extracted with alcohol. The oil is said to promote growth of hair.

40. *Cephaelis ipecacuanha* (Brot.) A. Rich.; (Syn. Psychotria ipecacuanha Stokes); Rubiaceae

Na : Eng.; Ipecac.

The drug ipecacuanha consists of dried roots or rhizomes of *Cephaelis ipecacuanha* (Brot.) A Rich. The roots are used as an emetic and expectorant and also in the treatment of amoebic dysentery and pyorrohea. The active principle is emetine. The plant is indigenous to Brazil and is cultivated in India at Mangpo near Darjeeling in West Bengal and in the Nilgiris. Its cultivation on experimental scale has also been undertaken at Nongpoh Umsaw (Assam) by State Forest Department. It is also cultivated in Malaya and Burma. Many strains of ipecac are known and named after the places of their origin. Of these, best drug comes from Motto Grasso. The plant is a low straggling shrub up to 30 cm high. The underground portion consists of a smooth rhizome bearing young smooth quadrangular roots and older annulated ones. Stem is angular, Leaves are few. Flowers are white in terminal solitary pendunde heads. Fruits are dark purple clustered berries, each bearing two seeds.

Cultivation: lpeacac thrives in shady situations in humid tropical climate with an annual rainfall of 2500-5000 mm preferably distributed during monsoon, later part of cold months and, spring. Absence of frost and a temperature ranging between 10° to 38 °C is most suitable. Ipecac grows best in more or less virgin sandy loam forest soils rich in humus, potash, manganese and lime with loose texture and especially treated with leaf mould and organic manure in proper proportion. At Mongpo it was raised at an elevation of 300-600 m. and in Assam at 100 m to 800 m.

Clearing of land: The site for ipecac cultivation should be more or less a level ground, if possible in a virgin forest land having a light loam and sandy soil rich in leaf mould, near or along river beds and between an elevation of about 300 and 400 m. preferably protected from storm and wind. The place must be easily accessible. After extracting timber and fuel the soil is given a good burn. The land is then slightly levelled to form terraces generally of 3.6 m width.

Nursery and preparation of soil: In the centre of each terrace, the ground is thoroughly trenched two feet deep and all stones and roots dug up. Then a MACHAN is built facing north-1.65 m high at one end and 75 cm at the lower end. The nursery is divided into plots, locally called KAMRAS of 1.8 m × 1.8 m. The raised beds are protected along the sides with pieces of timber. The roof of the MACHAN is then covered with a thick layer of thatch or split bamboos or veneer waste. The covering must be thick and strong enough to prevent damage from storm and rain-drip for three years otherwise the seedlings and even mature plants damp-off. The shed for ipecac should have its higher end towards the East or the North. The beds are then raked over and about four to six basket loads of newly collected leafs and one basket of sand are added to each KAMRA and dug in. The amount of leaf soil added depends upon the type of soil. The beds are then levelled off and thoroughly watered. After a few days the soil is raked over and a further layer of leaf soil placed about 2 to 3 cm thick on the top.

Seed sowing: The beds are now ready for seed-sowing which can be done from January to March depending upon the climatic conditions. Ipecac seeds have to be sown under drip proof thatched sheds. Seeds should not be buried deep under the soil. At Mungpo and Rongo in the district of Darjeeling in West Bengal 15th January to 15th February is considered a good time for sowing. About 85 to 140 grams of seeds are sown in a KAMRA of 3.24 sq.m. The sowing is done either by broadcast or in drills. During the process of germination, which takes four to five months, very careful watering is necessary and filamentous algae and fungi, slime-mould and mosses and germinating weeds of flowering plants, and ferns have to be removed. After sowing, covering the nurseries with a layer of pure fine sand 0.5 cm thick prevents the growth of algae and moulds. Germination trials conducted in Assam (Forest department) show that ipecac seeds need a post ripening period of about 40 to 60 days. Ipecac seed is sown in shaded beds during February obtaining germination up to 50% till the middle of July. Best yield of seed comes from 3-year and older plants. It has been observed in Assam trials that the seeds from 1 hectare of beds (in the third year) can produce plants only for about 2.7 hectare of new area. After germination, seedlings are transferred to another nursery prepared on similar lines, but with double the amount of leaf soil and sand added. The seedlings are planted two inches both ways. This is done in July-August and the seedlings are tended and watered carefully till March-April the following year when they are again transferred to another prepared nursery and planted with a 12.5 cm - 15 cm spacing both ways, where they will remain till they are dug out. Here again they are carefully watered and after the first year of transplanting a top dressing of well-rotted cow manure and sand is given annually. It may be given twice a year if manures are at hand. With the plants spaced 12.5 to 15 cm each way they are allowed to grow for three years (or four years from sowing the seeds) and then dug out. The usual weeding and attending to the beds are continued till harvesting time. The proper and careful application of water to the plants at different stages is a very important factor as over watering or water-logging may prove fatal to them. Another vital factor is exposure to sunlight. The leaves of the plants get etiolated and look yellowish and unhealthy when they get exposed to sunlight. Nitrogen yielding plants are to be grown alongside the beds to protect the KAMRAS from sunlight as ipecac is primarily a shade loving plant. At the end of the first year (12.5 cm. x 12.5 cm or 15 cm. x 15 cm planting) all seed is collected. During the second and third years, layers are taken and more

Cephaelis ipecacuanha **(Brot.) A. Rich**

seed is collected. Just before the final uprooting (i.e., in March after three years in the nursery) the layers are removed and all cuttings are taken out when the radix is uprooted. The radix is then thoroughly washed and dried, which will take two weeks. Production of dry radix from a KAMRA of 1.8 m × 1.8 m varies from 0.7-0.9 kg. and when grown on a large scale the average yield varies from 0.3 to 0.7 kg. The yield is highly variable. Trials conducted in Assam gave a yield of air dried roots from 400 kg to 1900 kg per hectare depending on edaphic conditions and the care that is bestowed.

Methods of propagation: It is advisable to grow ipecacuanha plants from seeds. The germinating power of the seed is rather limited and varies mostly from 20 to 30 per cent. Experiments and experience prove that vegetative methods of propagation give good results. In fact ipecacuanha plant can easily be propagated from – 1. root and stem cuttings 2. layering and 3. matured but not old leaf cuttings. The general orthodox practice followed in vegetative propagation is described:

(a) The root cuttings and leaves should be placed in pans not less than three inches deep containing very fine sand.

(b) The plants must be kept constantly damp but never saturated with water.

(c) The pans containing the plants must be kept under cover where they will have free ventilation but not exposed to droughts and protected from the direct rays of the sun and drip of water. Under favourable circumstances the cuttings from the tops and shoots may be expected to form callus in two to three weeks and will root and will be ready for putting off in another three weeks. The layers may be expected to form callus in about a fortnight to three weeks. After this a mass of fibrous roots grows out from the callus and then two, three even four young shoots make their appearance above ground. When these shoots are about 2.5 cm or 4 cm in height they may be separated and potted off. The root cuttings should be buried just beneath the surface of the sand and kept damp but not saturated with water. The plants raised from these cuttings are much slower in their development than those derived from either the shoots or leave. The young plants may be potted off when they are about 2.5 cm or 4 cm in height into pans not less, than 12.5 cm deep. These pans may have a layer of broken porous gravel 2.5 cm deep placed at the bottom. The gravels may be covered by a thin layer of coconut refuse, half-rotted leaves or some such material which prevents the soil being washed down etc. The pans should be filled with a mixture of two parts of thoroughly rotted leaf mould one of sand and one of old manure. The plants may be put out into the beds when they are about 10 cm to 15 cm high. The ipecacuanha is an unusually slow growing plant and while its requirements in regard to soil, temperature, moisture and shade are known to a certain extent, it must be acknowledged that hitherto the difficulties faced in its successful cultivation in India have deterred many from taking up as a commercial enterprise. In short, great skill is needed to grow the plant successfully.

Regarding the application of inorganic fertilizers, trials have been conducted in Assam by State Silviculture Division. Variable results were obtained. However it was observed that depending on the soil conditions, nitrogen and phosphorous fertilizers in combination with humus have a much favourable effect on root yield.

Storage: Prepared ipecacuanha roots should be kept in a well closed container and stored in a room which is cool and dust free, insect and fungus proof.

Diseases and Pests: Special care should be taken in the prevention and control of fungus and insect pests. The small red ants are an enemy of the plants and they build nests in the loose beds of soil and damage the roots of seedlings and mature plants. Mechanical and chemical means should be adopted to effectively check the pests. The soil is often found infected with Fusarium, Rhizoclonium and protonema. The cultivation trials conducted in Assam State indicated large scale mortality due to a wilting disease caused by a Fusarium species. Application of chestnut mixture, sprayed twice a month during rainy season was found suitable with much success in the germination beds. It will be clear from the method of cultivation of Ipecac that such a plantation is very costly. One of the most important items of expenditure is watering the nursery beds and soil working for which the best method of controlling the cost is to engage labour permanently. The other important item is the cost of preparing the beds by bringing humus soil from the forests, hoeing the beds deep, edging them with bamboos so that the bed surface stands some 30 cm above the ground level and provision of a thatched shed. If sheds are of thatch then the cost of shed making in experimental beds may be more. If the beds are made of split bamboos, locally known as "tarzas" in Assam then the cost of the bed may be less. It is found that seedlings found in one germination bed can be transplanted in about 7 beds. As a safe estimate, it may be taken that one germination bed is enough to plant up 4 transplanting beds and yet leave enough seedlings in the germination bed itself to give full stocking. Thus per square metre nursery cost can be spread over five square metres of plantations in all. For an economic unit of plantation with the staff and capital expenditure Ipecac beds will have to yield on an average, 150 gms of air dry roots per square metre, or 1500 kg of dry roots per hectare of net bed area. This target is well within the limits of achievement if the beds are prepared by filling them with humus soil, and ammonium sulphate and single superphosphate at 250 kg per square metre of bed area forked every year in April.

41. *Chlorophytum borivillianum*; Orchidaceae.

Na : Guj.: Gholi-musli, Janjaria; Hin.: Safed Musli; Mal.: Shedeveli; Mar.: Kuli, Safed Musli; San.: Musali; Tam.: Tanureta; Tel.:Kushelli.. Tubers are white and hence it is called as Safed mush.

Hab : Tubers available in the market are normally collected from hilly areas in tropical to sub-tropical regions in the country. The small seedlings of musli are found in forests during rainy season.

Botanical description : Leaves look like that of garlic but are slightly yellowish. Flowers are white with six petals arranged on the flowering stalk which emerges from the centre of the plant. About 2000 flowers appear on the flowering stalk in July. The seed is very small and black.

Geographical distribution : The plant is noticed in the sub Himalayan regions, Satpuda, Vindhyas, Arawali and in hilly areas of Bihar and Assam. It is also found in some parts of Maharashtra and Andhra Pradesh.

Soil and Climate : The loamy, gravelly soil with rich organic matter and well drained soil with pH ranging between 6.5 and 7 is suitable for growth. Good amount of soil moisture during growing season favours root development. It can survive at low temperatures. It can be grown in warm, humid climatic conditions. Temperature between 25 and 35 °C and 50-75 cm annual rainfall is suitable for its growth.

Propagation : Land preparation - The land should be ploughed and harrowed so that there should be proper aeration. Weeds like *Cyperus rotundus, Cynodon dactylon* should be killed completely. After land preparation, the bed, 90 cm broad and 15 cm in height should be prepared. Between two beds there should be water channels to drain out excess water. Well decomposed farm yard manure at the rate of 15-20 tonnes per hectare should be mixed in the soil properly before planting or the vermicompost can be used. Due to the use of organic manures, the water holding capacity of the soil increases.

Chlorophytum borivillianum

Mode of propagation : It is propagated by seed and tubers. The germination percent by seed is very less. It takes about 18 months for harvesting tubers, if raised by seeds. Hence the sprouted tubers are economical for planting. It takes about 6 months to mature in rainy season. Previous year's tubers are used in propagation.

The bunch of tubers should be taken out from soil in May and tubers are separated from the bunch in such a way that some portion of the crown remains attached to each tuber. The number of tubers varies from plant to plant and on an average there are 5-30 tubers per plant. Tubers separated should be stored in gunny bags filled with mixture of fine sand and silt under moist conditions under shade. In June, tubers sprout and these are used for planting. Planting is done with the onset of monsoon, preferable in June/ July. The soil should be moist at the time of planting. The planting should be done at a spacing of 6-8 cm between tubers while the distance between the rows maybe about 30 cm. Tubers should be planted in the soil and soil around should be pressed firmly. Nearly 6-7 quintals of tuber is required for 1 hectare of plantation.

Irrigation : Irrigation should be given as and when required during growth period to avoid any water stress. Sprinklers may be used for irrigation purposes. Weeding should be done as and when required. As the plants of musli look like grass, there is possibility of them getting removed instead of grass. However, the bed should be kept weed free.

Harvesting and Storage : Tubers are ready for harvesting after 6 months of planting. Harvesting is done in December/January when the matured leaves start yellowing and drop down. This is the best indication for harvesting tubers. The tuber bunches are removed by digging the soil. Bunches removed from soil should be thoroughly washed with water. Such clean bunches are dried in shade for 2 days. The skin/bark on the tubers should be removed by pressing the tubers by holding them between 2 fingers. Such peeled mush should be again washed with water and dried completely in shade. The dried musli is then marketed. Peeled tubers are not suitable for planting.

Yield : The yield of fresh tubers is 3-3.5 tonnes per hectare. (Dried tubers is 0.6-0.65 per hectare)

Chemical constituents : Tubers are of great medicinal value containing the steroid of sapogenine (1-2%), proteins (10-12%), calcium to some extent and some water soluble minerals. Tubers are fat free.

Medicinal uses : It is useful in *pitta and vata.* It is explained as *vrishya, dhatuvardhak, madhu, rasayana,* slightly bitter, healthier and energy giving in Ayurveda. It is aphrodisiac. It is effective in fatigue and weakness. It is also useful in blood purification and in certain diseases like renal calculus, leucorrhoea. It is lactating.

Cinchona calysaya Weld

Marketing and Trade : Safed musli roots and formulations are sold in the market. Nearly 35% supply of the herb has been fulfilled by wild resources (forests) of Maharashtra and M.P. while 45% from cultivated sources. The market of Safed musli and its products is highly volatile. The economics of cultivation is subject to fluctuations depending upon time, place and market trends.

42. *Cinchona calysaya Weld.*; and

43. C. *ledgeriana* Moens; Rubiaceae.

Na : Eng.: Quinine; Hindi: Cinchona, Quinine.

Hab : C. calysaya and C. ledgeriana are exotic species being grown in India, in Darjeeling district in West Bengal, Annamalai hills, Moyar valley in Nilgiris, Tinnevelly district in South India Khasi and Jaintia hills. Former is a large and stouter tree whereas the latter is a small tree. The species thrive in high rainfall areas of 2500 mm to 3700 mm with high atmospheric humidity and temperature ranging from 16-23 °C. *C. ledgeriana* Moens. is considered by some authorities to be a variety of *C. calisaya* Weld.

Propagation: Cinchona is propagated either by seeds or by vegetative methods. However in vegetative propagation selected strains can be ensured. Nursery raised seedlings when they are about 12 to 18 months old and 45 to 60 cm high are planted with the commencement of summer monsoon. The espacement adopted is normally 2 × 2 m. Younger plants require shade. Cinchona plants are prone to damage from insect pests and diseases. Hence some prophylactic treatment may be necessary.

C. ledgeriana **Moens**

Harvesting: Harvesting starts from 4th year. Maximum alkaloidal content is reached when the plants are 10-12 years old. Bark is removed from the felled trees and the cut branches. The stumps are coppiced and they throw out new shoots which will form the 2nd crop. During the process of thinning and cleaning operations also some bark is obtained. The harvested bark is dried in shade. The most important alkaloid of cinchona is quinine. The alkaloids exist in the form of salts. Greater part of the production of bark is employed in the manufacture of quinine. The bark of the stem of C.ledgeriana contains 5.79% of total alkaloid, quinine being about 4.14%. Root bark is rich in alkaloids while the bark of the branch and twig has less. C. calysaya has also about the same alkaloidal contents.

Uses: The most important use of quinine is for the treatment of malarial fever cough and cold but it is stated to have many side effects. The derivatives of quinine are also used in the preservation of fur feathers, wood, textiles felts etc. New and more effective anti-malarial drugs have been synthesised and as such the demand for cinchona has gone down.

44. *Cinnamomum zeylanicum* Bl.; Lauraeeae.

Na : Eng.: Cinnamon; Bengali, Gujrathi, Hindi, Kanerese and Telugu: Dalchini; Tamil: Caunalavanga pattai; Sanskrit: Tamulapatra.

Cinnamomum zeylanicum **Bl. Plant**

Hab : It is found in the Western Ghats in Karnataka, Kerala and also in Tamilnadu states.

The oil obtained from the stem bark is used in medicine as a carminative, antiseptic and as astringent against cold and diarrhoea and as a flavouring agent in medicines. The plant flourishes best and produces the finest quality of bark when it is growing on white sand overlying a good rich sub soil along the coast. It grows well in warm sunny climate with an average temperature of about 27 to 29°C and annual rainfall of 2160 to 3800 mm without long spells of dry weather. Cinnamon being a shade bearing species, shady places are preferred for raising plantations. The seeds do not retain viability for long and should be sown immediately after collection. The fruits ripen in May.

Bark

The seeds are sown in nursery beds immediately after collection. The nursery beds should be in shady situations. Germination occurs in 2 to 4 weeks. Usually one year old seedlings are transplanted in the field in pits at the onset of monsoon. The spacing adopted is about 2.5 × 2.5 m. The plantation is weeded 2 to 4 times a year. Saplings of about 3 years old are cut down to a few inches above ground level. Four to six shoots are allowed to grow from the stump and are kept straight by pruning. These shoots reach the exploitable size about 3 years later. Only young and tender shoots are harvested (1.25 to 6.5 cm. in diameter) since the bark from the older shoots is coarser and is of an inferior quality. The best time for cutting the shoots is the rainy season. The outer bark is carefully scraped off without injuring the valuable inner bark. A longitudinal incision, 15 to 20 cm long is then made into the soft, white inner bark, which is then carefully peeled off the hard core. Thus hollow tubes or quills which are 15 to 20 cm long are obtained, into which smaller or broken pieces are packed. Finally several of these quills are telescoped together to form tubes about 90 cm long. These are rolled by hand while fresh and soft and are very slightly pressed. The rolled quills are dried for about three days in the sun and for a similar period in the shade. During this period they are rolled and slightly pressed repeatedly or they will swell up and split. During the drying operation the original white colour of the quills turns to a yellowish brown. The quality of the bark improves with the age of the root stock and with the pruning. The best quality bark is of a uniform light, brownish yellow colour and has a characteristic delicate and sweet flavour. Small pieces and debris obtained during the preparation of quills are called "quillings" or featherings. Shavings consisting of the outer bark and particles of the inner bark are known as cinnamon chips. A hectare of plantation yields about 170 to 225 kg of cinnamon quills about 63 kg of cinnamon chips from 3rd coppicing cycle onwards. In the first two coppicing cycles the yield is low. The bark and leaves on distillation yield essential oil. Chief constituents of bark oil are cinnamaldehyde and eugenol. The chief constituent of the leaf oil is eugenol.

***Cissampelos pareira* L.**

45. *Cissampelos pareira* L.; Menispermaceae.

Na : Bengali and Hindi: Akanadi; Eng.: False pareira root; Guj.: Venivel; Kanerese: Padavali; Malayalam: Kattuvalli; Marathi: Paharvel; Sanskrit: Ambashtha; Tamil: Appatta; Telugu: Advibankateege.

Hab : It is a slender climber commonly found in the Punjab, Uttar Pradesh, Andhra Pradesh and other drier regions. The roots are used in indigenous medicine in diarrhoea, dysentery, cough and urinary troubles.

46. *Citrulus lanatus* (Thumb.) Mansf. Syn.*C. colocynthis* (L.) Schrad; Cucurbitaceae.

Na : Bengali & Hindi: Indrayan, Makal; English: Colocynth, Bitter apple; Gujrathi: Indrak, Indranan, Indravana; Kanerese: Pavamekkekai; Malayalam: Peykommutti; Marathi: Kadu vrindavana, Indrayan; Sans.: Mohendravaruni; Tamil: Peykkumitti; Telugu: Eriputccha; Urdu: Indrayan.

Hab : It is a trailing herb found in Northern India, Madhya Pradesh, Gujarat and South India. Fruits are used as a purgative. It is officinal in Indian and British pharmacopoeas. Local people use fruits, leaves and roots for skin diseases.

***Citrulus lanatus* (Thumb.) Mansf.**

47. *Cleome diandra* L.; Cleomaceae.

Na : Telugu: Tellavominta, Yetakura.

Hab : It is a herb growing to a meter in height found in waste lands in dry places. The plant is used by the local people for skin diseases. It is also said to be wormifuge.

***Clerodendrum inerme* (L.) Gaertn.**

***C. phlomidis* L**

***Cocculus hirsutus* L. Diels.**

***Colchicum luteum* Baker**

48. *Clerodendrum inerme* (L.) Gaertn.; Verbenaceae.

Na : Beng.: Banjai, Batraj; Hindi: Lanjai, Sangkupi; Kanerese: Kundali; Nuyitakkali; Malayalam: Nironchi; Marathi: Vanjai: Sans.: Kundali vanaja; Tamil: Anjali; Telugu: Takkolakamu, Etipisinika.

Hab : It is a small herb found wild in the eastern parts of Andhra Pradesh & South India. It is used for medicinal purposes as a febrifuge.

49. *C. phlomidis* L.; Verbenaceae.

Na : Sanskrit: Laghuagnirnantha; Guj.; Hindi & Marathi: Arni; Telugu: Takkali, Takkeda.

Hab : It is a shrub found in dry deciduous forests in Andhra Pradesh and other places. Local people use leaves, roots and stem bark in the treatment of venereal diseases.

50. *Cocculus hirsutus* L. Diels.; Menispermaceae.

Na : Beng.: Huyer; Gujrathi; Vevati; Hindi: Jamtikibel; Kanerese: Sogadiballi, Dusaribali; Tamil: Kattukkodi; Sanskrit: Chilipant, Patalagarudi; Telugu: Dusediteega, Dusariteega.

Hab : It is perennial climber. Leaves and roots are stated to have cooling effect on human system and also said to be used as having aphrodisiac properties.

51. *Colchicum luteum* Baker; Liliaceae.

Na : Eng.: Golden collyrium; Hindi: Hirantutiya.

Hab : It is a perennial herb found in the temperate regions of North Western Himalayas. The bulbs are used as a substitute for exotic *C. autumnale*. The bulbs have great medicinal value and are used in the treatment of rheumatism and gout.

52. *Commiphora caudata* (Wight & Arn.) Eng.: (Syn. *Balsamodendron caudatum* Wight & Arn; Burseraceae.

Na : Eng.: Hill mango; Kanerese: kondamavu: Tamil: Kiluvai; Telugu: Kondamamidi.

Hab : It is a tree or shrub with thick trunk and papery bark found in Andhra Pradesh, Tamilnadu, Karnataka and Kerala. The gum-oleo-resin obtained from the bark is used in medicine and embalming.

53. *Coptis teeta* Wall.; Ranunculaceae.

Na : Assam & Bengal; Tita; Beng.: Gold thread; Guj.: Haladiovachnag; Hindi: Mamira, Mamiran.

Hab : Plant is found in eastern Himalayas in Assam. Rhizome is a tonic and stomachic. It is found useful in fevers and debility.

54. *Corallocarpus epigaeus* Benth. & Hook. f.; Cucurbitaceae.

Na : Hindi: Akashgaddah, Rakasgaddah; Kanerese: Akashagarudagadde; Tamil: Akashgarudan; Telugu: Nagadonda.

Hab : It is a climbing herb found in the Punjab, Gujarat, Andhra Pradesh and Karnataka. The root is yellowish white marked externally with circular rings. Roots are used locally in dysentery and rheumatism.

Costus speciosus **(Koen. ex Retz.) Sm.**

55. *Cosmostigma racemosum* Wight.; Asclepiadaceae.

Na : Kanerese: Gharahuvoo; Malayalam: Vattu valli; Marathi: Marvel, Shendvel.

Hab : It is a climbing shrub found in the Western Ghats Karnataka and Andhra Pradesh. Leaves of this climber are used locally to cure sores.

56. *Costus speciosus* (Koen. ex Retz.) Sm.; Costaceae.

Na : Sanskrit: Kembuk; Telugu: Kevukinna.

Hab : It is an annual herb found in wet places in tropical forests like Andhra Pradesh and Madhya Pradesh. Local people use the rhizome in birth control measures.

57. *Crataeva adansoni* DC. S.sp. *odora* (Buch-Ham.) Jacobs; Capparidaceae.

Na : Telugu: Uskaman, Ulimidi

Hab : It is common in dry tropical regions. Bark in small doses is stated to be useful for ulcer in alimentary canal, piles and urinary troubles. In higher doses bark is toxic.

Croton tiglium **L.**

58. *Croton tiglium* L.; Euphorbiaceae.

Na : Eng.: Croton; Hindi & Marathi: Jamalgutta; Gujrathi, Kanerese & Telugu: Nepala; Malayalam & Tamil: Nervalam.

Hab : It is a small ever green plant found as an escape in Wastelands and forest areas. The seeds are the source of croton oil which is used as a purgative.

59. *Daphne oleoides* Scrrebe.; Thymelaeaceae.

Na : Hindi: Kutilal; Panjabi: Kanthan, Kutilal.

Hab : It is a shrub found in the Western Himalayas particularly in Kashmir. The roots are locally used as a purgative.

Daphne oleoides **Scrrebe.**

60. *Datura innoxia* Mill; Solanaceae.

Na : Hindi: Sada dhatura.

Hab : It is a herb commonly found in the Western Himalayas and South India. Dried leaves are used locally in the treatment of asthma.

Datura innoxia **Mill**

Cultivation: The plant can be grown directly from seeds or by transplanting seedlings. Germination of seeds is slow and irregular and may be hastened by alternate exposure to freezing and thawing to weaken the seed coat. Seeds are sown during spring. About 11 kg seeds are required for sowing one hectare of plantation. Liberal quantities of rotted cow-dung manure are applied to ensure healthy growth. Harvesting is done at the time of flowering when the alkaloid content is at its maximum.

Extraction: The dried leaves and flowering tops of *D.innoxia* along with those of *D.metel* are recognized in the B.P.C. under the name *Datura folium*. The alkaloid scopolamine is readily extracted from powdered material by percolation with acidulated water after soaking in a menstrum of 40% isopropyl alcohol, 48% water and 4% glacial acetic acid. The percolate is concentrated to syrup under vacuum, rendered alkaline with ammonia and refluxed with isopropyl alcohol.

D. metel **L. – plant**

61. *D. metel* L.; Solanceae.

Na : Hindi: Sadadhatura.

Hab : It is a herb occurring throughout India and sometimes grown in gardens. Leaves are triangularovate. Flowers are large white, violaceous or purple. Leaves are used in asthma. Leaves and flowering tops are recognized in British pharmacopoeia under "*Datura folium*". Cultivation and exploitation details are the same as for *D. innoxia* Mill.

D. metel **Fruit**

62. *D. stramonium* L.; Solanaceae.

Na : Bengali, Hindi, Gujrathi and Marathi: Dhatura; Eng.: Thorn apple, Stramonium; Sanskrit: Dhattura, Unmatta, Kanaka, Shivapriya; Kanerese, Malayalam, Tamil & Telugu: Ummatta.

***D. stramonium*. L.**

Hab : It is a common weed growing to a height of about 1 to 2 meters. Leaves are ovate or triangular ovate; flowers are large, white or violaceous. Leaves and seeds are used in indigenous medicine in the treatment of asthma. Chief alkaloids are hyoscyamine, scopolamine and atropine. Its uses are similar to other species of Datura and belladonna. In larger doses the seeds are highly toxic.

63. *Desmostachya bipinnata* (L.) Stapf. (Syn. *Poa cynosuroides* Retz.); Poaceae (Gramineae.)

Na : Bengali & Sanskrit: Darbha, Kasha; Hindi: Dab, Durbha, Durva, Kusa; Kanerese & Telugu: Kushadarbha, Darbha.

Desmostachya bipinnata **(L.) Stapf.**

Hab : It is a perennial grass found in many parts of the country. This grass is locally used in cases of dysentery.

64. *Dichrostachys cinerea* L. wight & Arn.; Mimosaccae.

Dichrostachys cinerea **L. wight. & Arn.**

Na : Hindi: Vurtuli; Kan.: Odatare; Marathi: Segam-Kati; Sanskrit: Veerataru; Telugu: Velthuru; Tamil: Vidottalai.

Hab : It is a thorny shrub or small tree, found in North Western India, Madhya Pradesh, Rajasthan, Andhra Pradesh and Karnataka.

The root is used locally for rheumatism and renal troubles. It is said to dissolve bladder stone.

65. *Digitalis lanata* Ehrh.; Scrophulariaceae.

Na : Eng.: Grecian foxglove; Austrian foxglove.

This plant is known as Austrian digitalis or Austrian foxglove. It is a biennial or perennial herb 60 to 90 cm high. It produces smaller and linear leaves than *Digitalis purpurea* L. and is also less robust in growth. It has high glycosidal content. It is the source of a drug which is used as a cardiac stimulant and tonic. The glycosides found are digitoxin; gitoxin and digoxin, the last one is a highly active crystalline glycoside which has not been found in other species.

Digitalis lanata **Ehrh.**

Distribution: The plant is cultivated in India in Kashmir at Tangmarg and Baramulla nurseries. It is indigenous to Austria and the Balkans and is cultivated in Great Britain.

Propagation: Propagation is done by direct sowing or transplanting. The seed can be sown direct in April in drills about 30 cm apart. About 2.25 kg seed is required per. hectare as it is very small and light. It should be mixed with fine sand to ensure even distribution. Another method which economizes seed is to sow in boxes or in a seedbed. The seedlings are pricked out and subsequently transplanted when large enough to handle preferably with the commencement of spring. Yield of leaf is lesser than that of *Digitalis purpurea* L.

Storage: Leaf should be stored in well-closed containers which prevent access to moisture and are protected from light.

66. *D. purpurea* L.; Scrophulariaceae

Na : Eng.: Digitalis, Foxglove.

It is a hardy biennial plant, which occasionally forms a rootstock and lasts for 3 or 4 years. Flowering shoots, which are 0.9 to 1.5 m high, come up during summer from a rosette of leaves. The leaves are the source of important cardiac medicines. Kashmir leaves have given excellent results in both biological and clinical trials. The dried leaves are officinal in Indian and British Pharmacopoeias. Their action is due to several unstable but very active glycosides, among which digitoxin is the most important.

D. purpurea **L.**

Distribution: The plant is widely distributed throughout Europe and is common in England and naturalised in North America. Not indigenous to India, but it has long been grown in gardens in different hill stations as an ornamental border plant. It was tried in Government gardens Saharanpur and Mussoori for a regular crop of leaves for medicinal purposes. The plant, however, did not flourish. It is growing freely in Kashmir Hills and has done fairly well in Sikkim and the Nilgiris.

Locality factors: This is a plant of the temperate zone. It neither likes excessive moisture nor extreme drought. The soil should be of open texture and well drained and should not have excess of lime. Good results are obtained on loamy soils if a balanced mixture of fertilizer is added. It does well under light woodland shade. In Kashmir, it has done very well under the light shade of Blue Pine, *Pinus wallichiana* A.B. Jackson. The climate of Kashmir seems to be very suited to the growth of digitalis. Most localities from 1500 to 2100 m in elevation with dry climate may be suitable.

Propagation and cultivation: Seeds should be sown in the early spring. About 2 to 2.5 kg seed may be sufficient to give seedlings to plant out a hectare. As the seed is very small and light, it should be mixed with fine sand to ensure even distribution in the nursery bed. Germination takes place within 2 or 3 weeks. Seedlings when they are 5 to 8 cm high are transplanted on ridges towards the end of rains at an espacement of 30 to 60 cm. Pricking out of seedlings before transplanting may be beneficial for getting better out turn of leaf. The land should be well prepared and manured beforehand. The seeds mixed with sand can directly be sown in drills in rows 10 to 40 cm apart. Later the seedlings, which can be transplanted readily in damp weather, are thinned 15 to 25 cm apart in the rows. The plant is cultivated somewhat like tobacco and is generally harvested and dried like belladonna.

Harvesting, storage and yield: Leaves provide the drug. Mature leaves are collected either in the first year or at the time of flowering in the second summer. With a good crop, up to 560 kg of dried leaves per ha may be harvested. The leaves must be dried as quickly and as carefully as possible and packed in airtight containers. Immediate and rapid drying at a temperature of 54° to 60°C may be necessary in order to avoid deterioration of leaves, which sets in rapidly. Drying must be complete as even air-dried leaves lose their potency. The leaves should be stored in dark sheds or in airtight containers.

67. *Dioscorea deltoidea* Wall and

68. *Dioscorea prazeri* Prain & Burkill; Dioscoreaceae.

About 30 species of Dioscorea grow wild in India. Of these only *D.deltoidea* Wall and *D. prazeri* are medicinally important. The tubers of these species are the source of diosgenin a starting material for the preparation of cortisone, Hydrocortisone and for the synthesis of steroid sex hormones e.g. Testosterone, progesterone, pre-gnenolone etc. Recently these hormones have been used as oral contraceptives. Similarly there are certain other exotic species viz. *D.composita* and *D.floribunda* of Maxico and *D. caucassica* of Russia, which are the source of Diosgenin.

Na : ***D.deltoidea*:** Hindi: Psingli mingli; Kashmir: Kildri; Punjab: Kins, Kirta.

***D.Prazeri*:** Lepcha: Kencheong; Hindi: Kurtorul.

Hab : Both the species are tender climbers. *D.deltoidea* occurs naturally in the temperate to sub-alpine regions. In Kashmir it occurs in Baramulla South East of Srinagar, Pahalgaun, and Jammu between Sera punch and Kishtwar. In Himachal Pradesh it occurs in Pangi, Chamba, Bilaspur, Suket, Kangra Distt., Dharmasala, Manali, Phulga (Parvati River), Simla Hills, Kasauli, Narkanda, Thoog, Kanawar and Bashr. In Uttara Khand it is found in Jaunsar hills, Mussoorie, Tehri Garhwal, Jhala (2400 m to 3000 m), Nainital and AImora Districts. *D.prazeri* is indigenous to eastern Himalayas in North Bihar, North Bengal, Abor hills, Naga Hills and Assam upto 1500 m and in Sikkim and Bhutan.

D.deltoidea, because of its commercial and industrial importance is continuously being exploited resulting in the depletion of the stock. It is understood that M/s John Wyeth Ltd., of Bombay are processing Dioscorea tubers. It is generally noticed that the tubers of *Dioscorea deltoidea* contain varying percentages of diosgenin depending on the climatic conditions of different regions. It is reported that the content varies from 2-3% in samples analysed from heavy monsoon areas in the eastern Uttara Khand to 3-8% in samples from Himachal Pradesh and Kashmir. In general, *D. deltoidea* compares favourably with any other exotic species with regards to its diosgenin content. Similarly *D. prezeri* contains 3.3% diosgenin.

Dioscorea deltoidea **Wall**

***Cultivation and exploitation*:** The species grows well under the shade of trees in the forest. It can be raised by planting rhizome cuttings, transplanting nursery-raised seedling and even by direct sowing. In case of rhizome cuttings about 30 to 35 gms of rhizome is planted at an espacement of 50 × 50 cm with the commencement of spring. Similarly in the case of direct sowing or raising seedlings, for transplanting, seeds are sown at the commencement of autumn or beginning of the spring in the field or nursery as the case may be. In the case of direct sowing in the field above spacing may be adopted. In the case of nursery sowing, seeds may be uniformly spaced and sown at about 7.5 cm to 10 cm spacing. If sown before autumn the seeds will remain dormant and germinate only with the commencement of spring. Seedlings are transplanted in the field with the onset of monsoon. Necessary weeding operation will have to be carried out at least twice a year. Any support occurring nearby may not be disturbed. The sub-aerial part of the plant dries with the approach of winter and sprouts again with the commencement of spring. The plant may be harvested after 3 to 4 years growth. The work carried out at Chakrata (Uttara Khand) and elsewhere has shown that the plants raised from seeds take about 3 to 5 years to grow to marketable size of tubers. Those raised from rhizome cuttings show an increase of about 5 to 6 times in weight after growing for about 3 years. A hectare of plantation raised through rhizome cuttings may yield about 5000 kg green weight of the raw materials after growing for 3 years. Considering the low yield of rhizome and price being offered for the raw material, cultivation of the species does not appear to be promising under the present conditions.

D.myoporoides R. Br.

69. *Duboisia leichhardtii* F.V. Muell &

70. *D.myoporoides* R. Br.

The genus Duboisia comprises of three species all endemic to Australia. Leaves of *D. myoporoides* & *D. leichhardtii* are the richest source of mydriatic alkaloids hyoscyamine and hyoscine occurring to the extent of 1.7% much higher than that in *Atropa* and *Hyoscyamus* species. Both the species are medium sized trees reaching to a height of 13 to 14 m though rarely, with leaves broad lanceolate to ovate, glabrous 7.5 to 10 cm long in *D.myoporoides*, smaller and narrower in *D. leichhardtii*. Flowers are small white, bell shaped, slightly bigger in *D.leichhardtii*, borne in cymose panicles in spring. Fruit is a small black globular berry.

Elettaria cardamomum (L.) Maton var. minor Watt. - Plant

Soil and Climatic requirements: *D.myoporoides* thrives on a variety of soils but prefers a rich basaltic type soil. *D.leichhardtii* thrives best on the red soils. In Australia *D.myoporoides* is confined to eastern coast in the moderate rainfall zone and where frost rarely occurs. *D.leichhardtii* is confined to a relatively small area of the high lands of south eastern Queensland. Both *D.leichhardtii* and *D.myoporoides* are difficult to propagate owing to low germination percent of seeds. Their viability is also short (6-8 months). Vegetative propagation is also difficult. Seeds from ripe fresh berries (May-June) germinate quickly. But in areas like Dehradun (Uttara Khand) where rainfall is much higher and weather too humid in rainy season, sowing prior to rains is not advisable as high mortality occurs in rains. However some of the ungerminated seeds germinate after the rainy season is over. Sowing at the end of monsoon is therefore advisable. Germination occurs after about 4-8 weeks and continues for a period of over three months. Seed can be stored in dry state for about 6-8 months after which these cease to germinate. In spring when the seedlings are about 10 cm tall they may be pricked out singly in pots or alkathene bags, before these are planted out in the field at the break of monsoon. Pricking out in larger sized pots (23 cm) and planting of seedlings with earth ball are more suitable. Plants of about 25 cm or more in height give better survival percent. Duboisias coppice well and harvesting can be done by cutting the stems when they are about two years old at 50 to 60 cm above the ground level in spring when the plants start flowering (alkaloidal content is said to be high at this stage). Depending on the growth of plants a next harvest of leaves may also be collected in early autumn otherwise only one harvest should be collected. The leaves should be dried in shade and stored in cool and dry place in suitable containers.

Elettaria cardamomum (L.) Maton var. minor Watt. Branch with capsules

71. *Elettaria cardamomum* (L.) Maton var. minor Watt.; Zingeberaceae

Na : Eng.: Cardamon; Hindi: Chotielaichi.

Hab : It is a herb being grown in Kerala, Karnataka and also in other parts in private estates. It is also grown in forests to a lesser extent. Seeds are used in medicine as a stimulant, a stomachic and a carminative besides being used as a spice.

72. *Embelia ribes* Burm. f.; Myrsinaceae.

Embelia ribes **Burm. f.**

Na : Bengali: Biranga, baibarang; Gujrathi: Vyvirang, vavading; Hindi: Baberang, Wawrung; Malayalam: Vizhal; Marathi: Kakannie, Vavidinga; Kanerese, Tamil & Telugu: Vayuvilanga; Sanskrit: Vidanga, Vrishanasana.

Hab : It is a shrub with elliptic lanceolate, gland dotted leaves found throughout India. Fruit is nearly black with a short pedicel. Fruits are used locally as an anthelmintic. They are stated to be used by the local people as a birth control measure. Fruits are available in September/October.

73. *Enicostema verticellatum* (L.) Engl.; Syn. E. littorale Bl.; Gentianaceae.

Na : Eng.: Indian gentian; Gujrathi: Mamijava; Hindi: Chota chirayata; Marathi: Kadavinayi; Tamil and Malayalam: Vellaragu, vallari.

Hab : It is a common herb with opposite, deccusate leaves. Flowers are white or bluish in whorled axillary clusters. Plant is bitter in taste. The herb is used locally as a stomachic, a tonic, a laxative and a blood purifier. It is said to be useful in abdominal ulcer, itches and insect poisoning.

74. *Ephedra gerardiana* Wall.; Ephedraceae.

Na : Bushahr: Rachi, Khadaphag; Eng.: Ephedrine plant; Hindi: Khanda, Phok, Somalata; Ladakh: Tse, Teapat, Trano; Punjab: Asmania, Budagur, Chewa.

Hab : It is a small nearly erect shrub with dark green striated cylindrical branches arising in whorls. It is found in the temperate to alpine regions in the Western Himalayas. The sub-aerial stems yield a drug which was known for its medicinal value even in Vedic age and was called Somalata. In modern times Ephedrine and Ephedrine hydrochloride are extracted from the stems. The extracts are used for bronchial spasm, for prevention of a fall in blood pressure; for the treatment of allergic conditions such as hay-fever, urticaria, serum sickness and for relieving the symptoms of acute nasal catarrh, sinusitis, rhinitis etc.

The fruits ripen from middle of June to the middle of July when they are collected for raising plantation. If the seeds are allowed to remain in the fruit for a considerable time after their ripening they germinate viviparously. The seeds are sown in the nursery in September/October. They start germinating in about a month's time. Germination is complete before the onset of winter. They remain dormant in winter and sprout up with the commencement of spring. One or two weedings may be necessary before transplanting. The seedlings may be transplanted in July with the setting-in of the monsoon. Two or three weedings may be necessary in the first year. In order to ensure good growth of plants no harvesting is done during the first year of planting. It can be harvested up to the 5th year. Since the total alkaloidal content as ephedrine among the plant is higher in October to November (when dormancy sets in), the crop is harvested at this stage. The plantation is productive till the 5th year after which it may be renewed.

Ephedra gerardiana **Wall.**

E. intermedia schrenk & Mey.

The harvested crop is dried completely, preferably in shade before it is stored or marketed. The plant is officinal in British and Indian Pharmacopoeias, the B.P. and I.P. limits being not less than 1.25 and 1.0 per cent respectively. The drug ephedrine and ephedrine hydrochloride are being imported into the country every year in large quantities as the supply from the indigenous source is limited. Therefore it is desirable to raise the plant artificially.

75. *E. intermedia* schrenk & Mey.; Ephedraceae

Hab : It is a herb found in Kashmir and Kulu valley in Himachal Pradesh. Locally this herb is used in the treatment of asthma and bronchial disorders but it is mostly used for adulterating the valuable drug *E. gerardiana* Wall.

76. *Eucalyptus fruticetorum* F. Muell. (Syn. *E.polybractea* R.T. Baker); Myrtaceae.

E. globulus Labill.- Trees

Na : Eng.: Blue Mallee.

Hab : It is an Australian shrubby plant and has been introduced at Dehradun. In India the medicinal eucalyptus oil is obtained as a by product from the leaves of *Eucalyptus globulus* Labill. where it is grown mainly to supply firewood. This oil is not acceptable to the British Pharmacopoeia as it yields less than 70 per cent cineole and the yield of oil is also low about 1 per cent. The essential oil has to be rectified to increase cineole content. *E.fruticetorum* F. Muell. grows wild in its natural habitat and yields 1.5 to 2.5 times the amount of oil obtained from *E.globulus* Labill. containing 77-84 % cineole. The results of analysis of sample of leaves of *E.fruticetarum* received from Araku valley in Vishakhapatnam district indicate that the yield of oil on zero moisture basis is 2.5930 on distillation for 3 1/2 hours. Cineole content is 80%. It grows well under summer temperatures occasional winter frosts. Mean annual rainfall should be 500 mm having its distribution mostly in winter months. The species needs trials in drier localities in the country. The oil is used in medicine for cold and cough.

77. *E. globulus* Labill.; Myrtaceae.

Na : Eng.: Blue gum.

E. globules Labill.-
Branch with leaves

Hab : Blue gum is a tall tree native of Australia. It is raised in the Nilgiri Plateau in less frosty localities at elevations between 1500 to 2200 metres with a good distribution of rainfall not exceeding 200 cm. It is also grown in other parts in the country like Pulney hills, Simla and Shillong. It is comparatively wind hardy, immune from grazing but frost tender when young.

Nursery technique: Soil is worked to a depth of 45 cm. and roots, pebbles etc., are removed. Raised beds (12 × 1.2 × 0.3 m) are formed. Broadcast sowing is done at the rate of 60 grams per standard bed in October-November. The seeds are evenly sown in the bed and covered with a thin layer of earth and layer of thin brushwood, bracken grass. Shade is provided by erecting a "Pandal" about 60 cm in height to protect the seedlings from sun and frost. Seeds germinate in about a fortnight and when the seedlings are about 7 to 10 cm shade is removed to avoid damage by drip action. Seedlings are pricked out into polythene containers when they are 7 to 10 cm in height in the month of December-January. Beds are watered twice a day by rose can on non-rainy days. Number of plants obtainable from one standard bed is 5000-6000 seedlings.

Planting: In grass lands in the Nilgiris in South India, aligning and staking are commenced in November-December and pitting is usually completed by March. The pit is 0.3 metre cube. The espacement adopted is 2 m × 2 m. Blue gum fertiliser mixture (6.7%N, 10% P205 and 10% K20) is mixed up with the dug up soil at 75 grams per pit and the pits refilled and kept ready for planting. Planting commences with the on-set of south west monsoon showers in May-June and polythene container seedlings are used for planting. Casualties are replaced periodically during the monsoon as and when they occur. Weeding and soil working operations are carried out as per the requirement. Casualties are replaced in second year if the stocking is less than 75%. Plantation is completely protected from fire.

Euphorbia nivulia **Buch.-Ham**

Uses: Leaves and branches yield essential oil on distillation. The leaves after harvesting are dried in shade for 2 to 3 days to drive out excess moisture. The yield of oil is 1 per cent on an air dry basis. The oil usually contains 60 per cent active principle cineole; the minimum acceptable to the British Pharmacopoeia is 70 per cent. As such unless it is rectified, to increase the cineole content it is not acceptable in trade. The oil is colourless when it is rectified. When it is old and poorly stored it becomes pale to dark yellow in colour. It has aromatic odour and spicy taste. It is mainly used in pharmaceutical preparations for colds and bronchial diseases. In perfumery also it finds its use. During 1986-87, 100 kg of oil valued at Rs. 6175 was exported. The wood of the species is used as timber, for pulping in paper manufacture and as fuel.

78. *Euonymus tingens* wall.; Celastraceae.

Na : Eng.; Spindle wood; Hindi: Barphalli. Kungku; Jaunsar: Bambeli, roini; Kumaon: Gwali; Simla: Chopra, marmakaol.

Hab : It is a small tree found in the temperate Himalayas. The bark is locally used for eye disease and constipation.

79. *Euphorbia nivulia Buch*.-Ham.; Euphorbiaceae.

Na : Beng.: Sij; Hindi: Katathohar; Kanerese: Dubbakalli; Malayalam: Illakalli; Marathi: Newranj; Sans.: Patrasnuhi, vajri; Telugu: Akujammudu.

Hab : It is a perennial herb common in waste lands. On injury it gives milky juice from all parts. Leaves and milky juice are said to be useful in cough and bronchial diseases.

E.tirucalli L.

80. *E.tirucalli* L.; Euphorbiaceae

Na : Sanskrit: Kanda snuhi; Telugu: Kadajummudi.

Hab : It is a herbaceous plant stem being green. It is found in dry places in wastelands. Milky juice and the plant extract are said to be used for constipation.

81. *Ferula assafoetida* L.; Umbelliferae.

Na : Eng.: Asafoetida; Bengali, Gujrathi, Hindi, Kanerese & Telugu: Hing; Malayalam & Tamil: Perungayam; Sans.: Balhika, Hingu.

Hab : It is a perennial herb grown in Kashmir. The gum oleoresin obtained from the carrot shaped roots is used in indigenous medicine in the treatment of bronchial diseases, coughs, colds and indigestion.

82. *F.galbaniflua* Boiss & Bushe; Umbelliferae.

Na : Eng.; Galbanum; Hindi: Gandha biroza.

Hab : The plant is found in North western Himalayas. The gum oleoresin obtained from the stem and root is used in indigenous medicine in the treatment of bronchial diseases.

Ficus hispida L. f.

83. *Ficus hispida* L. f.; Moraceae.

Na : Eng.: Dumoor, Kakodumar; Guj.: Janglianjir; Hindi: Daduri, Dagurin, Gobla, Kagsha, Katgularia; Marathi: Bhokoda, Bokria, Kalamber, Kharairi; Telugu: Bodamamidi, Brahmamedi.

Hab : It is a shrubby plant found in dry places in waste lands. Leaves are said to have useful effect in the treatment of leucoderma.

84. *Gardenia gummifera* L.f.; Rubiaceae

Na : Bengali, Hindi & Marathi: Dikamali; Guj.: Kamarri, dikamali; Kanerese: Chittubikke, Dikkemalli; Oriya: Gurudu, Bryodikoli; Sanskrit: Pindava, Nadihingu: Tamil: Kambilippicin, Dikamali; Telugu: Bikki, Chitamita, Dikamali.

It is a small tree commonly found in South India particularly in Andhra Pradesh and Karnataka. The gum oleoresin obtained from the bud is used medicinally as carminative, stimulant and in dyspepsia.

Gardenia gummifera L.f.

85. *Gentiana kurroo* Royle; Gentianaceae.

Na : Bengali & Hindi: Karu, Kutki Bombay: Pakhanbhed, Pashanveda; Eng.: Himalayan gentian, Indian gentian root; Kashmir: Nilkanth; Punjab: Nilakil, Nilkant.

Hab : It is a small perennial herb with a stout rhizome bearing decumbent flowering stem. The plant is found in Kashmir and other parts of North Western Himalayas. Roots are used as a tonic, a stomachic and for urinary ailments.

86. *Gloriosa superba* L.

Gloriosa superba L.

Na : San.: Haripriya, Kalihari, Langali ; Telugu: Adavinabhi, Nagaligadda, Pottidumpa.

Hab : Gloriosa is an ornamental slender climber available in tropical regions in central and south India, particularly in Andhra Pradesh, Maharashtra, Karnataka, Tamilnadu and Madhya Pradesh. The species is scarcely seen in the forest areas. It emerges with the commencement of monsoon and seeking the support of adjoining bushes grows and produces ornamental orange coloured flowers. The apex of leaves is tendrilar, which helps the plant to seek support. With the dry spell of summer, the sub-aerial part dries up and sprouts in the ensuing monsoon.

Soil and Climate: It grows in warm and well drained localities. It is not fastidious about soil and grows in a variety of soils. It is found to grow even in inhospitable localities. However the plant prefers neutral soils.

Method of planting: The plant can be raised either by rhizome cutting or by seed. The sub-aerial part of the plant dries up after producing seeds. The plant repeats the cycle for about 6-7 years. In the case of seedling planting, the seeds are collected soon after ripening and dried. They are sown/dibbled in beds at a distance of 7.5-10 cm at least 2 months before the onset of monsoon. The beds should be watered to keep them moist in summer. Alternatively, 2-3 seeds should be sown in polythene bags. The soil in polythene bag should be fertile. With the commencement of monsoon, the seedlings are pricked out from beds and planted at 60 × 60 cm. In the case of polythene bag seedlings, bottom portion of the bag is torn and separated and planted at the above spacing. Weeds smothering the seedlings should be removed. In the case of rhizome cutting, freshly collected rhizome is cut into small pieces of 30-40 grams each with 1 or 2 nodes and internodes and planted at site at the above spacing. In about a week's time the rhizome cuttings sprout and start growing. Full advantage of growing season of the plant should be taken and planted. In places where natural support is not available, artificial supports should be fixed nearby. In about 6 months time their fruits ripen. Then they are collected and dried. In the case of rhizome collection, plants are dug out after 4-5 years growth to have economic yield of crop. After harvesting, rhizomes are cut into pieces of suitable size and dried and marketed. A hectare of plantation may yield about 200-250 kg seed and chaff and 3 tonnes of dried rhizomes depending on fertility of soil and growth period when harvested after 5 years.

Glycyrrhiza glabra **L. Plant**

Pests and diseases: There is not much damage due to pests and diseases.

***Uses*:** Seeds and chaff as well as rhizomes are useful medicinally. Rhizomes are used to hasten up delivery. It is also used as insecticide and in skin ailments. Seeds yield colchicines which is used in gout, arthritis. As such it has good export market.

87. *Glycyrrhiza glabra* L.;Fabaceae (Papilionaceae).

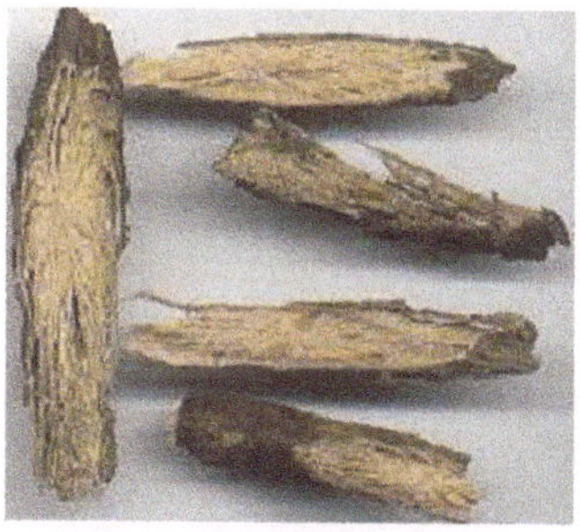

Glycyrrhiza glabra **L. Root**

Na : Bengali: Jashtimadhu; Eng.: Liquerice; Guj.: Jethimadha; Hindi: Jethimadh, Mulhatti; Kanerese: Jeshtamadh, Yashtimadhuka, Atimadhura; Malayalam: Irati madhuram; Marathi: Jushtamudha; Pers.: Bikhemahak;

Sans.: Madhuka, Yashtimadhu; Tamil.: Atimadhurani.; Telugu: Yasthimadhukam, Atimadhurram.

Hab : It is a small shrub native of the mediterranean region but is being grown in the Punjab, Jammu and Kashmir and Himachal Pradesh. The roots are officinal in pharmacopoeias. The active principle is glycyrhizine. The roots are used as a tonic, a laxative, for coughs and sore throats. Owing to short supply of the drug from indigenous sources, it is being imported in large quantities. During the year 2002-03 the drug in the raw form as well as powdered valued at Rs. 20.6 millions was exported..

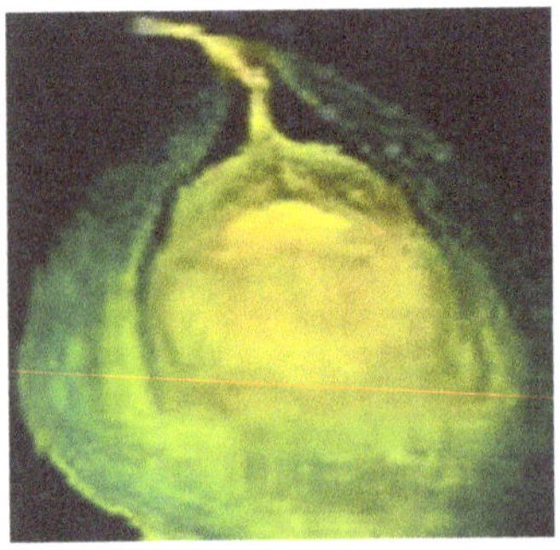

***Gnetum ula* Brongn. – fruit**

88. *Gnetum ula* Brongn.; Gnetaceae.

Na : Bombay: Kumbal, Umbli, Lolumbi; Kanerese: Kodakamballi, Navurukattee; Malayalam: Odul, Ula; Oriya: lolori; Tamil: Anapendu, Peiodal.

Hab : It is a large dioecious plant with thick scaly bark and swollen joints. It is found in the Western and some parts of the Eastern Ghats of India. Seed oil is locally used for rheumatism. The leaves are also used in the treatment of Alzamir (forgetfulness)

***Gymnema sylvestre* (Retz.) R.Br. ex schutt**

89. *Gymnema sylvestre* (Retz.) R.Br. ex schutt; Asclepiadacea

Na : Bengali: Merasingi; Guj.: Dhubli, Mardashingi; Hindi: Gurmar, Merasingi; Kanerese, Sannagerasehambu; Marathi: Kavali, Kalikardori, Vakundi; Sanskrit: Meshashringi, Madhunashini; Tam.: Adigam, Chemkurinja; Telugu: Podapatri.

Hab : It is a perennial plant. Leaves are opposite and ovoid. The roots and leaves are used by the local people in diabetes.

***Hemidesmus indicus*. R Br**

90. *Gynocardia odorata* R.Br.; Flacourtiaceae.

Na : Assam: Lemlem, Bonsha, Dieng-sohphailying, Umphu, Bailibu, Koilur; Lepcha: Tukkung; Nepal: Kadu, Bandrophal.

Hab : It is an evergreen dioecious lofty tree found in Khasi hills and Sikkim. Seeds yield oil which is used in the treatment of leprosy in indigenous system of medicine.

91. *Hemidesmus indicus* R.Br.;Asclepiadaceae

Na : Bengali: Anuntmul; Eng.: Indian sarasaparilla; Guj.: Sariva, Upalasari, Durivel; Hindi: Anantamul, Hindi: salsa, Kapuri, Magrabu; Kanerese: Anantamul, Sogade; Malayalam: Naruninli; Mar.: Anantamul, Upalasari; Oriya:Onontomulo; Sanskrit: Anantamula, Nagajihva; Tamil: Wanari; Telugu: Sugandhpala.

Hab : It is a twining shrub commonly found in Central and South India. The roots are aromatic and are used as tonic diuretic, diaphoretic and demulcent. The roots are officinal in Indian Pharmacopoeia and are used as a substitute for Sarsaparilla from Smilax species. (S.medica-Mexico, S. officinalis-Honduras and S.ornatus-Jamaica). The roots have export value.

***Hippophae salicifolia* D. Don.**

92. *Hippophae salicifolia* D. Don.; Elaeagnaceae.

Na : Hindi: Chuma, Kalabis.

Hab : It is a small deciduous tree found in the temperate Himalayas. Bark of this small tree has given promising results in the treatment of certain forms of cancer.

***Holarrhena antidysentirica* Wall**

93. *Holarrhena antidysentirica* Wall.; Apocynaceae.

Na : Assam: Dhulkari, Dudkhuri; Bengali: Kurchi; Bihar: Dudhiari; Eng.: Easter tree, Ivory tree; Guj.: Drowda, Kudakari; Hindi; Kurchi, Karva indarjau, Kutaja; Kanerese: Koodsaloo, Korchie; Malayalam: Kodagapala; Mar.: Kodaga; Oriya: Kherwa, Korwa; Kurwa, Pitakorwa; Punjab: Keor, Kewar; Nepal: Khuria; Sanskrit: Kutaja, Kalinga; Tamil: Veppalei, Kodagapalei, Indraban; Telugu: Pala Kodasha.

Hab : It is a small common tree found almost through out tropical deciduous forests in the country. Leaf, bark and the seeds are locally used in the treatment of dysentery. Bark and seed are also stated to be blood purifier. Stem bark (kurchi) is officinal in Indian pharmacopoeia.

94. *Hydnocarpus alpina* Wight; Flacourtiaceae.

Na : Hindi: Kastel: Kanerese: Torathi; Malayalam: Malamaravetti: Marathi; Kastel; Nilgiri: Maralatte; Tamil: Attuchankalai;

Hab : It is a large tree found in the Nilgiris and the Western Ghats. Seed oil is used locally in the treatment of leprosy.

95. *H.pentandra*; (Buch.-Ham.) Oken (Syn.;*H. laurifolia* (Dennst.)Sleumer;); Flacortiaceae.

Na : Bombay: Kanti, Kava; Hindi: Chaulmoogra, Kava; Kanerese: Saranti, Surti, Toratti; Malayalam: Kodi, Koti, Maravetti, Maroti, Niralam, Nirvetti, Tamana, Vetti; Marathi: Kadu Kavata, Kastal, Katukavatti, Kobasett; Sanskrit: Garudaphala, Tuvruk; Tamil: Maravattai, Maravetti, Niradimuttu; Telugu: Adibadamu, Nirndi.

Hab : It is a tall tree found in the Western Ghats, in Karnataka and Kerala. Seeds yield oil which is used in the treatment of leprosy and skin diseases in indigenous system of medicine.

96. *Hyoscyamus niger* L., Solanaceae.

Na : Bengali: Khorasani ajowan; Eng.: Henbane; Guj.: Khorasaniajmo; Hindi: Khurasaniajvayan; Kanerese: Khorasanivadaki; Marathi: Khorasanivova; Sanskrit: Dipyaparasikaya; Tamil: Kurasani yomam; Telugu: Khurasanivomam.

It is a branching, narcotic and poisonous annual or sometimes a biennial, 1 to 2 feet high, more or less hairy and viscid, with a nauseous smell. The leaves or leaves and flowering tops constitute the commercial drug which contains the alkaloid hyoscyamine together with smaller quantities of atropine and hyoscine. It is officinal in Indian Pharmacopoea and should contain not less than 0.50 per cent of alkaloids. The drug derived from the leaves or leaves and flowering tops is used as a sedative, narcotic and also in the treatment of asthma and whooping cough.

***Hyoscyamus niger* L.**

Distribution: It is found wild in the temperate Western Himalayas from Kashmir to Garhwal at altitudes of 1500 to 3300 m. Its cultivation has been tried with varying degree of success in Kashmir, Saharanpur, Nilgiris, and Dehradun. It is under experimental cultivation at Chakrata (Uttaranchal).

Locality factors: It is found naturally and grows most easily in a temperate climate, but can be cultivated successfully in the plains e.g., Saharanpur. It prefers sandy loam soil, and is fairly easy to cultivate. The cultivated plant is said to yield a better quality of drug than the wild plant. It has been found at Research Laboratory, Jammu Tawi, that *Hyoscyamus niger* leaves growing wild or cultivated in Kashmir at altitudes of over 1500 m give alkaloidal content well up to and even above B.P. and U.S.P. Standards. The leaf grown at lower altitudes or in the plains is said to have a lower alkaloidal content than the B.P. or U.S.P. Standards. It would be best to cultivate it in the Western Himalayas at altitudes of 1500 to 2400 metres.

Propagation and cultivation: In Europe two forms, annual and biennial, are said to spring from the same crop of seed; the one flowering and perfecting seed in the first summer and the other producing in the first season, only a tuft of radical leaves, which disappear in winter, leaving a thick underground root, from the crown of which a vigorous branched flowering stem rises in the following spring. There is no definite record of this type of biennial growth occurring in India. In plains at Saharanpur, it has been raised as a winter crop but at higher altitudes it could be raised in April-May and harvested in October-November. 3.5 to 4.5 kg of seed per hectare are required for direct sowing in rows 60 cm apart. A nursery may also be raised for transplanting seedlings at a spacing of 60 × 45 cm. The comparative performance of direct sowing and transplanting needs further study under Indian conditions. Plants may require frequent irrigation in early stages of growth. Seeds usually germinate in about 2 weeks but sometimes germination is erratic.

Manuring: Soil rich in lime appears to be better suited to this crop, in other soils a dressing of lime may be given before planting the crop. A light dressing of farm-yard manure supplemented by superphosphate may be given when preparing the land. It is important that the land should be thoroughly clean and a periodic change of land is an important factor, contributing to the success of the crop.

Collection: In plains, in October sown crop of annual type, flowering starts about the beginning of January or later and the leaves and flowering tops are collected when flowers appear. The crop should be rapidly dried. It is necessary to retain the green colour of the leaves. For purposes of getting seeds, some flowering tops should be left intact and a few leaves only should be collected from such plants. On an average about 670 kg of dried material can be obtained from hectare of plantation. In the higher altitudes April sown crop could be collected in the same way as described above when it flowers in October/November. The crop is reported to be adversely affected by some beetles. In such cases treatment with derris or D.D.T. dust insecticide may help.

97. ***Indigofera tinctoria* L..; Fabaceae (Papilionaceae.)**

Na : Bengali: Nil; Guj. & Hindi: Gali, Gari, Nil; Kanerese, Marathi & Tamil: Nili; Sanskrit: Nili, Nitika, Nilla, Rangapatri; Telugu: Aviri, Nili.

Hab : It is a small shrub found in wastelands. Flowers yield indigo dye. Roots are used by local people in dental diseases, and the leaves in venereal diseases.

Indigofera tinctoria L. Roth-Plant

98. ***Ipomea nil* (L.) Roth; (Syn. *I.hederaceae* Clarke); Convolvulaceae.**

Na : Hindi: kaladana.

Hab : It is found wild throughout tropical regions in India. The seeds are used as purgative and as a substitute for jalap.

Ipomea nil (L.) Roth-Plant

99. ***Jasminum malabaricum* Wight; Oleaceae**

Na : Hindi: Kusar; Kanerese: Dolle kusdiballi; Marathi: Kusur, Kusuri, Mogra, Ranmogra.

Hab : It is a large climbing or scandent shrub with fragrant flowers commonly found in South India. Juice of the stem is said to be used by the local people in the treatment of eye cataract.

100. ***Jatropha curcas* L.; Euphorbiaceae.**

Na : Bengali: Bagbheranda, Erandagagachh; Eng.: Physic nut; Guj.: Jamalgota; Hindi: Jamalgota, Ratanjota, Safed arund, Bagbherenda, Jangliarundi; Kanerese: Adaluharalu, Bettadaharalu, Maraharalu, Karnocchi; Malayalam: Kadalavanaka, Kattavanakka; Marathi; Mogalieranda, Ranaerandi; Oriya; Jahazigaba; Sanskrit: Kananeranda, Parvalaranda; Tamil: Kadalamanakku, Kallamanankku; Telugu: Nepalamu, Adaviamidamu.

Hab : It is a shrub or small tree native of tropical America, but now it is found growing wild in wastelands and along the fringes of forests in tropical regions. The seed oil is a drastic purgative.

Ipomea nil (L.) Seeds

101. ***Lallemantia royleana* (Wall. & Benth.) Benth.; Labiatae.**

Na : Hindi: Gharei, Kashmalu, Tukhmalanga.

Hab : It is a perennial herb mainly found in Himachal Pradesh, Kashmir and Kumaon. Seeds are used locally in flatulence and constipation.

102. ***Lasiosiphon eriocephalus* Dene; Thymelaeaceae;**

Na : Hindi: Rametha, Rami; Kanerese: Enujariga, Nukute; Malayalam: Nanea, Nangu; Marathi: Ramita, Rametta, Romi; Tamil: Nachinaar.

Hab : It is a large shrub or small tree found in the Western Ghats. The local people apply the leaves to swellings.

Jatropha curcas L.

Leea indica (Burm.) Merr. – branch with fruits

103. *Leea indica* (Burm.) Merr. (Syn. Staphylea indica Burm.); Vitaceae.

Na : Assamese: Kukurathengia; Bengali & Hindi: Kurkurjiwah; Kanerese: Andilu; Malayalam: Erattayani, Maniperandi; Marathi: Dino, Karkani; Tamil: Nalava, Nyekki, Ottanali; Telugu; Ankadosa.

Hab : It is a large shrub with flowers in short compact cymes and is found throughout India. Local people use the roots in diarrhoea and dysentery.

L. macrophylla Roxb. ex Hornem – branch

104. *L. macrophylla* Roxb.ex Hornem; Vitaceae.

Na : Bengali & Hindi: Dhola samudra, Hathikana, Khasi. Pharunbarne; Lepcha; Dampamtom-kung; Marathi: Dinda; Santal: Hatkan.

Hab : Root is used as a remedy for ringworm and in the cure of guinea-worm.

Leonotis nepetifolia R. Br. – flowering stalk

105. *Leonotis nepetifolia* R. Br.; Labiatae.

Na : Bengali & Hindi: Hejurehei; Gujrathi: Matijer, Matisul; Marathi: Dipmal; Telugu: Ranabheri.

Hab : It is a common ornamental herb found wild throughout tropical parts in India. Flowers are bristly and are orange scarlet in colour. They are arranged in verticelaster.

The villagers use the ashes of flower heads in the treatment of ringworm and itchy skin diseases. Powder of the flower heads is used by the local people in stomach ulcer and other ailments of stomach.

Leonurus cardiaca L.

106. *Leonurus cardiaca* L. (Syn. *Cardiaca vulgaris* Moench); Labiatae.

Na : Eng.: Common mother-wort.

Hab : It is a perennial herb found in Kashmir, Kumaon and the Punjab. The plant is used as a diaphoretic and stomachic.

L.sibiricus L.

107. *L.sibiricus* L.; Labiatae.

Na : Eng.: Siberian mother wort; Hindi: Guma.

Hab : It is an annual weed found in Assam, Bengal, Bihar and South India. Roots and leaves are used as febrifuge by the local people.

108. *Leucas cephalotes* Spreng.; Labiatae.

Na : Bengali: Barahalkura; Guj.: Gubo, Kubi; Hindi: Dhurpisag, Deldona, Goma, Gubbha, Motapati; Marathi: Devakhumba, Shetvad, Tumba; Telugu: Peddalumni.

Hab : It is a common herb which is pungent to taste. It is used as a pot-herb. Flowers are used by the local people for cough and cold.

109. *Limnophila indica* (L.) Bruce (Syn. *L. gratioloides* R. Br.); Scrophulareaceae.

Limnophila indica (L.) Bruce

Na : Bengali: Karpuri; Gujrathi: Turali; Hindi: Kuttra; Malayalam: Manganar; Mar.: Ambuli.

Hab : It is a common profusely branched herb found in damp places through out India upto 900-1200 m in the Himalayas. It is considered to be antiseptic. It is used locally with coconut oil in elephantiasis.

110. *Limonia acidissima* L. (Syn. *Feronia elephantum correa*); Rutaceae.

Na : Beng.: Kerit, Katbel; Guj.: Kavit, Kotha, Kothi, Kothun; Hindi: Kaveet, Kait; Kanerese: Belavi; Malayalam: Vila, Vilatti; Marathi: Kavith; Tamil: Velanga; Telugu: Velaga.

Hab : It is found throughout dry tropical forests. Leaves are stated to be used in reducing nausea. Local people use the bark in the treatment of venereal diseases.

111. *Litsea glutinosa* (Lour.) C.B. Robinson (Syn .L. *sebifera* Pers.); Lauraceae.

Limonia acidissima L.

Na : Bengali: Kukurchita, Garur, Ratan; Gujrathi, Hindi, Marathi: Maida lakdi; Hindi: Garbijaur, Medh; Oriya: Jaisanda; Tamil: Uralli; Telugu: Naramamidi, Narralagi.

Hab : It is an aromatic shrub or tree found in the Punjab, Assam, Bengal and South India. Oil obtained from the fruits is used by the local people in the treatment of rheumatism.

112. *Lobelia nicotianifolia* Heyne; Lobeliaceae

Na : Beng.: Badanala; Guj.: Nali; Hindi: Nala, Narasala; Kan.: Kaduhogesoppu, Kandele; Malayalam: Kattupokala; Marathi: Devanala, Thoradevanala; Tam.: Kattuppugaiyilai, Upparichedi; Telugu: Adavipogaku.

Hab : It is an herbaceous plant like tobacco and is found in Western Ghats from Bombay to Travancore, round about Mahableshwar, Lonawala and Khandala. The leaves and twigs contain an active principle Lobellin. The content of this active principle is 3 to 4 times more than the American drug plant *L.inflata* L. The plant is harvested when it is in full flower and dried in shade. The plant can be grown from seeds. The seed should be sown in March/April and the seedlings are transplanted in the field at 60 cm apart during monsoon.

Litsea glutinosa (Lour.) C.B. Robinson

Uses: The drug is used in respiratory affections especially in spasmodic cough, bronchitis etc.

Lochnera rosea (L.) Reichb

113. *Lochnera rosea* (L.) Reichb. (Syn. *Vinca rosea* L.; *Catharanthus roseus* G. Don); Apocynaceae.

Na : Eng.: Madagascar periwinkle or Periwinkle; Hindi: Sadasawagan, Sada bahar; Telugu: Billaganneru.

Hab : It is a small shrub native of the West Indies. It has_pink or whitish flowers found for the greater part of the year. Although it is grown as an ornamental plant in India it is found as an escape in waste lands in dry places. Recent researches have shown that the plant contains certain active principles like reserpine which are useful in the treatment of hypertension. It has also been found to be effective in the treatment of certain forms of cancer. Locally the plant is used in diabetes.

Mallotus philippiensis Muell. Arg.

114. *Luffa echinata* Roxb.; Cucurbitaceae.

Na : Bengali: Bindal; Hindi: Bindal; Ghagarabela; Gujrathi: Kukaravel; Marathi: Devadali, Devadangari; Sanskrit: Devadalika, Koshaphala.

Hab : The plant is found in Bihar, Bengal, Dehradun and Gujarat. It is a climber with slender hairy furrowed stem and 2-fid tendril. The fruit is intensely bitter and fibrous. Locally the fruits are used in the treatment of dropsy and also as a purgative.

115. *Mallotus philippiensis* Muell. Arg.; Euphorbiaceae.

Na : Assami: Jorat, Losan: Bangali: Sinduri; Guj.: Kapilo; Hindi: Kamala, Rohini, Sindur; Kanerese: Kunkumadamara; Mal.: Manjana, Kuramuddakku; Mar.: Shudri; Tamil: Kapli, Kungumam; Telugu: Kunkuma, Sinduri.

Hab : It is a small tree found commonly along the foot hills of Himalayas from Bengal westwards and in Orissa, Madhya Pradesh and Maharashtra, The red dye obtained from the glands found on the surface of the fruity is uses for removing thread worms and Ascaris.

Marrubium vulgare L.

116. *Marrubium vulgare* L.; Labiatae.

Na : Eng.: Hoar hound; Hindi: Pahadi gandana.

Hab : It is a robust herbaceous perennial found in Western Himalayas from Kashmir to Kumaon at altitudes of 1500 to 2400 m. Dried leaves and flowering tops are the source of a drug which is used in cough, cold and for rheumatism.

Matricaria chamomilla L.

117. *Matricaria chamomilla* L.; Asteraceae (Compositae.)

Na : Eng.: German chamomile, Chamomille; Hindi: Babuna, Suteigal.

Hab : It is a much branched aromatic annual herb native of Europe commonly found in Jammu & Kashmir, Himachal Pradesh and Uttar Pradesh. The essential oil obtained from the leaves and twigs of this herb is used medicinally as tonic and gastric stimulant.

118. *Melia azedarach* L.; Meliaceae:

Na : Beng.: Mahanim, Ghoranim; Eng.: Pride of India, China tree: Persian lilac; Guj.: Bakam limbodo; Hindi: Bakain, Drek: Kanerese: Arebevu, Hutchubevu; Malayalam: Karin vembu, Simaveppu; Marathi: Pejri, Pedrai; Tamil: Malaivembu; Telugu: Turakavepa.

Hab : It is grown as an avenue and in soil conservation and afforestation areas. Bark is dark grey. Flowers are fragrant. Juice of the leaves is used as an anthelmintic. The seeds are used in rheumatism.

***Melia azedarach* L.**

119. *Mentha arvensis* L.; Labiatae.

Na : Eng.: Japanese mint, "Corn mint" Field mint: Hindi: Pudina an important source of menthol.

Introduction: The chief use of the Japanese mint oil is for the preparation of natural menthol which is extensively used in medicine in the treatment of bronchial and nasal catarrh, as a carminative and analgetic in the treatment of neurelgia and rheumatism and also in confectionary, drinks and in mentholated cigarettes etc. In India menthol forms a constituent of high class supari powders. It is also used along with betel. The "dementholised oil" (Oil from which part of the menthol has been removed by freezing) finds use as a substitute for the true pepper-mint oil (from M.piperita L.) for flavouring purposes, in oral and other pharmaceutical preparations. Indian demand for menthol is still being met by imports.

Habit and distribution: It is an erect perennial herb 30 to 90 cm in height with an extensive sub-aerial growth and trailing suckers. The plant has strong aromatic odour. It is native of Japan and China and Brazil. It has been successfully cultivated in Dehradun and a few other places.

Soil and climatic requirement: It can be grown both in the temperate and tropical regions in India which are not excessively dry in summer and where the sub-soil is always moist but not water-logged. Heavy clay is not suitable but it prefers a light sandy-loam soil. It thrives well in areas where the rainfall varies from 1000 to 1500 mm with an average winter temperature of about 20°C and summer temperature of 25 °C. Where the summer is more severe the plant requires copious watering during such months.

Cultivation: Two to three inch cuttings of the rooted suckers (underground stem) are made and then pressed in the soil horizontally about 1.5 cm deep. The distance between end to end and also between rows may be 22 to 30 cm (9" to 1') depending on the availability of the planting material. Beds 1.2 m (4') wide and upto 12 m (40') lengths are quite suitable. If planting is done in the spring watering may be necessary. It can be done in the spring (mid. Feb. to March) or rainy season. In about ten days, plants sprout and send out new shoots. During summer season regular watering is necessary so that the plantation is kept moist. However, water-logging inhibits the growth and ultimately results in the decay of the plant. In the early stages of the plantation weeding may be done to avoid suppression. In about 3 month's time whole area is covered up and the weeds are also suppressed.

***Mentha arvensis* L.**

Treatments required: The soil should be thoroughly ploughed and manured prior to planting. Farm-yard manuring is necessary to increase the yield of oil. Green manuring may also be done before the mint is planted. Periodic manuring and liberal irrigation are necessary. Weeding in the early stages is also essential. Application of fertilizers may be done after each harvest at 168 kg Nitrogen, 56 to 90 kg phosphorus and some potash (if necessary) per hectare to have a good crop. The actual quantity of the fertilizers which will contain 168 kg of Nitrogen and 56 to 90 kg phosphorus can be worked out on the basis of percentage of nitrogen and P_2O_5 present in the nitrogen and phosphorus fertilizers respectively. 2% urea in the form of solution in water can also be sprayed after planting and after every harvesting.

Harvesting and yield of material: The crop is harvested with sickle or mower when the plants are in flowers or just before full bloom or when they are about 30 to 45 m. Harvesting should be done in the morning on a bright sunny day after the dew has disappeared. In the first year one to two harvests can be had and in subsequent years depending on the soil and climate of different regions, harvests may vary from three to four. The plant is harvested about 2.5 cm above ground level. In some places 5 to 6 harvests are also being obtained.

M. piperita **L.**

Handling of harvested material: The harvested material is spread in open or shade for a few hours to 2 days to drive out excess moisture. Excessive drying results in the reduction of yield of oil.

Yield of fresh material and oil: Annual average yield of fresh material as worked out at Forest Research Institute, Dehradun is about 23000 kg per hectare and of oil about 48-70 kg per hectare containing 18-26 kg of separable menthol. A yield of 61.5 kg to 67 kg of oil per hectare has been reported by the Regional Research Laboratory, Jammu.

Distillation: The harvested material after drying as suggested above is distilled by steam or water-cum steam distillation method.

Recovery of menthol: The oil contains 78 to 83 per cent menthol which is present both in the free and combined form. On repeated congealing, the oil in the refrigerator, about 40 to 50 per cent menthol which is in the free form separates out and is subsequently recrystalized. Menthol and the remaining oil are separated by decantation. Menthol crystals should be free from oil as otherwise they melt with high temperatures in summer and may make a mass. A centrifuge can be used for separating the remaining oil from crystals.

Dementholised oil is sold at Rs. 500-600 per kg.

120. *M. piperita* L.; Labiatae.

Na : Eng.: Peppermint oil; Hindi: Velaithi pudina.

Its method of propagation is quite similar to *M.arvensis* L. Oil is obtained by distillation. The best period is when plant is flowering. In most localities it is harvested in Spring/Summer. Two to three harvests are obtained. The yield of oil is comparatively less than *M.arvensis* L. It ranges from 0.3 to 0.4% which may sometimes go upto 1%. The yield of oil per hectare on an average is about 27-35 kg. The oil is poor in menthol content but yields peppermint oil of good quality.. The oil is used as a carminative in addition to its extensive use in flavouring pharmaceuticals, oral preparations like toothpastes, dental creams and mouth washes, in cough drops, chewing gums, confectionary and alcoholic liquors. It is also used as a stimulant and for allaying nausea, sickness and vomiting.

***Mimosa pudica* L.**

121. *Mimosa pudica* L.: Mimosaceae.

Na : Beng.: Lajjabati; Hindi & Sans.: Lajjawanti, Sharmeli, Chuimui; Kanerese: Lajjamachike; Malayalam: Tintaramani; Marathi: Lajalu; Tamil: Tottalwadi; Telugu: Althapathi, peddanidrakanne.

Hab : It is a common weed. Leaves are sensitive to touch. The plant is said to be useful in the treatment of piles.

***Moringa oleifera* Lamk.**

122. *Moringa oleifera* Lamk. (Moringa pterigosperma Gaertn.); Moringaceae.

Na : Bengali: Sajana; Eng.: Drumstick tree, Horse radish; Guj.: Saragavo; Hindi: Sahinjna, Soanjna; Kan.: Nugge; Mal.: Murimna, Sigru; Marathi: Shevgi; Tamil: Murangai; Telugu: Munaga.

Hab : It is a small tree grown in gardens for its edible fruits. The oil obtained from the seeds is used locally in gout and acute rheumatism.

123. *Mucuna pruriens;* (L.) DC. Fabaceae. (Papilionaceae)

Na : Hindi: Beej, Kavach, Kavanch; Telugu: Atmagupta, Dulagondi, Duradagundi, Kachu, Kapi, Markata, Vanari.

Hab : The plant grows naturally in moist and humid tropical regions with a temperature range of 27-42 °C in the country in Andaman and Nicobar Islands and Srilanka. It is a perennial climber, dries up in summer and sprouting in rains. It trails on to bushes and trees in the forests and waste lands and is conspicuous with compound trifoliate leaves and light violet coloured flowers. Fruits are bean like in shape with 3-5 seeds in each. Fruits are in bunches and bear hair like stinging glands. The seeds are white to black in colour.

***Mucuna pruriens* (L.) DC.**

Propagation: The plant thrives in a variety of soils preferring well drained neutral sites. Addition of compost or farm-yard manure in poor soils may increase the yield of fruits. Direct sowing of 2-3 seeds at each place keeping roughly 60-90 cm spacing is done near the bushes and trees. Since the fruits bear stinging glands wider spacing may be adopted. About 15 kg of seeds per hectare is required. Sowing is done with the first regular shower of monsoon, in about a week to 10 days time the seeds germinate. Crop of

Myristica fragrans **Houtt. – Branch with opened fruit**

fruits is ready in 3-3 and half months. With regular irrigation during dry periods a second harvest of crops can be obtained.

Weeding: Weeds smothering plants are removed periodically.

Prophylactic treatment: Larvae damage the crop as such neem oil or decoction of leaves of neem or any other suitable insecticide is sprayed to prevent damage. Pruning of plants when they are about 30 cm in height will help them to develop bushy crown. When the fruits are ripe they are collected. Rubber gloves should be worn to collect. Collection of fruits is made 3-4 times in the season. About 3-3.5 tonnes per hectare can be collected in a year. The seeds are thoroughly dried and stored in a dry place.

Uses: Seeds are used as a nervine tonic. The chemical L.Dopa obtained from the seed is used in Parkinson disease. It has good export value.

124. *Myristica fragrans* Houtt.; Myristicaceae.

Na : Eng.: Nutmeg; Bengali, Gujrathi, Hindi and Marathi: Jaiphal, Jatre, Jaipatri; Kan., Mal., Tamil, Telugu: Jajikai, Jadikae, Jadipattire, Japatri.

Myrsine africana **L.**

Hab : It is an aromatic evergreen plant native of Molucca Islands but is grown in the Nilgiris and Kerala and can be grown in forest areas in Karnataka, Kerala, Maharashtra, Andhra Pradesh and Tamilnadu. Nutmeg oil is used externally for rheumatism.

125. *Myrsine africana* L.; Myrsinaceae.

Na : Garhwal: Rikhdalmi; Hindi: Chapra; Kashmir: Gugil; Kumaon: Ghani; Punjab: Bebrang, Shamshad.

Hab : It is a small shrub found in the western temperate Himalayas. The fruits are used medicinally as an anthelmintic especially for tape-worm. The ointmtent prepared from the fruit is considered effective against ring-worm and other skin diseases.

126. *Nardostachys jatamans* DC. Valerianaceae.

Na : Bengali & Sanskrit: Jatamansi; Eng.: Spikenard; Guj.: Jatamasi, Kalichhad; Hindi: Balchhar, Jatamansi; Marathi: Jatamavshi; Kanerese, Malayalam & Telugu: Jatamamshi; Tamil: Jatamashi.

Nardostachys jatamans **DC.**

Hab : It is a perennial herb found in the alpine Himalayas in Kumaon, Sikkim and Bhutan. Leaves are radical, elongate & sessile. Short thick, dark grey rhizome is covered with reddish brown tufted fibrous remains of the petioles of radical leaves. Essential oil obtained from the roots is used in pharmaceutical preparations. It is used as a substitute for valerian. An infusion of the rhizome is reported to be useful in epilepsy, hysteria, palpitation of heart and chorea. The rhizome is used as an adjunct in the preparation of medicinal oils.

127. *Nepeta cataria* L.; Labiateae.

Nepeta cataria L.

Na : Eng.: Catnip, Catmint.

Hab : It is a native of Europe and is also found in the temperate region of western Himalayas. Dried leaves and flowering tops are used medicinally as carminative and stimulant. Leaves are chewed to relieve toothache.

128. *Nymphoides indicum* (L.) O. Kuntze (Syn. *Limnanthemum cristatum* Grisele.); Gentianaceae.

Na : Bengali: Panchuli, Chandmalla; Bombay: Khatara, Kumudini; Hindi: Cumuda, Ghainchu, Tagarmul; Marathi: Kolarechikale; Telugu: Anthara thamara.

Hab : It is a common aquatic herb with long floating stem rooting at the nodes found in fresh water lakes almost throughout India. It is used as a substitute for chiretta in fever and jaundice.

129. *Ocimum sanctum* L.; Labiateae

Ocimum sanctum L.

Na : Beng: Tulsi; Eng.: Sacred basil, Holy basil; Gujrathi: Tulsi; Hindi: Tulsi, Baranda; Kanerese: Vishnutulsi, Sritulsi; Mal.: Trillavu; Marathi: Tulasi; Sanskrit: Ajaka, Brinda, Manjari, Parnasa, Patrapushpa; Tamil: Thulasi; Telugu: Tulasi, Brinda, Krishnatulasi, Nallatulasi.

Hab : It is a much branched herbaceous annual found throughout the country upto 1800 m as an escape. The plant is held sacred by Hindus. The oil from leaves is reported to possess antibacterial properties. The juice of leaves possesses stimulating and expectorant properties. It is applied to skin in ring-worm.

130. *Operculina turpethum* (L.) Silva-Manso; (Syn. *Ipomoea turpethum* R. Br.); Convolvulaceae.

Operculina turpethum (L.) Silva-Manso

Na : Bengali: Dudhkalmi, Tohri; Eng.: Indianjalap; Guj.: Nashotar, Nahotara; Hindi: Nisotar, Nisoth, Pitohri; Kanerese: Bilitigade, Bangadaballi; Mal.: Chivaka, Trikdpakonna, Rochani, Tribhandi; Mar.: Nishottar, Phutkari; Oriya: Dudholomo; Tamil: Shivadai, Kumbum; Telugu: Tellategadda.

Hab : It is a common twining herb with milky juice and fleshy branched roots. A resinous substance obtained from the root bark of this twining herb is used as a pargative and acts as a substitute for jalap.

131. *Oxalis corniculata* L.; Oxalidaceae.

Oxalis corniculata L.

Na : Assamese: Changeritenga, Tengeshitenga; Bengali & Hindi: Amrul sak, Chuka tripali; Kanerese: Hulichikkai, Pullampurachi, Uppinasoppu; Mal.: Puliyarel; Marathi: Ambuti, Anjati, Bhinsarpati; Punjab: Ambika, Khattamitha: Tamil: Puliyarai; Telugu: Pulichinta.

Hab : It is a small herb with trifoliate leaves found in wet places. Leaves are sour to taste. They have carminative properties when eaten and also prevent loose motions.

Phyllanthus emblica L.

132. *Pedalium murex* L.; Pedaliaceae.

Na : Bengali: Badagokhru; Guj.: Mottogokhru; Hindi: Badagokhru, Kadvagokhru; Kanerese: Annegalugida; Malayalam: Kakamulla, Ananerinnil; Marathi: Hathicheratte, Mothegokhru; Oriya: Gokhara, Gokshura; Punjab: Gokhru Kalan, Tamil: Anainerinji, Perunerinji; Telugu: Enugupalleru.

Hab : It is a common succulent herb found in waste lands in dry places. Plant is said to be useful in urinary troubles and is said to dissolve bladder stone.

133. *Phyllanthus asperulatus* Hutch Syn. *P. niruri* L.; Euphorbiaceae.

Na : Sanskrit: Bhoodathri; Telugu: Nelausri.

Plant is said to be used by local people in certain forms of jaundice, skin diseases and kidney troubles.

134. *Phyllanthus emblica* L. Syn. *Emblica officinalis* Gaertn.; Euphorbiaceae.

Phyllanthus emblica L.
Scrapings of Fruit

Na : Bengali: Amla, Amlaki; Eng.: Emblic myrobalan; Gujrathi: Amah, Ambala; Hindi: Amla, Aonla; Kanerese, Malayalam & Tamil: Nelli; Sanskrit: Adiphala, Amalaka, dhatri; Telugu: Usri.

Hab : It is a small to medium sized tree commonly found in the mixed dry deciduous forests in India. It is also cultivated in home yards. One of the cultivated varieties bears larger fruit than the ordinary type.

The species is propagated by seeds and vegetatively by budding, cuttings, layering etc. The fruit is much esteemed for making pickles, preserves and jellies. The fruit is rich in vitamin C and also in iron. The fruit juice contains nearly 20 times as much vitamin C as orange juice and a single fruit is equal in antiscorbutic value to one or two oranges. Vitamin C can be preserved in the fruit either in dry or jelly form but is lost slowly on storage. Vitamin C in the form of amla is more readily assimilated than with synthetic vitamin C. The fruit is held in high esteem in indigenous medicine. It is cooling, diuretic and laxative. Dried fruit is useful in haemorrhage, diarrhoea and dysentery also in anaemia, jaundice, dyspepsia and also in cough. Amla is one of the Triphalas, the others being belleric myrobalan and chebullic myrobalan. Triphala is one of the important drugs in Ayurvedic medicine.

135. *Picrorhiza kurrooa* .Royle ex Benth.; Scrophulariaceae.

Picrorhiza kurrooa Royle ex Benth.

Na : Bengali & Hindi: Kuru, Kulki; Guj.: Kadu; Kashmir: Kaur; Marathi: Kutaki; Malayalam, Tamil & Telugu: Katukarogani, Kadugurolhini; Punjab: Karru.

Hab : It is a hairy perennial herb found in the alpine to subalpine Himalayas from Kashmir to Sikkim. Leaves are radical. The rhizome and roots are used as stomachic tonic and as a purgative. The active principles are glucoside, picrorhizin. It is officinal in Indian pharmacopoea.

136. *Piper longum* L.; Piperaceae.

Na : Bengali: Piplamor; Guj.: Pipli; Hindi: Pipli, Piplamul; Kanerese: Hippali, Lippali; Malayalam: Tippali, Pippali, Magadhi; Marathi: Pimpli; Tamil: Tippili, Pippili, Sirumulam, Kandan Lippili; Telugu: Pippallu.

Hab : It is a climbing or trailing aromatic perennial herb. The twigs contain aromatic oil. Its leaves resemble Beatle leaves but are small in size. Fruit heads are called catkins. It is a shade tolerant plant as such it can be grown under the shade of big trees.

Soil and Climate: The plant requires hot and humid climate with fertile moist soils. Water logging inhibits its growth. It is found in Karnataka, Maharashtra, Andhra Pradesh, Kerala, and Tamilnadu. It is occasionally found in the foot hills of the Himalayas in Uttara Khand and West Bengal.

Propagation: The plants can be raised by suckers growing around parent plants or by cuttings. In the case of cutting planting, stem pieces with 3-4 nodes are planted in polythene bags filled with good rich soil. This operation has to be done much in advance of planting at site so that rooted cuttings are available for planting. This is done at least 2-3 months in advance. Periodic watering of bag plants is essential. In about 10-15 days time cuttings strike roots and start growing. With the first regular shower of monsoon the rooted cuttings or suckers as the case may be are planted at site at 60 × 60 cm apart in 30 × 30 × 30 cm pits. During dry spells watering is required. Weeding may be done as and when necessary. As stated, the plant is shade tolerant and as such it can be grown as an under crop.

***Piper longum* L.**

Prophylactic treatment: In case of any insect, pest or disease insecticides and fungicides maybe sprayed.

Harvesting: After about 6 months of planting when the green catkins start turning to black they should be collected and dried in sun for about 5-6 days. Irregular or insufficient drying results in mould formation. Dried catkins loose 15-20 % moisture. The yield of catkins increases with the age of plants till about 3 years. In the first year the yield may be 25 kg, in the second year about 600 kg and in the third year it maybe about 1 tonne. Thereafter the yield declines when the crop is required to be replaced. Since the roots are also valuable they are dug out and dried and cleaned. These dried roots are called Modi in Indian medicine. About 5 tonnes of dried roots can be expected from a hectare of plantation.

Storage: Dried catkins should be stored in a dry and cool place.

Uses: Catkins or Pipli are used in blood purification, helps in digestion as diuretic in liver ailments, general debility to alleviate cough and bronchitis and as stimulant and expectorant. The roots are used as carminative in liver ailments and ladies ailments. They also have aphrodisiac properties.

Pistia stratiotes L.

137. *Pistia stratiotes* L.; Araceae.

Na : Bengali: Takapana; Guj.: Jalashankhala; Hindi: Jalkhumbi; Takapana; Kan.: Antaragange; Malayalam: Akasatamara, Kudapayal, Muttapayal; Marathi: Prasni, Gondala, Jalamandvi; Oriya: Borajhanji; Tamil: Akashtmarai, Kodilamarai; Telugu: Antharaidhaman, Nirubuduki, Antharathamara.

It is a hydrophyte. The plant is said to be useful in the treatment of urinary affections and helps in the disintegration of bladder stone. It is also stated to be useful in bronchial diseases.

Plumbago zeylanica L.

138. *Plumbago zeylanica* L.; Plumbaginaceae.

Na : Bengali & Hindi: Chita, Chitrak, Chitra; Guj.: Chitaro, Chitrak; Kanerese: Chitramula, Vahni; Malayalam: Tumbakoduveli, Vellakoduveli; Marathi: Chitramula, Chitraka; Oriya: Chitamulo, Chitapro, Krisanu, Ogni; Tamil: Chitramulam; Telugu: Agnimula, Chitramoolam.

Hab : It is a small perennial shrub found in drier parts.

Roots are used as a carminative by the local people.

Podophyllum hexandrum Royle

139. *Podophyllum hexandrum* Royle (Syn. P.emodi Wall. ex Hook. f. & Thoms.); Berberidaceae.

Na : Eng.: Indian podophyllum; Hindi: Bakrachimyaja, Bakrachimyaka, Bhavanbakra, Nibishi, Nirbash, Papra, Papri, Pilijati.

Hab : It is an herb growing wild in the temperate and alpine regions in the central and western Himalayas. It is an important pharmacopoeial drug plant. The rhizomes and the roots are the source of podophyllum resin which is used as a cholagogue purgative. Recently the resin from this plant has been found to affect the growth of malignant tumours and it has also been shown to be effective in controlling some skin diseases.

It can be raised either by direct sowing of seeds or by planting of rhizome cuttings (from tops bearing buds and roots). In the case of direct sowing the seeds may be sown with or without fruit pulp immediately after their collection in July/August. At least seeds may be dibbled at the specified spacing to ensure uniform stock as some of the seeds may fail, to germinate. Planting of rhizome cuttings may be done at the end of June or beginning of July at an espacement of 30 × 30 cm. Rhizome cuttings start sprouting in about a week's time after planting. Germination of seed, in the case of sowing, starts only in the following spring. They remain dormant in winter. Germination is complete in about a month's time. During winter the sub-aerial parts die back due to frost and snow and with the commencement of spring the plants again sprout. Generally one or two weedings are required to be done during the rains. The plants are dug out for roots and rhizomes at the age of 5 years preferably in early autumn.

Pongamia pinnata (L.) Pierre (tree)

140. *Pongamia pinnata* (L.) Pierre (Syn. *P.glabra* Vent.); Lcguminoseae.

Na : Assamese: Karchaw; Bengali, Hindi, Gujrathi, Marathi: Karanj, Kanarese: Honge; Mal.: Pungu, Punnu; Oriya: Koranjo; Tamil: Pongam; Telugu: Pungu.

Hab : It is a glabrous tree with spreading crown. Bark is greyish green imparipinnate. It is commonly found in moister places particularly along river banks, in central western and southern India and also in Bengal. The oil obtained from the seed is used locally in the treatment of skin diseases and rheumatism.

***Pongamia pinnata* (L.) Pierre Branch**

***Pongamia pinnata* (L.) Pierre Fruit**

141. *Prosopis cineraria* L. (Druce); Mimosaceae.

Na : Beng.: Shami; Guj.: Semi, Semru, Khijado, Hamra, Kandi; Hindi; Jand, Chaunkra, Khar, Khejra; Knnerese: B.mni, Shame; Malayalam: Parampu, Tambu; Marathi: Shemri, Saunder, Oriya: Shami; Tamil; Jambu, Perumbay; Telugu: Jammichettu.

Hab : It is a small tree found in dry tropical forests in Andhra Pradesh and other places. Stem, bark, leaves and fruits are said to be used for removing hair.

***Prosopis cineraria* L.**

142. *Psoralea corylifolia* L.; Fabaceae (Papilionaceae.)

Na : Bengali: Bavachi, Kakuch, Latakasturi; Guj.: Babchi, Bavchi; Hindi: Babchi, Bavanchi, Bukchi; Kanerese: Bavanchigida, Karbekhiga; Malayalam: Karpokkari, Kaurkoalari; Marathi: Babachi, Bavanchya; Oriya: Bakuchi; Sans.: Bakuchi, Kushtanashini, Sugandha Kantak; Tamil: Karboka arisi, Karporgam; Telugu: Bavanchalu, Bapunga, Bavanchi.

Hab : It is an annual plant with alternate leaves. Leaves are serrate. Seeds when fresh are yellowish and become black on drying. Seeds are the source of an oleoresinous substance which is used locally in the treatment of leucoderma, leprosy, kidney troubles and other skin diseases.

***Psoralea corylifolia* L.**

143. *Pterocarpus marspium* Roxb.; Fabaceae (Papilionaceae)

Na : Bengali: Pitshal; Gujarati: Biyo, Hiradakhan; Hindi: Bijasal, Bija; Kanerese: Honne Bange; Malayalam: Venga; Marathi: Asan, Bibla, Dhorbenla; Oriya: Byasa. Tamil: Vengai; Telugu: Yegisa, Peddagi; Trade: Bijasal.

Hab : It is found in tropical forests in Western Central and South India. The gum obtained by tapping the stem is used locally in diarrhoea and for toothache. The wood is said to contain curative effects on diabetes and also helps in reducing the frequency of urination. In such cases local people drink the water which is kept overnight in the bowls made from the wood.

***Pterocarpus marspium* Roxb.**

144. *Punica granatum* L.; Punicaceae.

Na : Bengali: Doom; Eng.: Pomegranate; Guj.: Dadam; Hindi: Anar; Kanerese: Dalimba; Malayalam: Matalam; Marathi: Dalimba; Tamil: Madulai; Telugu: Danimma.

Hab : It is a shrub or small tree mostly cultivated. It is found growing wild in Kulu valley. The peel from the fruit is used by local people in diarrhoea and dysentery.

***Punica granatum* L**

145. *Rauvolfia serpentina* Benth.; Apocynaceae.

Na : Assamese: Arachoritita; Bengali: Chandra; Hindi: Chandrabhaga, Chota-chand, Sarpagandha; Kanerese: Sarpagandha, Sarpagandhi, Shivanabhiballi, Sutranavi, Patalagandhi; Khasi: Todong-pait-parao; Malayalam: Churannavilpori, Suvapavalporiyan; Marathi: Harkaya, Harki; Oriya: Patalagarur, Sanochado; Sanskrit: Sarpagandha, Chandrika, Patalagaruda; Tamil: Chevanamalpodi; Telugu: Patalaguni, Patalagaruda, Sarpagandha; Trade: Rauvolfia, Sarpagandha.

Hab : It occurs practically throughout India from Kerala to the Himalayan foot hills upto an altitude of about 1200 m except Rajasthan. Although the range of distribution of this species is very wide, it is not common and the plant grows only sporadically. Increased tempo of modern life with stresses and strains has brought about an increasing incidence of mental and physical ailments. Hypertension is generally considered as a major threat to active and useful life after 50 years. Sen and Bose reported the drug of Rauvolfia serpentina Benth. valuable and safe in the treatment of this disease. It has now acquired great importance in modern therapeutics and is extensively used in reducing blood pressure in hyperpiesis and as a sedative in the treatment of insomnia; hypochondria, mental disorders and certain forms of insanity. It is also used in numerous other diseases like psoriasis, excessive sweating and itching, gynaecological ailments such as menopausal syndrome, in old-age afflictions, angina pectoris and rapid or irregular heart action and toxic goitre. As high blood pressure is one of the commonest constitutional diseases of the present day, the drug is considered a great boon. The drug is administered in the form of whole root powder or its extract reserpine.

Rauvolfia serpentina Benth.

The plant prefers soil with plenty of humus or rich in nitrogenous and organic matter with good drainage. It grows in tropical or subtropical regions, having the benefit of monsoonal rains preferably the south west. It grows luxuriantly well where the rainfall is 2500 mm or more. In localities with moderate rainfall with no irrigation facilities, the plant could be grown under the partial shade of tall or medium sized trees. The areas having more equable climate seem to be more suited than the areas having larger climatic variations. The plant can easily be propagated through seeds sown in nursery as well as by vegetative means i.e. stem as well as root cuttings but the yield of roots of the plant raised through seeds is far more than the yield obtainable from plants raised vegetatively. The seeds are sown in nursery beds of convenient size during March/April. Before sowing, the seeds may be soaked in water for about 24 hours. They are sown in drills at 1 cm depth. The distance between the seeds may be kept at 2.5 cm. The beds are kept moist during dry summer months. Germination starts in about 25 days. Good viable seeds give about 30 percent germination. About 20 kg of seed is required for raising one hectare of plantation. After germination frequent and regular weedings are essential. Transplanting is done by end of June or beginning of July depending on the break of monsoon. Seedlings are about 7 to 10 cm in height at the time of transplanting. They are planted in holes made with a stick. Spacing between the seedlings is kept at 30 × 30 cm. Two or three weedings may be necessary in the first year. Plants are dug out for roots after 2 or 3 years growth. At the time of digging soil must be moist to facilitate digging and separation of roots from the adhering soil. Two year old Plantation can give about 1120 kg of dried roots per hectare. After digging, the roots are thoroughly dried in the sun otherwise they get mouldy and become useless. The present market price of dried roots is Rs.15 per kg. Indian pharmacopoeia prescribes that the foreign organic matter in the roots should not be more than 2 per cent and the total alkaloids not less than 0.8 per cent in the roots.

Rheum emodi **Wall.**

146. *Rheum emodi* Wall.; Polygonaceae.

Na : Beng.: Bangala revanchini; Bhutan; Tuzha; Gujrathi: Gamnirevanchini; Himachal Pradesh: Ladu, Chuchi; Hindi: Revandchini; Kanerese; Natrevachinni; Kashmir: Chutial; Kumaon: Archu; Ladakh: Lachu; Marathi: Mulkacha revalchini; Punjab: Chutial, Chuki, Khabium, Khandaul, Lachu, Pumbash, Alsu, Ribas (stalks), Revandchini (roots); Sanskrit: Gandhini, Pitamulika, Revatchini; Tamil: Nattuirvelchinni, Nattumanjal chinakhizangu; Telugu: Nattupasupu-chinnigadda, Nattu revalchini.

Hab : This herb is found in the sub alpine Himalayas. It is a good substitute to Chinese rhubarb R.officinalis which is imported. The roots and rhizome of the rhubarb are the source of a drug which is used as a stomachic in dyspepsia, laxative, tonic and purgative. Small cuttings of rhizome are planted at about 1m espacement at the commencement of monsoon. At the beginning of winter the rhizomes are dug out.

147. *Rhus succedanea* L.; Anacardiaceae.

Na : Eng.: Japanese Wax tree; Bengali & Hindi: Kakrasingi; Kanerese: Karkata shringi; Marathi: Kakada-shing; Tamil: Kharkhaadagasurgi; Telugu: Kakeerasringi.

Hab : It is found in the temperate Himalayas from Kashmir to Khasia hills. A liquid obtained from the mesocarp of the fruit is used locally in ointments.

148. *Rivea hypocrateriformis* (Desv.)Choisy Syn.*R.ornata* (Roxb.) Choisy; Convolvulaceae.

Na : Bengali: Kalmilata; Hindi: Phang; Marathi: Kulnilata, Phanji; Tamil: Budthi, Kiray; Telugu: Bodditeega, Niruboddi.

Hab : It is a perennial climber with grey coloured orbicular leaves. The plant is locally used in the treatment of piles.

149. *Saussurea lappa* C.B. Clarke. (Syn. Aucklandia costus Falc.); Asteraceae (Compositae).

Na : Bengali & Hindi: Kut, Kur, Pachak; Gujrathi: Upleta, Kur; Kanerese: Kashtra; Kashmir: Kuth, Chob-i-qut, Post khai; Malayalam: Sepuddy; Marnthi: Kushta; Sanskrit: Kushta; Tamil: Kostum, Putchuk, Goshtain; Telugu: Changala, Kustam, Trade: Kuth.

Rhus succedanea **L**

Hab : It is a perennial shrub attaining a height of 1.8 to 2.1 m. It is found wild in Kashmir at altitudes of 2400 to 3600 m and is cultivated in Himachal Pradesh (Chamba, Kulu and Lahaul Valleys), the Uttara Khand (Garhwal Hills) and in Sikkim.

Soil: The plant flourishes in rich, well-drained loamy soil.

Propagation: Kuth is propagated from seeds sown in well-prepared nursery beds in March-April soon after the snow melts. The beds are weeded and irrigated as and when required. The seedlings are transplanted in intensively prepared and well manured beds, Kuth being a root crop this aspect is very important. In Sikkim, the practice adopted is to prick out the seedlings in beds at a spacing of 5 × 7.5 cm in September or October before snow-fall. In the second year the seedlings are transplanted in their final situation in well- prepared pits 2 m apart.

Harvesting: In Sikkim, the plants are reported to be harvested after three years in the field. If the harvesting is delayed the roots tend to become hollow. In Garhwal the best time for harvesting the roots is at the end of October or early November. The harvested roots are cut into short lengths of about 5 to 10 cm, and partially dried on the spot after which they are transported to a drying godown at a lower altitude for further drying.

Saussurea lappa **C.B. Clarke.**

Chemical constituents: Roots contain essential oil (1.5-2.5%), alkaloid saussurin (0.05%) and a bitter resin (6%) and tannins. The constituents of oil are costus lactene ($C_{15}H_{20}O_2$) 11, Costusic acid ($C_{15}H_{20}O_2$) 14, dihydrocostus lactone 15, bicyclic sesquiterpene alcohol costal 7, sesquiterpene hydrocarbon, aplotoxin 20, £-costene 6, B-costene 6, phellandrene 0.4 camphene 0.4 and a terpene alcohol 0.2 per cent. The root also contains inulin 18 % fixed oil, potasium nitrate, sugar etc. A liquid constituent named Kusthin ($C_{20}H_{26}O_3$) has also been isolated from the roots.

Action and Uses: It is antiseptic, disinfectant, relaxes the involuntary muscle tissue, cardiac stimulant, carminative, expectorant and diuretic. Alkaloid saussurine has a depressant action on the vagus centre in the medulla as well as on the involuntary muscle fibres of the bronchioles and gastrointestinal tract. It produces a slight but persistent rise of blood pressure and increases the force of contraction and amplitude of the ventricles. The drug has an effect in controlling bronchial asthma, especially those of vagotonic type. The paroxysms are cut short by the combined action of the essential oil and the alkaloid. The essential oil, during this operation in the lungs, not only relaxes the bronchial muscle but has a marked expectorant action which relieves turgecence of the mucosa. It is also useful in persistent hiccup.

Dried and powdered root is astringent and stimulant and is applied to severe ulcerations. The powder of dry root is used in hair wash. The root is universally employed by shawl merchants to protect Kashmir fabrics from the attack of moth and other vermins due to its insect repellent property. It is an agreeable fumigatory and yields excellent pastilles which burn fairly well.

The essential oil is used as a fixative and for blending high class perfumery for which it has a high reputation in countries like U.K. France and U.S.A. where it is sold at fabulous prices.

Standards: Crude Drug I.P.C (Indian Pharmaceutical Codex):

Foreign organic matter not more than 2%, Volatile oil not less than 1.6% Costus oil. (Ess.oil Ass.,U.S.A.) Spefic gravity - At 25 °C 0.995 to 1.039

Refractive index - At 20 °C 1.512 to 1.523

Optical Rotation - + 10° to + 36°

Solubility in alcohol-Soluble in 0.5 vol. of 90% alcohol, cloudy upon further dilution with occasional separation of paraffin crystals.

Acid value-Not more than 42 (usually 8-40)

Ester value-90 to 150

Descriptive characteristics:

1. Solubility- Benzl benzoaic soluble in all proportions.
   Diethylophthalate-Soluble in all proportions.
   Fixed oil-Soluble in all proportions.
   Glycerine-Practically insoluble.
   Mineral oil soluble in all proportions.
   Propylene glycol-Practically insoluble.
2. Stability acid-Relatively stable in presence of weak organic acids.
   Alkali- + Relatively unstable in presence of alkali.
   Storage: May be stored preferably in tight glass or aluminium containers and kept in cool and dark place.

Schleichera oleosa **(Lour.) Oken**

150. *Schleichera oleosa* (Lour.) Oken. Syn *S.trijuga* Willd.; *Pistacia oleosa* Lour.; Sapindaceae.

Na : Bengali: Kusum, Kosano; Gujrathi: Kusum, Kosumbe; Hindi: Kusum; Kanerese: Sagade, Kendala, Jendala, Chakota; Malayalam: Puvam; Marathi: Kusumb, Koon, Kohan; Oriya: Kusamo, Kusam, Swad, Kussum; Punjab: Sama, Jamoa, Gausam, Kussani; Tamil: Puvathipuvam, Pulachi, Karanchi, Kula, Pumaratha; Telugu: Posku, Pusku, Botanga, Sagade pusuku, Mavita vitiki.

Hab : It is a tree found in the Sub-Himalayan tract, Bihar, the Punjab, Madhya Pradesh and South India. The powdered seeds are applied to ulcers of animals for removing maggots.

151. *Seseli indicum* Wight & Arn.; Umbelliferae.

Na : Bengali: Ban-jowan; Hindi: Banajawan; Kanerese: Kinumanji ajwana, Kaduajwana; Marathi: Kirminji ajvan; Sanskrit: Vanayamani.

Hab : It is a herb commonly found in Uttar Pradesh, Bengal and Assam. The fruits are used as stimulant, stomachic and are also used in expelling round-worms.

Sida cordifolia L.

Siegesbeckia orientalis L.

S. nigrum L. – branch and fruits

S. Virginianum L

152. *Sida cordifolia* L.; Malvaceae.

Na : Bengali: Swetberela, Berla, Bala; Gujrathi: Bala baldana, Mahabala, Khapat; Hindi: Kungyi; Kanerese: Hettuthi, Hettugigida, Kisangi, Chittu, Chitu haralu; Malayalam: Kurunthotti, Velluram, Kathuram; Oriya: Badianaula, Bisvokopari; Punjab: Kowar, Simak; Tamil: Nilatutti, Paniartuthi; Telugu: Tellantisa, Tellagorra, Chirwabenda, Survarnamu.

Hab : It is a small shrub which occurs in the plains of India. Ephedrine has been found in small quantities in this plant.

153. *Siegesbeckia orientalis* L.

Na : Assam: Gawalbahalgani, Soh-barthud, Lib; Bihar: Latlatia, Narangkal megh; Garhwal: Lichkura; Gujarat: Pilibadkadi; Marathi: Katampam; Tamil & Telugu: Kalampam, Kadambu.

It is a large erect, pubescent annual herb upto 1.3 m high found very commonly on the edges of fields and streams or in damp fallow lands throughout the Dun valley abundant in the Himalayas upto 1500 m. It is cosmopolitan in warm localities. It is conspicuous by the large spreading linear-spathulate outer involucrate bracts which by means of their viscous glandular hairs catch on to clothes. The leaves (bitter) are used as febrifuge.

154. *Smilax zeylanica* L. (Syn. *S.macrophylla* Roxb.); Smilacaceae.

Na : Hindi: Ramdataun, Jangliaushbah.

Hab : It is a shrub found in Kumaon, Bihar, Madhya Pradesh and Assam. The roots are used as a substitute for sarasparilla in the treatment of venereal diseases. They are also used in dysentery and rheumatism.

155. *Solanum nigrum* L.; Solanaceae.

Na : Assam; Pichkati; Bengal: Gurkamai, Kakmachi, Tulidan; Marathi: Kamuni, Ghati, Mako; Gujarathi: Piludi; Hindi: Makoi; Sanskrit: Kakamachi; Tamil: Munatakli. Telugu: Kachchipandu Kachi, Kamanchi, Gajju chettu.

Hab : It is found in wastelands in dry places. Extract of leaves is said to be used in urinary and skin diseases, as a blood purifier and in jaundice.

156. *S. virginianum* L. (Syn. *S. xanthocarpum* Schrd. & Wendt.); Solanaceae.

Na : Hindi: Kateli, Ringani, Katai; Guj.: Bhoyaringani; Marathi: Bhulrungani; Mal. & Tamil: Kandankathiri; Sanskrit: Kantakari, Nidigadhika; Telugu: Nelamulaka, Vankuda.

Hab : It is a common shrub. The roots are used in cough, asthma and the leaves for rheumatism.

157. *S. viarum* Dumal (Syn.S. *khasianum* C.B. Clarke Sen Gupta); Solanaceae.

Na : Va.-Khasianum

Hab : It is a shrub found in high rainfall areas in Bengal & Assam along the foot hills of Himalayas.

When the berries of this shrub are just ripe they are collected to obtain active principles which are used in synthesis of sex hormones and as birth control devices.

Soymida febrifuga A. Juss

158. *Soymida febrifuga* A. Juss.; Meliaceae.

Na : Bengali: Rohan, Rohina; Guj.: Rohina; Hindi: Rohan, Rohunna, Rakat; Kanerese: Suani; Trade: Indian red wood, Bastard cidar;

Hab : It is a common deciduous tree found in the tropical forests in western peninsula, Rajasthan, Bihar and central and South India.

The bark is used in local medicine as a stimulant, in diarrhoea and dysentery.

159. *Spilanthes acmella* Murr.; Asteraceae. (Compositae)

Na : Assam: Pirazha; Beng.: Paracress; Hindi & Marathi: Akalkara, Pipulka; Kanerese: Vanamugali; Hemmugalu; Malayalam: Kuppamanjel; Punjabi: Akarkarha, Pokarmul; Telugu: Maratimogga, Maratiteega.

Hab : It is an herb found in the temperate Himalayas. The leaves are used to relieve toothache.

Spilanthes acmella **Murr.**

160. *Stevia rebaudiana* : Asteraceae, (Compositeae)

Na : Eng.: Stevia

Hab : It is a small sized perennial herb native of Paraguay and southwest Brazil. It is also grown in China and Japan. Though it is not native to India and does not grow in Indian forests, considering its economic importance and usefulness as a calorie free substitute for sugar to diabetic patients and in preparation of sweetmeats it is recommended that it may be grown in nurseries as undergrowth in forest plantations. It can conveniently be grown in tropical agro climatic conditions in cooler and moister localities.

Chemical constituents: There are 8 glycosides in the leaves which have sweetening properties, the chief constituent being Stevioside having the sweetness of 250-300 times the sucrose. Stevia of Paraguay is 300 times sweeter than sugar. Another constituent Rebaudioside found along with Stevoside in the leaves of Stevia is 400 times sweeter than sugar. Stevioside concentration ranges between 3 and 10 % of the leaf dry weight. While Rabaudioside is less concentrated, ranging from 1-3 %. It is non fermentable sugar substitute.

Stevia rebaudiana

Cultivation: Stevia is easy to grow. The plant produces flowers and seeds. Seeds are small. Rooted cuttings give quicker yield. 15-20 cm cuttings of stems of better strains are dipped in 500 PPM indolbuteric acid (root hormone solution) for about 15 to 20 minutes and planted in polythene bags, taking care to see that they are not planted upside-down. Plants are watered regularly. In about 15 to 20 days they start rooting and after another 10 to 15 days they can be planted in the field at an espacement of 60 × 60 cm. Planting should be done with regular shower of monsoon. Adequate organic manure as a basic dose with NPK fertilizer of 100,50,50 kg per hectare. And sufficient irrigation will boost up the yield of the leaves. In about 3 month's time, flowering takes place. The first harvest of leaves can be had before flowering. While harvesting, the branches with leaves are pruned and dried. The dried leaves are separated. Annually about 4 to 5 harvests can be had. Depending upon irrigation facilities and application of fertilizers about 100 quintals of leaves can be obtained. The plant can be grown in one's backyard for daily use.

Uses: It is non calorific, non fermentable sugar substitute and does not become brown like sugar. It is used in soft drinks, cakes, pies, ice-creams, brewing tea. Diabetic patients can use Stevia leaves extract as a sweetening agent. No ill effects on the body system by the use of Stevia has been reported.

Strychnos nux-vomica L.

161. *Strychnos nux-vomica* L.

Trade and Indian names: The drug is known in the trade as nux vomica. The common Indian names of the plant are Bengali: Kachila, Kuchila, Thalkesur; Gujarat: Kuchla; Hindi: Bailewa, Chibbenge, Kajra, Kuchla; Kanerese: Hemushti, Ittangi, Itti, Kajavara, Kasara, Kasaragade, Kasaraka, Kasarakai, Kasarika, Kasarakana, Kasaragida, Khasea, Kujarra, Mushti, Nanjina, Nanjinakordu, Nanjunda, Yetti Yettica; Konkani: Kazro; Koya: Mushti; Madhya Pradesh: Kuchla; Malayalam: Kanhiram, Kanjera, Kanjirap; Marathi: Jharkatchura, Kajrakar, Kara, Karo; Tamil: Eddikanchera, Kanchurai, Yetti, Yettikanjuri; Telugu: Musadi, Mushadi, Kushti; Oriya: Karan, Kerra, Kora, Korra, Kuchila. In English it is known as Crowfig, Nux-vomica tree, Poison nut, Snakewood, Strychnine tree and in Sinhalese as Goda-Kaduru.

Introduction: Nux-vomica, to be more specific strychnine, is obtained from the seeds of Strychnos nux-vomica L. It is one of the important pharmacopoeial drugs. Strychnine is used as a stimulant in nervous disorders and also as a stomachic and general tonic. It is extensively used to kill rodents and pests. Nux-vomica does not appear to have been used in early Sanskrit medicine. In the more recent Sanskrit compilations we find it mentioned under its local name kuchila. The Muhammadans knowledge of the use of nux-vomica seems to have been derived from the Hindus as Makhzan-el-Adwiya concludes the description of the drug by saying that, much information will be found about the drug in Hindu medicine. In European medicine it was first known about the middle of the 16th century when Valerius Cordus who wrote in German gave a very accurate description of the seeds. However it was not used in medicine till a century later, since Parkinson (1640) remarks that its chief use was for poisoning cats, dogs, crows etc., and it was not till the beginning of nineteenth century that its value as a nervine tonic was recognised by European practitioners. It is a curious fact that the origin of the name nuxvomica is not traceable. The seed was never used as an emetic as the word "Vomica" would imply and it is likely that the earlier writers might have possibly been confused with some other seed that possesses emetic properties possibly *Strychnos potatorum* L. since Ainslie says that this species is reckoned in its natural state as an emetic by the Tamil doctors in South India.

Botanical characters: It is a slow growing small handsome deciduous to semi-evergreen shade bearing tree growing frequently in shade. It attains a height of 9 to 12 m its growth and development depending largely on the extent of rainfall. Where it occurs in moist deciduous forests it attains a fairly large size of about 30 m height with a girth of about 350 cm at Breast Height. Branches are opposite. Bark is Yellowish grey to blackish grey, thin, smooth, covered with minute tubercles and containing chlorophyll tissue. It has dullish green white flowers. Unlike the flowers, the fruit is bright looking, brownish yellow berry about the size of a small orange and containing a large quantity of gelatinous pulp in which 1 to 5 seeds are embedded. The seed is flat disc-shaped 15 to 30 mm in diameter and 4 to 6 mm thick, hard, sometimes irregularly curved, the edges rounded or subacute. It is depressed in the centre. The abundant endosperm in it is horny and translucent. It is odourless. It is light greyish or greenish satin appearance which is due to the presence of an outer coat of hairs. 60 to 90 seeds weigh about 100 gms. The wood when freshly cut is light cream white in colour but turns to light brown on exposure. The wood is hard and close grained. Weight of wood is 987 kg per cu.m.

Phenology: In moist types of forests the tree is semi evergreen to evergreen but in dry types on lateritic soils, it loses its leaves for a short time in the hot season. In South (Tamilnadu and Kerala) large tufts of dullish green white flowers appear from March to May and the fruits ripen in the cold to hot season from December to June.

Distribution and Availability: Nux-vomica plant is widely distributed throughout the tropical regions in India particularly in South India and the world's supply of the drug is almost entirely met from here. In the hills the species ascends up to 1200 m altitude. In economically exploitable form it occurs in Calicut and in Travancore and Cochin hills in Kerala state; in Madurai, Tanjore, Trichinapalli, Coimbatore, Vellore in Tamilnadu; in Bellary, Shimoga and Kanara districts in Karnataka state; in Warangal, Khammam, Rajamundry, Nellore, Godavari districts, Kurnool, Anantapur, Cuddapah in Andhra Pradesh; in Ganjam and Berhampur in Orissa state; in Raipur in Madhya Pradesh, and in South Chanda, Bhandara and Berar in Maharashtra. In the North it is found in Gorakhpur in Uttar Pradesh; in Chota Nagpur in Bihar; in Midnapur and Mayurbbang, in West Bengal; in the North Cachar hills and Nowgong in Assam state but the quantities of seeds obtainable in these places are negligible in comparison with the production in South India. In Sri Lanka it is found in the northern jungle districts between Kurunegalla and Jaffna. It is also found in Malaya.

Climate and Soil: In its habitat the absolute maximum shade temperature varies from 35 to 48 °C and the absolute minimum from 4 to 18 °C. The normal rainfall varies from 1245 to 3556 mm. (35 to 150 inches) or more. Although it prefers sandy loam soils it also occures on lateritic and clayey loam soils.

Propagation: The plant is a good coppicer. As such when it is felled new shoots come up from the periphery of the stump and start growing. It has been observed that when trenching all round the tree is done, the injured roots send out root suckers which in turn get established. No systematic work on the artificial regeneration of the species has been done so far. However from some trials it appears that the seeds sown in December-February have germinated after 70 to 120 days. Because of the long period and difficulty in getting the germination, transplanting appears to be more suitable than direct sowing for which the usual procedure is to sow the ripe seeds in the nursery soon after their collection. They should be soaked in hot water for 24 hours before sowing. It takes 2 to 3 months for germination and sometimes even longer. In the nursery the seedlings may be protected from sun and heat by having a moderate shade. When the seedlings are at least 30 cm in height, preferably two year old seedlings are transplanted in pits in the field at an espacement of about 9 m with the onset of regular rain of monsoon. The size of the pit maybe 30 × 30 × 30 cm. At the time of planting, care should however be taken to see that the collar of the plant is kept just above ground level and also the soil around the plant is pressed firmly so that the root is in complete contact with the soil. Two or three weedings may be necessary in the first two years. Weedings may be done as and when required in the initial stages of growth.

Rate of Growth: Little is known regarding the rate of growth of the plant. It starts bearing fruits after about 10 to 12 years. In about 30 years it attains a girth of about 60 cm depending on the locality factors.

Collection and Processing: The right to collect the fruits is sold by a system of leases. The areas are put up to auction by the forest department in the month of June/July. The lease period is generally for one year. The lessee having taken the area on lease engages his agents who in turn go to the villages and approach the people to collect and supply the seeds to them. Wages are paid to the collectors on a quantity basis. The actual work of seed collection is done by the forest tribes – Gonds, Santals and Mahars and other villagers to whom this is a secondry occupation. The collection of the produce should very carefully be made. The fruits may be collected on the trees as soon as they ripen or immediately after they fall. If they are left on the ground they are devoured voraciously by the cattle. However in practice very slip-shod methods of collection are adopted and even the seeds damaged by wild animals and birds are not left uncollected although these fetch considerably lower prices. The way to get the best seeds is to shake the mature fruits off the branches. The pulp is removed by retting or by rubbing on a strong sieve preferably in running water. The seeds are dried in the sun by spreading them on clean mats and agitated twice or thrice a day. It takes roughly about seven days for the seeds to dry. They are then sold to the agents of the lessee. The crop usually begins to arrive in the market in good quantities in December and gathering season extends through the cold weather upto end of April/May. It has been observed that although there is a long period of collection, the effect of seasonal variation on, the alkaloidal content is insignificant. Therefore it is immaterial if the seed is collected in winter or summer.

Yield of Seed: The yield of seed is generally proportional to the size of the tree. The most interesting feature is the amazing variation in yield from year to year and the failure of many trees to yield any seed at all.

Storage: It has been observed that the seeds stored in damp places no doubt lose their original brightness and get covered with a surface layer of mould but the alkaloidal content remains almost unaltered. The fungal attack is only on the outside silky coat and does not penetrate into the horny portion of the "Button". However in order to retain the original brightness and consequently to get a better price, damp places are avoided for storage. Storage of seed for year and a half has not affected the quality of seed.

Constituents and standards: Alkaloids are present in both the seeds and pulp even in the very' early stages of development of the fruit. The seeds contain much more alkaloids than the pulp. During the process of ripening the pulp loses its alkaloidal content and rather rapidly towards the end so much so that only traces of it are left in December/January collection. Therefore seed is the only product which not only has high alkaloidal content but also retains it throughout its stages of development and storage. Nux-vomica seed contains the alkaloids strychnine and brucine together with traces of strychnicine and vomicine and of a glycoside loganin. It also contains fatty matter and chlorogenic acid. The total alkaloid varies from 1.8 to 5.3 per cent. According to the British Pharmacopoeial standard nux-vomica should not contain less than 1.20 per cent of strychnine.

Adulteration: The seeds of *S.potatorum* L. a species found widely in the tropical forests and S.colubrina L. found in the Western Ghats in India and a Burmese species of *S.nuxblanda* A.W. Hill are used for adulterating nux-vomica seeds. The seeds of *S. potatorum* L. are smaller thicker and also free from bitterness. These are often known as clearing nuts. *S. colubrina* L. has varying strychnine content in its seeds. Its seeds are ovoid, slightly compressed brown and glabrous. The seeds of S.nuxblanda A W. Hill contain no strychinine whatsoever and have no bitterness. This plant was being confused with Indian Nux-vomica till recently when it was identified as a distinct species. Powdered vegetable ivory *Phytolephas macrocarpa* Ruiz el Pav is also employed to adulterate powdered nux-vomica. Adulteration of the drug is bound to lower the alkaloidal content and its efficacy and potency is reduced.

Medicinal and other uses: The drug nux-vomica is widely used in general prescriptions. However it is a virulent poison in large doses. The dry extract is combined with cascara sagrada and other laxatives in the form of pills and given for the treatment of constipation. It is bitter in taste and increases gastric secretion and appetite. Strychnine is also used in the treatment of nervous disorder and paralysis. Brucine is used in analytical chemistry and its salt for denaturing alcohol used in cosmetic preparation. The seeds are used by the hill tribes in the Nilgiris as a fish poison. The bark is used as a tonic and as an external application for skin diseases in Srilanka. The wood is said to be a febrifuge. Strychnine is commonly used in European countries to kill rodents. The seeds yield a dye which produces light brown shades on cotton cloth. The seeds are also used in the distillation of native liquors in some places in the country in order to add to its power of intoxication. The wood is not eaten by white ants probably on account of its bitterness. It is used in the manufacture of fancy and cabinet articles, cart wheels and other agricultural implements. Leaves are used as a green manure.

Pharmacological preparations: There are a number of nux-vomica preparations but some of them are as follows:

1. Elixir of Nux-vomica, B.P.C.
2. Extract of Nux-vomica, Dry, B.P.C.
3. Extract of Nux-vomica, Liquid; Nux-vomica Liquid Extract, B.P.
4. Mixture of Gentian with Nux-vomica, Acid, B.P.C.
5. Mixture of Gentian with Nux-vomica, Alkaline, RP.C.
6. Mixture of Nux-vomica, Acid, RP.C.
7. Mixture of Nux-vomica, Alkaline, B.P.C.
8. Mixture of Potassium Bromide and Nux-vomica, B.P.C.
9. Tablets of Cascara, compound, B.P.C.
10. Tincture of Nux-vomica; Nux-vomica Tincture, B.P.C.

Trade: The trade in nux-vomica seed and strychnine is practically the monopoly of India and Srilanka the only competitor is the Philippine Islands which produce *S. ignatii* Berg. the seed of which contains reasonable amounts of strychnine and brucine. These are also used as a poison for rodents and as an insecticide. The other producing countries of *Strychnox nux-vomica* seed are Burma, Malaya and Srilanka. Hitherto India used to export all the seed but with the recent developments in the country strychnine is being manufactured. The Governament has placed a ban on the export of seed. Only a limited quantity of seed is being exported. However the extracts are finding their way in the export trade. At present there are 3 or 4 manufacturers in India who are extracting nux-vomica alkaloids and salts showing their requirement as 1500 to 1600 tonnes of seed. In fact the actual consumption seems to be far less. As a result there is an abrupt decline in demand so much so that even after meeting the requirement of Indian manufacturers large quantities are remaining unexploited in the forests. During the year 2002-03 India exported Nux-vornica seeds valued at Rs. 329614.

Conclusion: It has already been stated that the species is sporadic in nature and very seldom gregarious. This makes the collection of seed more difficult and costly. Since there is a considerable demand for the seed in the trade, attention may have to be paid to raising the species artificially in a concentrated manner. In afforestation area, where the object of management is to conserve soil by growing trees and where production of timber is not aimed at, this species may be given preference over economically less important ones. This will result in increasing the production of seed and reducing the cost of collection. Thus in the years to come a steady and monopolistic trade of this drug could be built up.

***S.potatorum* L.**

162. *S. potatorum* L.; Loganiaceae.

Na : Bengali: Nirmali; Bombay: Nirmali, Gajrah; Hindi: Nirmali, Nelmal, Neimal; Kanerese: Chilu, Chilladabeeja; Malayalam: Tetranparal, Tetranparel. Oriya: Kotaku; Sanskrit: Kataka ambuprasada; Tamil: Tetankotai, Tettran, Tctta: Telugu: Indupachettu, Chillachettu.

Hab : It is a medium sized plant. Occasionally found throughout dry tropical forests. The seeds are said to be useful in cough and bronchial troubles and in the treatment of eye diseases particularly conjunctivitis.

163. *Swertia chirata* Buch.-Ham; Gentianaceae.

Na : Eng.: Chiretta; Bengali, Hindi, Gujrathi & Marathi: Kiryat, Charayatah; Kanerese: Nelabevu; Malayalam: Nelavepa; Sanskrit: Kiratatikta, Bhunimba; Tamil: Shiratkuchchi, Nilavembu; Telugu: Neelavemu, Neelaveru.

Hab : It is an annual or perennial herb 0.6 to 1.12 m in height found in the temperate Himalayas from Kashmir to Bhutan and Khasia hills, Sikkim and Darjeeling. The lower leaves are much larger. Flowering occurs in cold weather. The plant dies back in winter. The entire plant, branch stem and roots are used as a tonic, stomachic and laxative. During the year 2002-03 the material valued at Rs. 13098965.00 was exported.

***Terminalia bellirica* (Gaertn.) Roxb**

164. *Taraktogenos kurzii* King (Syn. *Hydnocarpus kurzii* (King) Warb. H.heterophyllus Kurz.); Flacourtiaceae.

Na : Assam: Lamtem. Dieng-soh-cap, Balibu, Bowaithing; Bengali: Daulmugri, Chaulmugra; Eng.: Chaulmogra; Hindi: Chaulmogra.

Hab : This tree is found in parts of Assam, Karnataka and Kerala. The chaulmogra oil obtained from the seeds of this plant has long been used in India for leprosy and skin diseases. Recent investigations carried out have also shown that the oil has curative properties of leprosy.

165. *Terminalia bellirica* (Gaertn.) Roxb. (Syn. *Myrobalanus bellirica* Gaertn.); Combretaceae.

Na : Beng.: Bhairah; Eng.: Bahera, Belleric myrobalan, Bastard myrobalan; Hindi: Bahera; Vibhitaka; Mal., Tamil & Telugu: Tani; Marathi: Baheda; Oriya: Bhara.

***Terminalia bellirica* (Gaertn.) Fruits**

Hab : It is a large tree growing wild practically throughout the tropical forests in the country. Stem is buttressed when old. It is one of the Triphalas of Indigenous medicine. The kernels possess narcotic properties and in

T. chebula Retz. - branch

Konkan they are sometimes eaten with betel nut and leaf for the treatment of dyspepsia. The ripe fruit is used as an astringent usually in combination with chebulic myrobalan. The part used in indigenous medicine is the pulp. It is used in dropsy, piles, diarrhoea and occasionally in fever. When half ripe it is used as a purgative. The oil is considered a good application for the hair.

166. *T. chebula* Retz.; Combretaceae

Na : Assam: Silikha; Bengal: Haritaki; Eng.: Yellow myrobalan, Chebulic myrobalan; Guj.: Hardo; Hindi: Harir, Harda, Haritaki; Marathi: Harda; Oriya: Haridra; Punjab: Harar; Tamil: Kadukkai; Telugu: Karakkaya.

T. chebula Retz. - fruits

Hab : It is found practically throughout dry, tropical forests in the country. The fruit of this plant is one of the Triphalas of Ayurvedic medicines. It is recognised by Indian Pharmacoepoea. The fruits ripen during December/ January and are credited with laxative, stomachic tonic and alterative properties. In combination with Emblic myrobalan, Phyllanthus Emblica. L.and belliric myrobalan Terminalia belerica (Gaertn.) Roxb. under the name Triphala, the fruits of T.chebula are extensively used as adjuncts to other medicines. The purgative principle in the fruit is a glycoside. The fruit pulp is used as a dentifrice to cure bleeding and ulceration of gums. It also has anthelmintic activity.

167. *Tinospora cordifolia* (Willd.) Miers; Menispennaceae.

Na : Bengali: Golancha; Guj. & Marathi: Gulvel: Giloe, Gullbel Gurcha, Jiwantika; Gulancha; Kanerese: Amrutaballi, Madhuparne, Uganiballi; Malayalam: Amrytu, Chittamritam; Oriya: Gulochi; Sanskrit; Amrita, Guluchi, Jwarari; Tamil: Amudem, Chindil; Telugu: Tippateega.

Tinospora cordifolia (Willd.) Miers

Hab : It is a common climbing shrub found in the dry tropical forests in the country. This climbing shrub is used in indigenous medicine as a tonic, in diarrhoea and chronic dysentery. Local villagers also use it as febrifuge.

168. *Tribulus terrestris* L.; Zygophyllaceae

Na : Bengali & Oriya: Gokhua, Gokshra; Gujrathi; Betagokhru, Mithagokhru; Hindi: Gokhru; Kanerese: Sannaneggilu; Ladakh: Rasha, Kokulla; Malayalam & Tamil: Nerenji, Nerinjeekai; Marathi: Lahangokhru, Sarala, Sharalte; Punjab: Lolak, Bakhra; Rajasthan: Gokhatri, Baragokhru, Kanti, Gokhrudesi; Sanskrit: Gokshura, Ikshugandha; Telugu: Chinnipalleru, Chirupalleru, Pallerukayalu.

Tribulus terrestris L.

Hab : It is a small plant with thorny fruits.

The plant is one of the constituents of Dasamula drug of Ayurveda and is stated to be useful in bronchial and kidney troubles.

169. *Tylophora indica* (Burn. f.) Merr. Syn. *T.asthmatica* Wight & Arn.; Asclepiadaceae.

Na : Bengali: Antomul, Ananthamul; Guj.: Damnivel; Hindi: Janglipikyam, Antamul; Kanerese: Adumutadhagida; Aiitmula, Nayelate, Nepalada beru; Malayalam: Vallipaala; Marathi: Khodikiraasna, Pitkari; Oriya: Mendi, Mulini; Tamil: Nach-churuppam, Nanjamurich-chaam, Mirkkurinja, Nay.

Hab : This herb is commonly found in Bengal, Assam and South India.

The leaves and roots are used as diphoretic and expectorant. Small quantities of leaves are stated to be efficacious in controlling and preventing bronchial diseases particularly eosinophlia.

Valeriana officinalis **L.**

170. *Urginia indica* (Roxb.) Kunth (Syn. *Scilla indica* Roxb.); Liliaceae.

Na : Bombay: Kochinda; Eng.: White squill, Indian drug squill; Gujrathi; Jangli kanda, Rankanda; Hindi: Banpiyaz, Janglipyaj; Kanerese: Adaviirulli; Marathi: Rancha kanda, Rankanda; Punjab: Kachwasal; Sans.: Vanapandan, Kolkanda; Tamil: Narivangayam; Telugu: Nakka Ulligadda, Advitellagadda.

Hab : It is a small glabruous bulbuous herb found in the subtropical regions of Western Himalayas, Bihar, and South India. The bulbs are the source of a drug which is used as an expectorant and stimulant. It is gaining importance in the treatment of cardiac trouble.

171. *Valeriana officinalis* L.; Valerianaceae.

Na : Eng.: Valerian, Gardenheliotrope; Hindi: Mushkwalee, Tagar;

Hab : It is an herb native of Euorasia and found in Kashmir. The rhizomes and the roots yield about 1% essential oil known as valerian oil which is used to relieve hysteria, coughs and is also used as a stimulant and tonic.

172. *V.wallichii* D C. Syn.*V.jatamansi* Jones; Valerianaceae.

Na : Beng.: Mushkbala, Tagar; Bombay: Tagarganthada; Eng.: Indian valerian; Garhwal: Sumaiya; Hindi: Muskbala, Balatagra; Kashmir: Mushkbala; Punjab: Balamush, Ba1a, Mushkwalee, Chargodar.

Hab : The herbaceous plant is found in the temperate Himalayas from Kashmir to Khasi hills. The rhizomes and roots yield about 0.3 to 1.2 % essential oil known as valerian oil. The roots and rhizome collected in spring gave 2.12 % essential oil. The drug is used in medicine as a substitute for Valariana officinalis L. to relieve hysteria and coughs. They are also used as a stimulant and tonic. The plant is recognised as Indian valerian in the Indian Pharmaceutical codex. The plant is propagated through seed or rhizome cuttings.

V.wallichii **D C.**

Vetiveria zizanioides (L.) Nash

173. *Vernonia anthelmintica* Willd.; Asteraceae. (Compositeae).

Na : Hindi: Kaliziri

Hab : It is a common herb in dry deciduous forests in Andhra Pradesh, Madhya Pradesh etc. Leaves are alternate and serrate.

The seeds are small black and pubescent and are, locally used in the treatment of urinary troubles, for leucoderma and other skin diseases. They are also used as an anthelmintic.

174. *Vetiveria zizanioides* (L.) Nash; (Syn. *Phalaris zizanioides* L.); Poaceae (Gramineae)

Na : Beng.; Khas-Khas; Eng.: Vetiver; Guj.: Valo; Hindi: Khas, Gandar; Kanerese: Vattiveeru, Laamanche, Kadukaridappa sajje hullu; Mal.: Ramaccham. Vettiveeru; Marathi: Vala; Sanskrit: Reshiru, Sugandhimula; Tamil: Vettiver: Telugu: Vettiveellu, Vettiveeru.

Hab : It is a perennial grass found along the river banks and in marshy places in South India in warm damp climate. It is also cultivated in Rajasthan, Uttar Pradesh, Punjab and the West coast. Local people use the paste of the roots for swellings

Viola odorata L.

175. *Viola odorata* L.; Violaceae.

Na : Bengali: Banafsha, Banosa;; Eng.: Sweet violet; Gujrathi: Banaphsha; Hindi: Banafsha; Kanerese: Violethu; Marathi: Bagabanosa; Tamil: Vialettu.

Hab : It is a herb found in the temperate regions of Western Himalayas.

The flowers are used medicinally as demulcent and in biliousness and lung troubles.

176. *Vitex negundo* L.; Verbanaceae.

Na : Assam: Pasutia, Aaglachita; Bengali: Niginda, Samalu, Nirgundi; Eng.: Chinese chaste tree; Guj.: Nagoda, Nagaol; Hindi: Nirgandi, Shambhalu, Nisinda; Kanerese: Lakkigida, Nakkilu, Nekki; Kumaon: Shiwali; Malayalam: Vellanocchi; Mar.: Nirgundi, Nisind, Nigudi; Tamil: Vellanocchi, Nirkundi, Venmochi.

Hab : It is a large aromatic shrub or small tree with quadrangular whitish tomentose branches found in waste lands and also grown for reclamation of erodable lands. Leaves are 3-5 foliate. They are used locally for swellings. They are reported to possess insecticidal properties.

Vitex negundo L.

177. *V. pedencularis* Wall.; Verbenaceae.

Na : Assam: Ashoi, Silaosai; Bengal: Boruna, Goda, Ashot; Cachar: Jadhgach, Jharua, Shelong-phang; Garo: Ashoi, Rangri, Shilangri; Hindi: Minjurgorwa, Nagpheni; Kanerese: Navalai; Oriya: Mado, Chulia, Erataka; Santal: Marakata, Bhadu.

Hab : It is a small tree found in South India, in lower eastern Himalayas, Bihar, Orissa and eastern parts of Andhra Pradesh. Leaves are generally tri-foliate. Leaflets are lanceolate. At one time it was thought that the drug obtained by the infusion of leaves cured the black-water fever, but recent reports have not been promising.

178. *Withania somnifera* Dunal; Solanaceae.

Na : Bengali; Ashvagandha; Bombay; Asgund, Asvagandha; English: Withania radix, Withania root; Gujrathi: Asan; Hindi: Asgand, Funir; Marathi: Askandha; Punjabi: Asgand; Sanskrit: Ashvangandha; Tamil: Asuragandi; Telugu: Asvagandhi, Penneru; Urdu: Asgandanagaori.

Specification of the Drug: *Ashvangandha consists of* dried roots of Withania somnifera Dun.

Characters

(a) *Macroscopic:* Roots occur in small pieces of 10-17.5 cm long and 6-12 mm in diam, uniform in appearance, smooth and tapering. External colour is light brown and that of internal is creamy. Odour is pungent. It tastes bitter and acrid.

Withania somnifera Dunal

(b) *Microscopic:* Root shows a narrow crumpled cork a moderate cortex (20-25 layers of cells) and a large wood. Periderm consists of phellem, phellogen and phelloderm. Below the perioderm is the secondary cortex of 20 to 25 rows of thin walled parenchyma cells mostly, filled with starch grains. No pigments and crystals occur in the cortex. Vascular system consists of a large number of parenchyma cells; vessels and trachieds are pitted and scalariform and with generally wall perforations, a few wood fibres are found scattered in the peripheral portion of wood; the parenchyma of wood contains plenty of starch grains.

Constituents: Ashvagandha contains an essential oil, ipuranol, a crystalline alcohol withaniol, C25H3505 hentriacontane, phytosterol and fatty oils. It also contains three separate alkaloids of which one, somniferin, C12H15N2, is crystalline. The total alkaloidal content in roots in India has been reported to vary between 0.13 and 0.31 percent. Some strains are reported to yield higher content of alkaloids in the roots. In addition to the alkaloids the roots are reported to contain starch reducing sugars. The green berries contain a high proportion of amino acid, cystine and cysteine.

Soil and Climate: The plant grows in a variety of soils in red, lataratic, loamy and BC soils with pH between 7.0 and 8.5. It grows in a temperature range of 28-38 °C and as low as 10 °C. If the temperature increases more than 38 °C, the plant dries up. The plant grows from sea-level to 1500 m altitude.

Propagation: About 5 kg of seed is required for raising 1 hectare of plantation. The seeds are treated with carbendazim to control wilt of seed borne diseases. They are sown in nursery beds at least 1 month in advance. A thin layer of sand is spread after sowing on the bed. The beds are watered with a rosecan. In about a week's time, germination takes place. They are ready for transplanting in a months time, when they are pricked out and planted in the well prepared site at an espacement of 60 × 60 cm. If the land is not fertile application of farm yard manure or fertilizer will increase the yield. Although the plant does not flourish in waterlogged areas, it grows in moister areas and responds to irrigation and watering during drought periods.

Harvest: The plant can be harvested after about 6 months. The roots are dug out or the land ploughed and roots collected. The roots are cleaned and cut into small pieces of about 7-10 cm and graded. Stout and long roots get premium price.

Yeild: A yield of 1-1.5 tonnes of roots per hectare can be obtained.

Zanthoxylum alatum **Roxb.**

Action and Uses: Ashvagandha is an important Ayurvedic drug and is now widely used in allopathy. It is regarded as a tonic, alterative and aphrodisiac and is used in constipation emaciation of children, debility from old age, rheumatism, etc. It has also narcotic, diuretic and deobstruent properties. The ground root is employed as a local application to carbuncles, ulcers and painful swellings. Roots are used for treating general and sexual weakness in human beings.

Standard and Test: Foreign Organic matter not more than 2 percent.

Preparation: Elixir Asvangandha, Extractum asvagandha, Liquidum pulvis Asvagandha.

179. *Zanthoxylum alatum* Roxb.; *Rutaceae*

Na : Bengali: Gaira, Nepalithanya, Tambul; Garhwal: Tezbal, Tezmal; Hindi: Darmar, Nepalidhaniya, Tejbal, Tumru; Oriya: Tundopoda; Punjab: Timbar, Timal, Timru; Nepali: Sungurukung, Baletimar; Sanskrit: Tumburu, Dhiva, Gandhalu.

Hab : It is a small tree found in the Punjab, Kumaon and the Khasia hills. The aromatic seeds are used as tonic.

5

Essential Oils

5.1 Introduction

The primitive man in search of food and clothing might have been fascinated by the fragrance emitted by some of the plants around him. With the passage of time he started making use of such of the aromatic plants and produced essential oils. The essential oils or volatile oils as they are often called are distinguished from fatty oils by the fact that they evaporate rapidly in contact with air and also have a pleasant taste and a strong aromatic odour. Essential oils are very complex in their chemical nature. Their physiological function in the body of the plant is not fully known. It is however understood that they are by products of metabolic processes rather than reserve foods. However a pleasant odour and aroma which they impart attract insects and animals for pollination or the dispersal of the fruits and seeds while unpleasant odours repel the enemies. The essential oils are found in about 60 families but the more important ones are Rutaceae. Leguminoseae, Lauraceae, Myrtaceae, Umbellifereae, Labiatae and Asteraceae (Compositae). The oil may be found in the internal glandular tissue or hair like structures of flowers, fruits, seeds, leaves, bark, root and wood. Sometimes it comes out along with gummy or resinous exudations. In some cases as in *Gaultheria* spp. (winter-green) the oil is not present in the tissue or hair like structures but develops as a result of chemical reaction when the plant tissue is ground up and extracted with water. The oil may be confined to any one particular part of the plant as in the case of *Michaela champaca* L. or may be found in more than one. The yield and nature of oil vary according to the species. In the same species they differ considerably in different strains and different parts of the plant. Very often they depend on various other factors like climate and soil.

At one time perfumes and attars of Indian origin were famous all the world over. But now production of many of the highly priced essential oils has dwindled partly because they require more attention than even the food crops. The perfumery industry cannot in all cases depend upon plant material growing wild and scattered since odoriferous principles are delicately balanced and any deterioration that may set in during transport over long distances adversely affects the quality and quantity of oil. Hence to produce aromatic oils economically on scientific lines it may be desirable to cultivate them. Excepting for a few essential oils the source of raw materials in the country is still the forest. Of the essential oils from forest origin turpentine from *Pinus roxburghii* Sar., Sandal wood oil from *Santalum album* L., palmarosa oil from *Cymbopogon martini* (Roxb.) wats, var. motia, agar wood oil from *Aquillaria agallocha* Roxb. stand out prominently.

The turpentine oil industry in the country is well established and the oil is used in paint and varnish industry, manufacture of synthetic camphor and also for medicinal purposes. L (alpha) pinene content in turpentine oil of *Pinus roxburghii* Sar. is low about 40 per cent. As such cost of production of camphor can be higher. There are however other species in the country like *P.wallichiana* AB. Jackson (Kail) and *P. insularis* Edle. (Syn. P. khasya Royle ex part.), which yield turpentine oil containing 85 to 90 percent of pinene. The former occurs in the interior ranges in the Himalayas and offers many difficulties in the extraction of resin; the warm and tapping season being limited, the yield of resin is low. Whereas the latter species has scattered and limited distribution and hence it is uneconomical to exploit. For the production of camphor, trials on the cultivation of *Cinnamomum camphora* (L.) T. Nees & Eberm were made but without much success. *Ocimum kilimandscharicum* Guerke which is another natural source for camphor was introduced under natural conditions in teak plantations in Tamilnadu and other states as undergrowth. The results have been very encouraging and it grows almost as a weed. But no systematic steps have been taken to exploit it and extend its cultivation. Palmarosa oil from *Cymbopogon martini* var. motia and *C. martini* var. Sofia are of forest origin and are of great commercial importance particularly the former. At one time India used to produce large quantities of vetiver oil from the roots of *Vetiveria zizaneoides* (L.) Nash (Khas) grass but ever since it was cultivated in Java most of the trade has dwindled. Another important essential oil is the sandal wood oil from *Santalum album* L. The inherent superiority of the Indian oil and the scientific management has enabled it to retain its position. Agar wood oil is obtained from *Aquillaria agallocha* Roxb. a species found in the wetter lower parts of Eastern Himalayas. The oil has great commercial value but the production is limited. Eucalyptus oil is being produced from *Eucalyptus globulus* Labill. in the Nilgiris in Tamilnadu. The cineole content being low is not recognised by British Pharmacopoea. Now, by fractional distillation, the cineole content is being increased but this will add to the cost of production. There are certain other Australian shrubby species like E.australiana Baker & Smith, *E.fruticetorum* F. Muel. Syn. *E.polybractea* R.T. Baker which contain high cineole content and can be introduced in the country. Although we are producing enormous quantities of essential oils in the country yet there are many others for which we have to depend on imports like citronella oil, clove oil, linaloe oil, etc. *Eucalyptus citriodora* Hook., *Cymbopogon nardus* L. Rendle (Ceylon variety), C.winterianus Jowitt (Java variety) all yield citronella oil and can conveniently be grown under forest conditions. Linaloe oil is obtained from *Bursera penicillata* Sesse & Moc. ex Dc. Engl. Syn. B.delpechianum Poiss. This species grows well in dry and even on shallow and skeletal soils. Efforts were however made to grow the species in forest areas in Andhra Pradesh and Karnataka. It is showing promising results and needs extensive cultivation.

Man's greed now-a-days is not confined to natural products alone, he has gone further to prepare more and more pleasant and lasting synthetic perfumes and essences. It is often said that the artificial essential oils are replacing the natural ones. It is not possible however to say that the artificial essential oils will replace the natural ones altogether. In some cases the artificial ones are inferior to the natural because small traces of highly odorous substances (esters etc.) occur in the natural ones without which the product lacks quality. Synthesis of compounds artificially is highly complicated and expensive and may not compete with the products from natural sources. However, at a period when science and technology are progressing so rapidly it is difficult to foretell the future position. But according to the present trend there is still a great demand for natural oil, and consumption of these products is increasing.

5.2 Extraction of Oil

Essential oils are extracted from the plant in various ways depending on the nature of the plant and its oil. There are three principal methods : distillation, expression and extraction by solvents. Extraction of oil must be preceded by preparation of the material for the purpose. Insufficient or defective method of preparation of the material raises the cost and decreases the yield of oil. Herbs, leaves and fresh roots are cut if they are big enough. Barks and dry roots and fruits are ground up or broken to small pieces. Wood is rasped and seeds are crushed. In the case of most grasses the material is spread in the shade for a short period to drive out the excess moisture.

If however, the material is dried excessively the volatile oil contents are lost. Therefore utmost precaution is necessary in preparing the material. The flowers contain the highest yield of oil just before the buds are fully open. Therefore they should be collected at this stage and subjected to extraction immediately.

1. *Distillation*

(a) *Water distillation:* Simplest method of distillation is boiling the material in water. The mixture of steam and essential oil is cooled in a condenser and the oil is separated by decantation.

Very often this method is used for distilling palmarosa oil in villages. It is a crude method and the oil obtained is of comparatively inferior quality. A better method would be to keep the material separate from water. (Fig. 5.1).

(b) *Steam distillation:* In this case distillable material is placed in a still and steam introduced. The steam softens and breaks the tissue and releases the essential oil. The vapourized oil along with steam passes into a condenser where the mixture is cooled. On cooling the oil gets deposited on the surface of the water which is then collected by decantation. In this method a separate boiler for generating steam, a copper still and a set of condensers are required but such a plant is costly and is inconvenient and uneconomic for cottage-scale purposes.

(c) *Water-cum-steam distillation:* In the steam distillation, costly equipment and the services of experienced personnel are required and it is inconvenient and uneconomical for cottage scale purposes. Water-cum-steam distillation unit is small and can be operated on a cottage scale basis. It is quite effective and could be used for distilling essential oils from palmarosa grass as well as camphor from camphor yielding *Ocimum kilimandscharicum* Gurke. It could be operated in forest areas utilizing stream water for cooling the condensers without lifting it to reservoirs or tanks using a pump etc. The unit consists mainly of (i) a still made from an empty oil drum of 200 litre capacity by removing the original top and fitting it with a suitable clamped top having a swan neck and a perforated false bottom about 8 inches above the real one, (ii) a double surface condenser Fig. 5.1 made of galvanized iron sheet and (iii) an oil separator made of galvanised iron sheet so designed as to serve the purpose of florentine flask. In order to run the unit, the still is placed over the fire place (chulla). For the first charge about 35 litres of water is poured in each still and the water level is kept just below the false bottom. For the subsequent charges the volume of water steamed off, is made up every time. The still can take about 18 to 20 kg of distillable material which is placed over the false bottom in each still every time a charge is run. The lid with the swan neck is placed over the still after arranging a 5 mm thick layer of a gasket or asbestos or soft cotton rope on the rim of the drum (still) in order to make it leak proof against the escaping of the vapours, during distillation. The top is then tightened with G bolts and fly nuts, provided for it.

The swan neck of the still is connected to the condenser. The Florentine flask is placed below the outlet of the condenser. The fire is now started below the still. Water is circulated in the condenser to keep it cool. The steam from the boiling water at the bottom of the still rising through the distillable material carries the camphor or essential oil vapours as the case may be with it and enters through, the adaptor into the annular space of the condenser. Because of the circulation of cool water, the steam laden with camphor or oil vapours get, condensed in the condenser and part of this mixture flows to the receiver of florentine flask, leaving behind most of the congealed camphor on the walls of the condenser. However if it is only essential oil it flows into the receiver. It takes about two hours to complete the distillation. Three charges of each unit could be run during the day. For

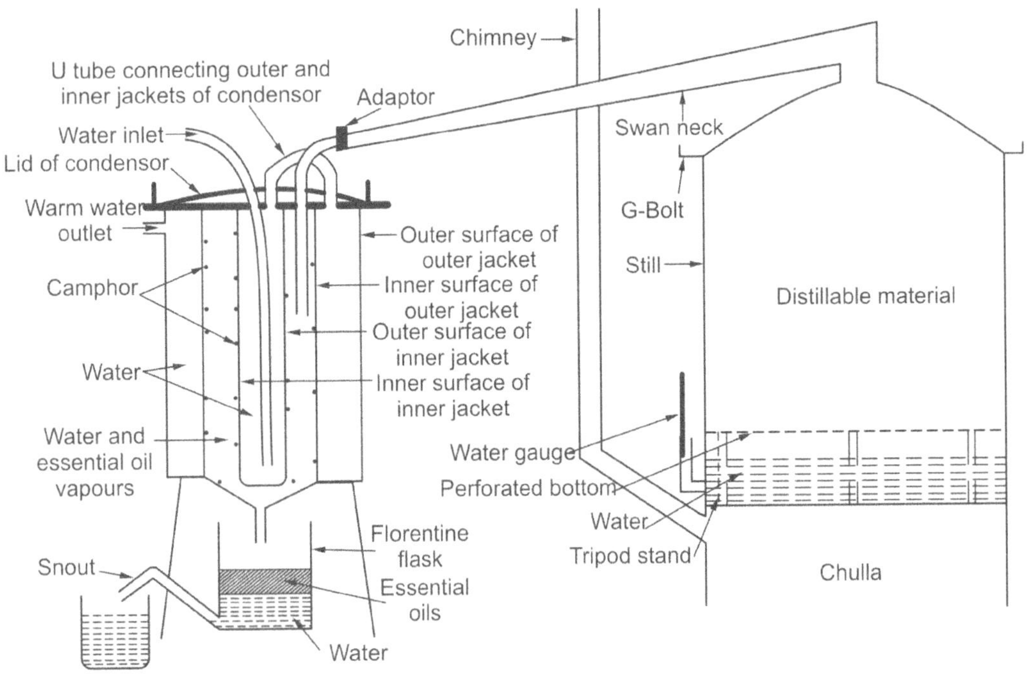

Fig. 5.1 Portable Unit for Camphor Distillation

distilling essential oils instead of camphor, the double jacket condenser could be replaced by an ordinary worm-condenser Fig. 5.2. A battery of three units could be operated conveniently by one man.

2. *Expression :* Oils contained in the fruit peels of citrus fruits etc., can be obtained by machine or Ghani pressing. Expressed oil usually contains a certain amount of non-volatile matter.

3. *Extraction by solvents:* This method is usually used for extracting oil from flowers. Both non-volatile and volatile solvents can be used in the extraction process. Among the non-volatile solvents, oils or fats are used to absorb the odours when they are called pomades. There are again two types in this process enfluerage and maceration. In enfluerage the extraction is carried out under normal temperature conditions. In this method non-volatile solvents are used repeatedly until a high concentration of aroma is obtained. In maceration the odorous plant material is digested with hot oil or melted fat. In this process essential oil bearing cells are broken up to liberate the oil. The liberated essential oil is readily absorbed by the hot oil or melted fat. After absorption the pomades are treated with ethylalcohol which dissolves out the essential oil and yields the floral extract.

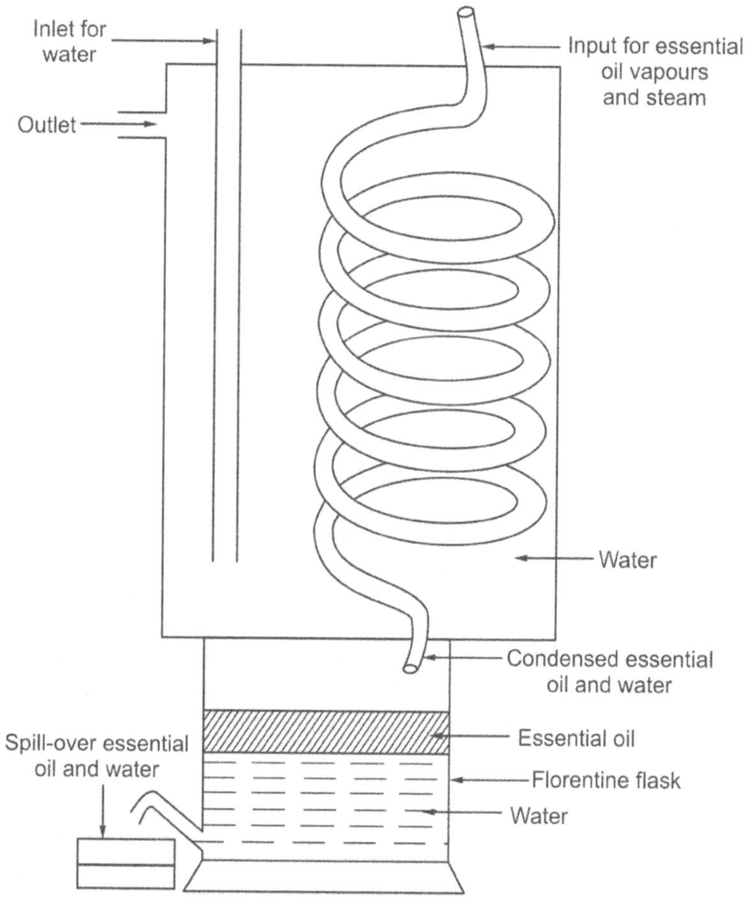

Fig. 5.2 Worm-condensor

In the volatile solvent process petroleum ether is generally used. It dissolves the odorous oil from the flowers. When the flowers are exhausted the solvent is distilled off leaving a semi-solid residue. The residue consists of the odorous oil and insoluble plant waxes. Thereafter alcohol is mixed with the residue. Alcohol dissolves the essential oil leaving the waxes. Waxes are removed by filtration or by freezing. The essential oil is obtained by fractional distillation. The resulting essential or perfume oil is highly concentrated. It is blended with other essential oils and fixatives. Essential oils with a low rate of evaporation like patchouli oil, sandal wood oil, palmarosa oil are used as fixatives as they contain both fixative and perfumery qualities. In this volatile solvent process dried material, resinous substances and even plant juices can be used. Modern perfumery is mostly based on skilful blending and requires a great variety of essential oils for preparation of newer notes.

5.3 Storage

On storage for a long time the oils usually deteriorate chiefly due to oxidation. Since they are volatile under normal conditions they should be stored in good airtight containers and kept in a cool and dark place.

5.4 Uses

Essential oils are used for various purposes. They are used in perfumes, cosmetics, soaps, flavouring agent or essence in culinary preparations, chewing gum, tobacco, tooth-paste, liquors and beverages and as disinfectants. Some have medicinal properties while others are used as solvents in the paint and varnish industry like turpentine oil. They are also used as deodorants to mask offensive odours at a time when personal cleanliness is overlooked. Quite a few of them are put to more than one use for e.g. clove oil is used as a flavouring agent in foodstuffs, in perfumery, medicine, as a source of synthetic vanilla and in the preparation of botanical sections.

A very common use of the perfumery material in India is in the manufacture of Agarbatti or joy-stick. It is a typical Indian cottage industry. It is a development of the age-old practice of burning dhup, an aromatic powder or paste, as a fragrant fumigant in religious and social functions. Agarbatties have now invaded almost every home and several varieties are in the market to suit every purse. Almost any type of aromatic material-woods, roots, barks, leaves, etc. is used in their preparation. In addition to the natural essential-oil-bearing materials, resins, balsams, and several synthetic aromatics are nowadays employed, and some manufacturers claim to use a blend of as many as about a 100 different aromatic materials. Some of the important materials used are amber, musk, civet, vetiver oil, kuth, rose, screwpine essential oil, deodarwood, rosin, cloves, cardamoms, cinnamon bark, saffron, jatamansi, etc. In addition to these, a large proportion of sandalwood powder is also used and it is said to be a "must" in forming the base. Agarwood (black) is also said to be used but its use is naturally restricted owing to its high cost.

A suitable formula for making agarbatti using agar wood (*Aquileria agalocha Roxb.*) or partially exhausted agar wood, is as follows: Exhausted agarwood powder 50 parts, Gur (Jaggery) 25, Lobhan 5, Safed ral (white gum) 5, Musk xylol 3, Bornel 2, Sandalwood oil I, Other aromatic materials (natural and synthetic) 4, Charcoal and other ingredients like saltpetre, colour etc.

For making agarbatti in a cottage-scale the above constituents, excepting the perfumes and Gur (Jiggery) are separately powdered to a fine stage (about 80 meshes) and mixed together in the proportions indicated. To the mixture the requisite quantity of jaggery and water is added so as to form dough and the whole mass is ground well on a stone mortar with a pestle till the size of the particles is reduced to further fineness. To the paste thus formed the requisite quantity of perfumes, both natural and synthetic, including borneol, musk xylol and sandalwood oil, are added and incorporated in it by a thorough mixing with hand. The quantity and nature of the perfumes to be added naturally depends on individual taste and the market demand. The next step is the incorporation of an adhesive into the paste. The best known adhesive for this purpose is the mucilage present in the bark of *Litsea sebifera*, commonly known as maida lakdi which is found in the forests of Madhya Pradesh. For this purpose the bark is finely powdered. The powder, about one-third of the weight of the paste, is mixed well with the paste. The final stage namely application of the paste to bamboo sticks, is a tricky operation which requires a lot of skill and experience. This is usually done by women labour in several places. A small amount of the paste is held between the fingers of the left hand. While rolling the bamboo sticks on a smooth horizontal wooden plank with the palm of the right hand, the paste held between the fingers of the left hand is moved from one end of the stick to the other so that a uniform layer is formed around the surface of the bamboo stick. The final finish is given by rolling the paste coated stick on the plank with the help of the entire fore-arm of the right hand. By this operation the paste sticks to the bamboo stick and assumes evenness, smoothness and shine. During rolling the desired aniline colour is sprinkled on the plank so that it sticks on to the paste. The rolled sticks are dried in shade and packed in small bundles.

***Abelmoschus moschatus* Medik – Branch**

5.5 Essential Oil Yielding Plants

Following are some of the plants which yield perfumery materials.

1. *Abelmoschus moschatus* Medik Syn. *Hibiscus abelmoschus* L.; Malvaceae.

Na : Eng.: Mallow, Ambrette; Hindi: Musk.

Hab : Abelmoschus moschatus is a tropical species valued for its scented seeds. It is native of India. (Names and Hab. Pl. see medicinal plants).

The plant is propagated from seeds collected from ripe fruits (capsules). Sowing may be done by broadcast directly in prepared field in rows made 60 cm apart with a plant to plant final spacing of 60 cm. Sowing may be done in March-April. The seeds start germinating in about a week and is completed further in 2 to 3 weeks. The field is to be irrigated and weeded as and when required. The plants start flowering in September and the fruits ripen from November to January and harvested when fully mature. The capsules are dried in sun when they burst open and the seeds fall out. Application of fertilizers improves the growth of plant and also the seed yield.

***Abelmoschus moschatus* medik Pod and Seeds**

Under Dehradun conditions experimental plantations of the species with direct sowing and a plant to plant final spacing of 60 cm × 60 cm gave yield of seeds up to 540 kg per hectare. Ambrette seeds possess a delicate musk-

Acacia decurrens Willd

like odour and are valued for the volatile oil present in seed coat. The oil is extracted by steam distillation of crushed seeds and used in high grade perfumery, to perfume hair and blending perfumes. The crude oil contains a high percentage of fatty acids and for use in perfumery, the fatty acids are removed by treatment with dilute alcohol or by precipitation as calcium or lithium soaps. The oil is also obtained by extraction of seeds with benzene and petroleum ether and alcohol.

2. *Acacia decurrens* Willd. var. *dealbata* F. Muel. (Syn. *A.dealbata Link.);* Mimosaceae.

Na : Eng.: Silver wattle.

Hab : It is an evergreen tree introduced into India and is now being grown in the Nilgiris. Bark of younger trees is generally olive green but is dark grey in older trees (Pl. see tanning materials). A perfume of good quality can be obtained from flowers.

Acacia farnesiana (L.) Wild.

3. *Acacia farnesiana* (L.) Wild.; Mimosaceae.

Na : Eng.: Cassie, Sweet Acacia; Hindi: Gandh Babul, Divanababul.

Hab : The tree is native to West Indies and is found throughout tropical and sub-tropical regions in India in cooler places. The fragrant yellow globose flowers are the source of cassie perfume.

4. *Acorus calamus* L.; Acoraceae.

Na : & *Hab :* Pl. see medicinal plants.

The essential oil obtained from the rhizomes is used in perfumery in addition to its use in medicine. It finds use in spice blends and flavouring of alcoholic beverages.

Acorus calamus L.

5. *Alpinia galanga* (L.) Wild. (Syn. *Languas galanga* (L.) Stuntz.); Zingiberaceae. .

Na : Eng.: Greater galangal; Bengali, Sanskrit & Hindi: Kulinjan; Kanerese: Dumbarasme; Marathi: Acorus calamus L. Kosht-Kulinjan; Tam.: Perurattai; Telugu: Peddadumparashtram.

Hab : This herb is found in the Eastern Himalayas and South India.

The rhizome is orange-brown in colour, aromatic, pungent and bitter and is a source of an essential oil.

Alpinia galanga (L.) Wild.

6. *Aquilaria agallocha* Roxb.; *Thymelaeaceae.*

Na : Assamese: Sasi; Eng.: Agar, Aloe Wood. Eagle wood, Calambac; Hindi: Agar; Sanskrit: Agaru.

Hab : It is a tree found in the Eastern sub-tropical Himalayas in Bengal, Assam,Manipur,Meghalaya, Garo and Naga hills and Mizoram.

Little is known about the formation of essential oil in the body of the plant.

Irregular patches of dark wood usually infected with fungus are charged with essential oil. These patches may not be perceptible from outside. They are usually found in and around old wounds and hollows. It is considered that formation of essential oil in the wood is due to fungal or insect damage. Three fungi viz. Aspergillus sp. Penicillium sp. and Fusarium sp. have been isolated from agarwood. The oil can be obtained by water-cum-steam or only by steam distillation. The yield of oil varies from 0.08 per cent from yellow to 2.49 per cent from black agar wood. Agarwood is graded into four qualities as per the local market.

1. 1st class agar
2. Bangtong
3. Bhoota doom
4. Doom.

The agarwood of the 1st category is very heavy and looks like almost black wood. Agarwood is heavier than water. There is considerable demand for the wood as well as essential oil in Egypt, Saudi Arabia, Middle East and the far eastern countries.

Aquilaria agallocha **Roxb.**

7. *Artemisia dracunculus* L.; *Asteraceae (Compositeae).*

Na : Eng.: Estragon plant, Tarragon plant.

Hab : It is an exotic species now introduced at minor forest products nursery at Chakrata in Uttara Khand. Its branches and leaves are the source of the important essential oil of estragon (tarragon). While the oil is used both as a flavouring agent and spice the leaves are used only as a spice. The species thrives in temperate climate. The seedlings raised in nursery beds or the suckers made available from older plants may be planted either in the beginning of spring i.e. March/April or when the first regular monsoon breaks i.e. July. But as the drought season follows planting in spring, it is preferable to plant during the rains. Suckers/seedlings 5 to 7 cm height are transplanted at an espacement of 30 × 30 cm. The plants get established in about 2 weeks. Weeding may be done as and when necessary. In the first year of planting one harvest can be had during October/November. From 2nd year onwards to 4th year two harvests can be obtained every year one in May/June and the second in October/November. After 4th year's harvest the yield of raw material and the essential oil declines when the plantation needs renewal.

8. *Blumea balsamifera* (L.) DC. (Syn. *Conyza balsamifera* L.); Asteraceae. (Compositae.)

Na : Eng.: Nagai camphor; Hindi: Kakaronda, Kakonda.

Hab : It is a small plant found in the sub-tropical Himalayas, Assam and the Khasia hills. Leaves are aromatic and have camphoraceous smell.

9. *Bursera penicillata* Sesse and Moc. ex DC Engl. (Syn. B. *delpechianum* Poiss.); Burseraceae.

Na : Eng.: Linaloe.

It is a medium sized deciduous Mexican tree growing to a height of 9 to 12 m with aromatic properties. It grows even under adverse conditions and in skeletal soils. In Mexico the "Linaloe" oil is extracted from its wood, although every other part of the tree including its berries, leaf yields the essential oil and the easiest way of identification of the plant is through the fine smelling oil which can be bled from any part of the plant by making a cut. In India the oil is also known as Indian lavender oil. It may not be correct to call it Indian lavender as it is neither Indian nor lavender plant. It was introduced into India near Bangalore by Anderson brothers in 1920. After they left the country in 1937, the plantations came into the possession of the famous film star Smt. Devika Rani Roerich. The value of the essential oil was rated so high at that time that the owners of the plantation treasured everything connected with it as a very guarded secret. Anyone wanting a cutting of the plant had no other alternative except smuggling it. The sweetness of the species naturally spread round, in spite of all the vigilance of its possessor. It is being grown in Karnataka, Andhra Pradesh and Tamilnadu and to a small extent in other states.

***Bursera penicillata* Sesse and Moc. ex DC Engl.**

Extraction of Oil: In Mexico the "Linaloe" oil is extracted by steam distillation of husk of the berries as well as the heart-wood from 40-60 year old trees. The bulk production is mainly from the distillation of heart-wood. Ripe berries are collected after they drop to the ground and dried in shade on perforated plates with free circulation of air. It is generally observed that the quality of fruits produced varies from tree to tree and from year to year on the same tree. Difficulty is experienced in collection of husk in September coinciding with heavy rains coupled with strong monsoonal gales. The yield of husk is reported to be strongly correlated with the morphology of the plant and sex-ratio (male and female ratio) in the plantation. A ratio of 1:9 (Male: female) is considered optimum. Compact bushy type plants with red coloured branches are observed to yield more than tall-erect growing, green coloured plants. On an average 100 kg of berries give 15 kg of husk. The husk is subjected to distillation of oil. The oil content varies from 2.5% to 3% in the wood to 10-19% in the husk on air dry weight, basis. In case of distilling heart wood, the young trees are sometimes wounded to increase the yield of oil. The heart wood is the most aromatic part of the tree and yields superior oil.

Properties and uses of oil: It is a light coloured mobile oil with a fine lily rose tone. It has high ester content and good fixative properties. The main constituent of the wood oil is linaloe (Linalylacetate 60-70% or more) with smaller quantities of geranial and £ or DX terpineol. Husk oil closely resembles wood oil in chemical composition except for the fact that the wood oil usually contains dextrorotatory linalool as the chief constituent.

The essential oil is used in perfumery, in cosmetic industry as a fixative in perfuming lily, lilac lavender, linden blossom cananga etc. and also employed in transparent soaps where colouration has to be avoided on keeping and also for flavouring foods and beverages. It has export potential, bulk of the Indian product being exported to United Kingdom and United States.

Method of Propagation: All the plants do not flower and fruit; some do not yield at all. Two varieties of berries may be recognised, the dark red and the greenish red. The morphological characters also differ visibly. The former variety appears to be a more prolific bearer than the latter. The plant may be propagated either by seed or by shoot-cuttings. The tree flowers in the 5th year of the planting and from 10th year it flowers and fruits profusely. It starts flowering in April and the berries will be ripe for plucking in August. Plant cutting appears to be faster in growth and surer way of propagation and also it starts bearing very much earlier than the seedling. It also ensures parental properties. Cuttings 15 to 22 cm in length are taken from high yielder of berries in April and raised in pot culture, preferably after treatment with any of the patent root hormones for planting out in July at the out-break of monsoon. Propagation through seed can be done in the usual way. The seeds are obtained from berries. They are said to be viable for about a year. Germination is said to be poor and takes place in about 15 days. Nine months to one year old seedlings are transplanted in pits 4 to 6 m apart. In the case of seedlings it may take as many as 10 years for fruiting. It is obvious that this will be reflected in the economics of the plantation.

10. *Cedrus deodara* Loud. (Syn. C. *libani* Barrel var.deodar Hook. f.; *Pinus deodara* Roxb.); Pinaceae.

Na : and *Hab :* Pl. see medicinal plants.

Cedrus deodara **Loud**

Himalayan cedar wood oil obtained from Cedrus deodara is gaining considerable importance in perfumery and soap industry. This species is abundantly found from central to Western Himalayas at elevations from 1500 to 2500 m. It yields a valuable timber which is extensively used for building purposes. During the process of conversion a lot of waste sawdust chips and wood shaving of the heart wood are obtained. These, because of their high oil content (4 to 6 per cent), can be employed for obtaining the essential oil. The oil obtained from this species conforms to the Indian Standard Institution specification of cedar wood oil and closely resembles the oil of Atlas cedar wood distilled from the wood of Cedrus atlantica. It is also a good substitute of imported cedar wood oil obtained from Juniperous virginiana. Forest department, Uttara Khand got a sample of Cedrus deodara oil analysed and they have found that the oil conforms to the I.S.I specification. Cedrus deodara Loud. is so abundant and economically available that with proper development of the industry good export trade could be built up.

It is therefore felt that there is a good prospect of starting the production of this oil on cottage-scale industry basis.

Cinnamomum camphora (L.) T. Nees & Eberm. – Tree &

Cinnamomum camphora (L.) Leaves

Cymbopogon citratus (DC) Stapf

11. *Cinnamomum camphora* (L.) T. Nees & Eberm.; Lauraceae.

Na : Eng.: Camphor tree; Hindi: Mushkapur, Karpur.

Hab : It is a tree native of China, Japan and Formosa and is also found in Assam and other places. Heartwood and leaves on water-cum-steam or steam distillation yield camphor which is much needed in a variety of industries. The yield of oil from wood is higher than that from leaves.

12. *Cymbopogon citratus* (DC) Stapf. Syn. *Andropogon citratus* DC. &

13. *C. flexuosus* (DC) Stapf. (Syn. *Andropogon flexuosus* Nees); Poaceae Syn. Gramineae.

Na : C.citratus - Bengali: Gandhabena; English: West Indian Lemongrass; Guj.:Lilicha; Hindi: Gandhatrina, Aginghas; Kanerese: Majjigehullu; Mar.: Hirvacha, Olecha; Tamil: Vasanapillu; Telugu: Nimmagaddi; Sanskrit: Bhustrina *C. flexuosus* - English: East Indian Lemon grass, Malabar or Cochin Lemongrass.

Introduction: Lemongrass oil is one of the most important essential oils. Citral, chief constituent of lemongrass oil, is the starting material for a number of aromatic chemicals and vitamin A. In trade two principal types of lemon grass oil are recognised viz.; West Indian oil derived from *Cymbopogon citratus* (DC) Stapf. and East Indian oil derived from C. *flexuosus* (DC) Stapf. The two oils, though almost identical having 75 to 85% aldehydes (chiefly citral), differing their solubility in alcohol, the West Indian oil has lower solubility. The species grows wild in many tropical and sub-tropical parts of Asia, Africa and America. However, the oil is extracted from cultivated lemongrass as it is more economical and also the quality from cultivated stands is reliable. C. *citratus* is cultivated in Central and Southern America, Madagascar, West Indies and Indo-China. C. *flexuosus* is cultivated in Southern India, mainly in Malabar in Kerala.

There are two types of East Indian lemongrass (C. *flexuosus*) viz., the red stemmed grass and white stemmed grass. The red stemmed grass yields the normal East Indian lemongrass oil while the oil from white grass has low aldehyde content and poor solubility in alcohol.

Cultivation: Soil and climatic requirements -The soil affects both the grass yield and oil content considerably. The grass grows best on well drained sandy loam soils, thriving even on light but fertile sandy soils giving relatively more oil with higher citral content than plants from very fertile soils. It does not grow in compact and poorly drained soils.

Grass grown on dry sandy soils yields oil rich in citral, whereas, in humid sandy loam soils the citral content tends to be lower. The grass requires warm tropical climate with plenty of sunshine, and well distributed but not excessive rainfall. Warmth and sunshine help in the increase of oil. In regions of abundant rainfall more harvests can be made but the citral content in oil tends to be low.

Planting: Lemongrass is propagated by separating slips from the clump. One mature clump may yield as many as 50 slips (stools). Lemongrass does not ordinarily produce seed unless grown in a suitable climate without being cut for a long duration. For vegetative propagation 2 to 4 stools with their tops cut off leaving about 1/3 of the total height of the plant, may be planted at an espacement of 60 cm × 60 cm (27,889 plants/ha)or 75 cm × 75 cm (17689 plants/ha) at the commencement of regular shower of monsoon. For propagation through seed, sowing is done in nursery beds in April and later transplanted in the field. The seeds germinate in 5 to 6 days. Approximately about 1.5 kg of seed is required for raising sufficient seedlings for raising a hectare of plantation. Transplanting is done with the break of monsoon, when the seedlings are about 45-60 days old. For broad-cast sowing, the seed is sown directly in the field in June in South India where monsoons start early, in Northern India the sowing may be done in June/ July with the commencement of regular monsoon. The grass comes up with the first regular shower of monsoon. Systematic weeding is necessary to promote good growth and also to prevent weeds from getting into distillation still. Two to three weedings are done during the crop season. Earthing the clumps may also help the crop.

C. flexuosus **(DC) Stapf**

In case of C. *citratus* propagation is done vegetatively by planting stools as it rarely flowers and seeds. However, C. *flexuosus* can be raised both vegetatively and from seed. In Kerala which is the main area of lemongrass cultivation in India, direct sowing is generally practised.

Harvesting: For harvesting, the grass is usually cut 15 to 25 cm above ground level with a sickle or a hatchet. In case of C. *citratus*, raised vegetatively the young plants, depending upon the climatic & soil conditions, require 4 to 6 months to develop sufficiently to have the first harvest. After the first harvest the grass grows rapidly and subsequent harvests may be collected at intervals of 3 to 4 months. Thus 3 to 4 cuttings can be had every year. Under Dehradun conditions, where the grass remains dormant during winters due to low temperature and heavy frost, 4 cuttings per annum can be collected in a growing period of nine months. A plantation normally lasts for 4 years and when it is raised on fertile soil with proper care it may last even up to 9 years.

In Kerala in case of C. *flexuosus* when raised from seed, the first harvest is made when the grass is about 1½ to 2 months old and from then on, the harvests are made at intervals of 40 to 50 days; the plants grow up rapidly after each cutting. In South India particularly the Western part in dry season lasting from February to April, the leaves and stems dry up. But in Northern India reverse is the case, the leaves dry up in winter season due to low temperature and severe frost. Fresh growth begins with the commencement of spring. Plantations of C. *flexuosus* are renewed every 5 to 8 years.

Lemongrass is quite a soil exhausting plant and requires nitrogen, potash and phosphate fertilizers. After a rotation of lemongrass, leguminous plants such as peas, gram and green manure may be planted to enrich the soil again. Thereafter these green manure crops are ploughed back.

Distillation: In India, lemongrass (C. *flexuosus*) oil is generally extracted by distillation with water, in indigenous types of stills (capacity 80 to 86 kg of fresh grass). The distillation requires 2 to 2 1/2 hours for completion. These stills are not very efficient. Now improved steam distillation plants with larger capacity stills are available and are actually in use in some areas. With these units the distillation period can be reduced up to 1 hour. In Madagascar and Comoro Islands the distillation is carried out in direct steam stills holding about 800-1000 kg of material and with a pressure of 7 kg in the steam generator. The distillation lasts about 1 hour. Where installation of such plants is not possible, smaller portable type water cum-steam distillation plants may be used. These are more advantageous than the crude water-distillation units. The capacity of such cottage-scale distillation units is about 20 kg of grass per charge. The unit may have a double jacket condenser or ordinary worm or multitube-condensor. It is however advisable to have distillation unit stills having a capacity to hold large quantities of distillable material at a time, as with smaller capacity stills, the cost of distillation is higher. It is advisable that prior to distillation the grass may be spread thinly in shade for one or two days or dried in open for few hours (actual period depending on the local conditions) to drive off the extra moisture. Prolonged drying should be avoided as it may result in substantial oil losses due to evaporation and the oil may develop grassy odour thereby rendering it inferior in quality. Chopping of leaves into small pieces of 2 to 5 cm. length, may be done in hand or power driven fodder chopping machine, prior to distillation. This will facilitate rapid and more efficient distillation of oil from the leaves and thereby the actual period of distillation can be reduced by 1/2 to 1 hour.

Handling and marketing of the oil: The oil obtained after distillation generally contains suspended impurities and water. The impurities can be removed by filtering the oil through fine cloth or filter paper. For removal of moisture the oil may be shaken with anhydrous sodium sulphate and decanted after being centrifuged or filtered directly. This sodium sulphate can be used again after removing its moisture by heating it in an oven. Anhydrous sodium sulphate should be stored in moisture proof airtight containers or in a dessicator over anhydrous calcium chloride. The oil should be stored in air-tight aluminium or tin or galvanised iron containers in a dry, cool and dark place. The aldehyde content of oil declines on storage and, therefore, prolonged storage should be avoided. Owing to this reason, the exporters of lemongrass oil standardize their lots to a minimum of 77% per cent citral against a minimum trade standard of 75% per cent citral content which would allow for the loss of some citral during the transport abroad.

Yield: The yield of grass, and thereby the oil, is small during the first year. It increases with the second year reaching to a maximum in third and fourth years, after which it declines progressively. The oil content of leaves depends on season and condition of the plant material. It is relatively higher during the dry season. In C.*citratus*, the yield of oil varies from 0.2 to 0.4 per cent (green weight basis) and one hectare of the grass yields from 50 to 80 kg of oil per annum in Madagascar and Comoro Islands depending upon the age of plantation. At Dehradun, where the species was tried on an experimental scale C.*citratus* gave (an average of 2nd, 3rd & 4th years yield) about 18 M.T. of grass and 51 kg of oil per hectare per annum with 4 harvests a year collected in April, June, Sept., & Dec. In India in the case of C.*flexuosus* the oil yield is approximately 0.2 to 0.3 per cent (green weight basis) and on an average about 30 to 40 kg oil is obtained per hectare per annum with four cuttings.

Utilization of spent grass: With the use of modern heating-appliances, utilization of spent lemongrass as fuel during distillation is declining. Studies carried out at the F.R.I. Dehradun have indicated that the exhausted grass is suitable for making paper. For spent-grass milder cooling conditions are required than with raw grass. Both soda and sulphate cooking give satisfactory yields of pulp. Since lemongrass yields short fibred pulp, it is necessary to mix it with long fibred pulp for obtaining paper with requisite strength and folding endurance.

14. *C martini* (Roxb.) Wats. (Syn. *Andropogon martini* Roxb.); Poaceae (Gramineae.)

C martini **(Roxb.) Wats**

Na : Eng.: Palmarosa/Ginger grass; Hindi: Ganchej or Rosha or Rosa grass.

Introduction: Palmarosa oil is obtained from the flowering tops and leaves of rosha grass Cymbopogon martini var. motia by steam, water-cum-steam or water distillation. This grass should be distinguished from the closely related C. martini var. sofia which is the source of the less important ginger grass oil. The two varieties are morphologically not easy to be distinguished though experienced persons can distinguish them in the field from their appearance. In odour however, they differ from each other.

Palmorosa is a perennial plant attaining a height of 1.8 to 2.4 m. The aerial parts die in winter. It is susceptible to frost. The root stock sprouts in spring and by the middle of October the grass is in blossom. Burkil describes his investigations regarding the two varieties. The odour given out by the two is certainly different. He states that the leaves of Sofia variety make an acute angle with the culm while those of Motia right or obtuse angle. The chromosome number in the two varieties is also reported to differ. In Motia 2N number is 40 where as it is 20 in Sofia variety.

Uses: Palmarosa oil is a colourless to pale greenish yellow liquid, with a soft sweet odour recalling that of rose. It is the source of the finest commercial, geraniol which the oil contains 75 to 95 per cent. It is extensively used in perfumery for imparting a rose odour and also as a fixative. It blends well with other perfumes. Ginger grass also has similar uses but is considered inferior.

Distribution: It occurs sporadically in forest areas as well as in wastelands and along berms of fields in several parts of India, particularly in Madhya Pradesh, Maharashtra, Andhra Pradesh and Karnataka. Small quantities are also found in Orissa, Gujarat, Tamilnadu, Bihar and Uttar Pradesh. Rosha grass is not gregarious in habit and occurs in patches. It grows in open forests or clearings, on dry sunny slopes. Bulk of the production comes from natural sources. However, in recent years plantations have been raised in some forest divisions in Madhya Pradesh, Andhra Pradesh and in some other places.

Soil and Climate: Rosha grass prefers a dry, well drained soil of basaltic composition, relatively bare and open forests or clearings. It thrives in regions with an annual rainfall ranging from 80 to 90 cm, but does not withstand stagnant water.

Cultivation: The plant can be propagated from seed and vegetatively by planting well rooted stools collected from the older clumps. Seeds can be sown direct by broadcast in well prepared fields (seed rate 4-5 kg per hectare) at the break of monsoon or earlier where irrigation facilities are available. Plantation can also be raised by nursery raised seedlings. In such a case seed may first be sown in well prepared nursery beds in April-May. About 1 to 1 1/2 kg seed will produce sufficient seedlings to plant one hectare with spacing of 75 cm × 75 cm between the plants. In very poor soils planting at spacing as close as 45 cm × 45 cm may be adopted when sowing of about 2-2.5 kg seed in nurseries for a hectare of plantation, will be required.

The seed may be mixed with double the quantity of chaff or sand to ensure its uniform distribution. Transplanting of the grass may be done with first regular shower of monsoon rain. Depending on the intensity of weeds one to two weedings may be done in first year of plantation. Hoeing once in the winter in each year of the plantation loosens the soil and helps the growth of plants. Depending on the nutrient conditions of the soil, application of suitable manures is apt to increase the growth of plants and thereby the yield of distillable material in poor soils.

Harvesting: The experiments conducted with the species at Forest Research Institute, Dehradun showed that though oil of palmarosa is distributed in different parts of the grass, the 1/3rd flowering tops contain much more oil (0.63-1.52%) than the 2/3rd lower leafy portion. In the 1/3rd flowering tops, the oil content is higher at full bloom stage (0.81-1.52%). The yield of flowering tops, which is about 17-28% of the total aerial parts of the grass, is maximum in 2nd and 3rd years and declines gradually. However, for exploitation of the grass the plantations are well productive up to 6-7 years of age. The grass may be harvested once a year at full bloom stage (October to mid November, depending on the climatic conditions and locality factors) but before advanced full bloom stage. In some places one more crop of leaf is harvested.

Distillation: The oil is best obtained by steam distillation but considering the cost of such plants and their being less handy for transportation in interior regions, a water-cum-steam distillation unit may be used. A cottage-scale unit has been evolved at Forest Research Institute, Dehradun which has double jacket condenser, especially designed for distilling camphor from camphor basil but has satisfactorily been used for distilling rosha and other aromatic grasses. Its double jacket condenser may be used as such or after replacing it by an ordinary worm-condenser. Its condenser can also be cooled by utilizing stream water in which case it has to be kept at a lower level than stream water. It takes about 2-2 1/2 hrs to complete one distillation.

Yield: At full bloom stage, under Dehradun conditions of soil and climate it is possible to get annually in the first three years of plantations a yield of about 25 kg of palmarosa oil per hectare per annum. With regular irrigation periodically and maintaining the crop throughout the year and harvesting mainly leafy crop higher yields of 45-50 kg of oil per hectare per annum can be expected.

Marketing: The principal producing centres in India are Akola, Ellichpur, Amravati, Nagpur, Khandesh, Nimar in Maharashtra and Betul in Madhya Pradesh, Zaheerabad and Mahboobnagar in Andhra Pradesh and Bidar, Gulbarga and Raichur in Karnataka State. Small quantities are also produced in Orissa, Tamilnadu, Bihar, Gujarat and Uttar Pradesh States. In the year 2002-2003 India exported palmarosa oil to the tune of about Rs. 2,07,22,364. In the year 2003 export price of palmarosa oil depending on the geraniol content was about Rs.700 per kg.

***C. nardus* (L.) Rendle**

15. *C. nardus* (L.) Rendle. (Syn. *Andropogon nardas* L.) &

16. *C. winterianus* Jowitt.; Poaceae (Gramineae.)

Na : C. *nardas* Bengali: Kamakher; Eng.: Ceylon citronella; Hindi: Ganjni; Kanerese: Ganda-Kanchi, Khaddi; Malayalam:Kamakshipulla; Marathi: Usadhana; Punjab: Khavai; Tamil: Kamochipillu; Telugu: Kamakshikasuvu.

C. *winterianus*: Eng.: Java citronella; Hindi: Mahapingri.

Citronella grass belongs to the genus *Cymbopogon* (Family Poaceae Syn. Gramineae) to which certain other so called aromatic grasses belong such as, lemongrass, palmarosa grass and ginger grass. Although botanically there are several varieties of citronella grass, from the economic and cultivation point of view, only two viz., *Cymbopogon nardus* (L.) Rendle, the Ceyion citronella, known as Lena Batu and *Cymbopogon winterianus* Jowitt. Java citronella known as Maha pingri, are important which yield the Citronella oil of trade. Both grasses are cultivated in Ceylon. Of these Lena Batu is grown more as it is hardy and grows on even poor soil and does not require much attention. Maha Pingri is the principal cultivated citronella grass of Java, Farmosa, Gautamala, Honduras, and Haiti. At present Java citronella oil has a greater demand in trade and industry due to its high citronellal and geraniol content. It contains between 80 to 90 per cent of total geraniol including about 35 per cent citronellal. In India three varieties viz Jorlab C2 , Bio-3 and Manjesh giving better quality and yield of oil have been isolated and cultivated. Ceylon citronellal oil contains only 54-65 per cent of total geranial including 8 to 15 per cent citronellal. The Ceylon citronella oil is used chiefly in mosquito repellent creams, in scenting of soaps, spray, detergents and polishes and in cheap perfumes. The Java oil is used for isolation of geraniol and citronellal which can be converted into some of the most widely used aromatics viz. citronellal, hydroxycitronellal, synthetic menthol and esters of geraniol and citronellal etc. India produces only small quantities of oil and the demand is mainly met by imports.

Soil and Climate: Citronella requires a hot moist climate and flourishes on rich soils. Sunny days alternating with moderate rain are ideal weather conditions. C. *nardus* in Ceylon thrives in dry regions, on hard gravelly soils which are not retentive of moisture; hence in Southern Ceylon, it is grown in high land and slopes of hills below an elevation of 300 m with an average rainfall of about 1500 to 1800 mm. C. winterianus however requires a richer soil and more rainfall and is less hardy than C. nardus.

Planting: Citronella is propagated vegetatively from stools (slips) obtained by separating them from the clumps. Regional Research Lab, Jorhat has evolved a new variety Jorlab C2 which gives higher yield of grass and oil content is also higher. A one year old clump generally gives about 20 or more stools. The stools are planted in the prepared field in rainy season. The optimum spacing for Ceylon Citronella is 90 cm × 90 cm. However, this spacing may be reduced to 75 cm. × 75 cm in poorer soils when intensive harvesting is practiced and the plantation has to last for a shorter duration. Java citronella, accordingly, may be planted at 75 cm × 75 cm. or 60 cm × 60 cm spacing. The number of plants per hectare will be 27,656 at 60 cm × 60 cm 17, 689 at 75 cm × 75 cm and 12321 at 90 cm × 90 cm spacing. The grass stools are usually planted by digging holes with a stick or dibbler. One or two slips preferably two may be put into each hole.

C. winterianus **Jowitt.**

Hoeing and Weeding: Weeding and loosening of the soil, in early stages of plantation and after each harvesting are advisable to improve the yield of grass and oil.

Manuring: The usual practice of manuring is to use the exhausted grass from the distillation stills as such and sometimes furnace ashes are applied mixed with waste grass. Adding of cattle manure, particularly in soils poor in organic matter, have beneficial effects. In Ceylon, artificial fertilizer mixture consisting of groundnut cake (150 kg) sulphate of ammonia (50 kg), kainit (potash fertilizer) (100 kg), super phosphate (50 kg) has been recommended for Ceylon citronella. About 220 to 330 kg of the mixture may be sufficient for one hectare. The CIMPO, southern zonal office, Bangalore recommend application of urea at 125 kg/ha (Split into 4 equal dosages and applied during rainy season) and superphosphate at 80 kg/ha just when the monsoon starts for Java citronella grass.

Harvesting: Slips planted in June-July may be harvested after 9 to 10 months of growth and thereafter the cuttings may be made regularly at intervals of 3 to 4 months, collecting 3 to 4 harvests per annum. However in areas where frost occurs in winter the first harvest in the first year of planting may be made in November or before commencement of frost. Experiments conducted at Forest Research Institute, Dehradun (growing season of 9 months) on Ceylon citronella 3 cuttings per annum (in June, September and early December) were optimum. Leaf blades were found to contain more oil than leaf sheaths and hence harvesting is to be made about 15 to 25 cm or still higher above ground level eliminating the sheath portions. Though the oil content in leaves was maximum when first new leaf just unfolded in the plant, but taking the weight of leaf material into consideration, optimum oil yield was obtained when the fifth leaf is well developed, i.e., the plants should be harvested when most of the slips in clumps have attained five leaf stage; The same harvesting schedule may be followed safely for Java citronella also. In case of Ceylon citronella maximum productivity was recorded in 3-year old plantation after which it diminished progressively. However oil content and quality of oil (geraniol content) up to 6th year compared favourably with that of good quality ceylon oils. Economic life of Java citronella plantation under normal conditions is about 4 years, after which it has to be replaced as both the oil yield and quality get diminished considerably. Citronella is rather soil exhausting, when plantations have to be renewed. Restoration of soil fertility by periodically adding sufficient quantities of organic matter and artificial fertilizers or growing of green manure crops is essential.

Distillation: The leaves are usually harvested in the morning and then partially dried for 2-3 hours in sun. Where facilities are available, it is advisable to spread the grass thinly in shade for 1-2 days. In case of shade drying, oil losses due to evaporation are lesser. Drying of material results in driving off excess moisture from the grass, increasing the quantities of distillable material per charge and thereby lowering of the distillation costs. Prolonged drying results in oil losses. During periods of continuous rain, distillation may be avoided, Chopping of the leaves into small pieces (2.5 to 5 cm long) prior to distillation results in a more efficient recovery of the oil. For chopping, the chopping machine used for preparing cattle fodder is quite suitable. For distilling the grass, water-cum-steam distillation or steam distillation may be employed.

To minimize the distillation cost it is advisable to have larger capacity distillation stills. The actual period of distillation depends on the capacity of the still and quantity of raw material filled in it.

Operating a still holding about 450 kg grass, by water-cum-steam distillation requires 3 to 4 hrs; direct steam distillation requires about 1 1/2 to 2 hrs. For smaller capacity units the time required is still lesser. For water-cum-steam distillation, stills of 1000 to 1200 kg capacity (distillation period 5 to 6 hrs) may also be satisfactory if the distillations are managed properly. The oil received in containers over water is separated in a separating funnel and filtered to remove suspended impurities. Even by ordinary method of decantation, the oil can be separated. Then the oil is treated with anhydrous sodium sulphate to remove traces of moisture.

Yield: The yield of oil depends upon climate, soil fertility, age of plantation, method of distillation etc. Under Dehradun conditions Ceylon Citronella, on an average of 8 years, produced about 50 M.T. of grass and 114 kg of oil per annum per hectare. In Java the yield of citronella oil (Java type) varies from 45 to 84 kg per hectare per year (oil content from 0.75 to 1.0%). In India the yield of Java citronella oil on simple steam distillation is about 50 kg per hectare but as reported by Regional Research Laboratory, Jorhat Asaam higher yield of 200 kg of oil per hectare per annum has been obtained with Jorlab C2 variety. The present market price of oil is Rs. 200 per kg.

Cyperus articulatus L.

C. longus L

C. rotundus L.

Eucalyptus citriodora Hook. Tree

17. *Cyperus articulatus* L.; Cyperaceae;

Hab : It is a perennial herb commonly found in Bengal. Tripura, Manipur and Gujarat.

Tuberous roots yield an essential oil which could be used in perfumery.

18. *C. longus* L; Cyperaceae.

Na : Eng.: English galingale.

Hab : It is found mainly in Rajasthan.

Rhizomes are used in perfumery.

19. *C. rotundus* L. Cyperaceae.

Na : Eng.: Nut grass; Bengali & Hindi: Motha, Mutha; Gujrathi & Marathi: Motha; Kanerese: Tungegadde; Sanskrit: Mustaka, Musta; Tamil: Korai; Telugu: Tungamaste.

Hab : The herb is found in Bengal, Uttar Pradesh and Eastern and Southern parts of India. The dried tuberous roots yield an essential oil which could be used in perfumery.

20. *Eucalyptus citriodora* Hook.; Myrtaceae.

Na : Lemon scented gum;

Hab : It is an Australian tree which has been grown in small patches in Karnataka, the Nilgiris and other tropical to subtropical regions in India.

The species is not fastidious about soil and grows in about all types of soils except clay soils. It can be grown from 300 to 1500 m elevations in tropical to subtropical regions and requires equable climate. Although the species can grow in regions with rainfall as low as 500 mm to as high as 2500 mm or more but localities with about 2000 mm rainfall are more suitable. In Shimoga district of Karnataka large-scale plantations of this species have been raised. The crop can periodically be harvested and distilled for oil.

The seeds are very small and are not viable for more than 6 months. They ripen by the end of December to mid January. About 55 to 85 gms of seed per bed of 3 × 1.2 m may be sown by about February/March and lightly pressed with hand. Straw or grass is then evenly spread on the nursery bed. Watering is done by a fine rose-can during drier period. Germination is complete in about two weeks. The young seedlings are badly attacked by white ants and cut-worms and sprinkling of some insecticides is necessary. As soon as the seedlings are about 7.5 to 10 cm tall they are pricked out into alkathine bags and-are kept in partial shade in the nursery. The seedlings are transplanted in pits at an espacement of 2.5 × 2.5 m or 3 × 3 m with the commencement of regular shower of monsoon. They are planted along with alkathine bags. However before planting the bottom of the bag is cut or torn. Two to three weedings may be necessary in the first year. To prevent the attack of white ants small quantities of gamaxene or hexidol powder are mixed with soil.

The leaves can be harvested for distillation of oil in the 4th year and again at an interval of every 3 years. For utilising the species as a commercial source of oil it is necessary to raise it as a special crop. It is pollarded at a height of 60 to 90 cm from the ground level allowing twigs to develop from the stump. Thus foliage can be collected from the branches for distillation. Young mature leaves give higher yield of oil than old leaves. The yield of oil varies depending on the locality factors, time of harvest and the age of the plantation. The yield of leaves (fresh weight) per tree may vary from 8 to 15 kg. When the trees are vigorous higher yields of leaves can be expected. In addition 25 to 65 kg of firewood is obtainable. Percentage of oil on fresh weight basis is 1.4 whereas on zero moisture basis it is 2.6. The oil contains citronellal 65% to 80%, citronellol (15% to 20%) and esters. The yield of oil per hectare of plantation may vary from 110 to 180 kg.

Glandular Leaves

Uses: The oil is employed in soap industry and perfumery and as a source of citronellal used in the manufacture of citronellal, hydroxyl citronellal and menthol. As a source of citronellal this oil has good demand and can compete with commercial citronella oil obtained from citronella grass.

21. *Eucalyptus globulus* Labill.; Myrtaceae.

Na : Eng.: Tasmanian blue gum.

Hab : and other details are given under medicinal plants.

The oil is used in cheap perfumery.

Eucalyptus globulus **Labill**

22. *Fagaria budrunga* Roxb. (Syn. *Zanthoxylum rhetsa* (Roxb.) DC.); Rutaceae.

Na : Hindi: Badrang.

Hab : It is a fairly large tree found in West Bengal, Western ghats and Andhra Pradesh. The seeds are the source of an essential oil.

23. *Ferula assafoetida* L., Umbelliferae.

Na : Eng.: Asafoetida; Hindi: Hing.

Hab : It is a perennial herb commonly grown in Kashmir (Pl. See medicinal plants).

The gum-oleo-resin in addition to being used as a flavouring agent in food products and in indigenous medicine, is used in perfumery.

24. *F. galbaniflua* Boiss. & Bush; Umbelliferae.

Na : Hab : and other details Pl. see medicinal plants.

Gum-oleo-resin obtained from the stem is used in perfumery.

F. galbaniflua **Boiss. & Bush**

F. jaeschkeana Vatke

25. *F. jaeschkeana* Vatke; Umbelliferae.

Hab : The herb grows in Kashmir.

Roots and the seeds yield an essential oil.

Gaultheria fragrantissima Wall

26. *Gaultheria fragrantissima* Wall.; Ericaceae

Na : Eng.: Fragrant wintergreen; Hindi: Gandapuro; Khasi hills: Jirhap Sohling-thrait

Hab : It is an aromatic shrub with orange brown bark found in Khasi hills, the Nilgiris, The Phulney hills and Kerala. It is gregarious in the Nilgiris.

The leaves are gland-dotted and are the source of an essential oil which is a good substitute for commercial wintergreen oil obtained from an American species *G. procumbens*. The yield of oil has been varying from 0.23 to 1.2 % on air dry basis. The colour of the Indian oil is pale yellowish or pinkish. It is used as a flavouring agent in non-alcoholic beverages and in toothpastes. The oil is used as an external application for rheumatic afflictions.

Glycyrhiza glabra L. – Root

27. *Glycyrhiza glabra* L.; Fabaceae (Papilionaceae)

(*Na : Hab :* and other details Pl. see medicinal plants)

An extract from the roots is used in confectionary and as a flavouring material in tobacco industry.

Hedychium spicatum Sm.

28. *Hedychium spicatum* Sm.: Zingiberaceae

Na : Eng.: Spiked ginger lily; Hindi: Sitruti; Bengali, Gujarati, Hindi & Marathi: Kapur Kachri; Kanerese: Gandhashati ; Sanskrit: Karpurakachli, Gandhashati, Tamil: Shimai-Kichchilik-Kishangu.

Hab : This herb is found in the sub-tropical to temperate Himalayas particularly in Chakrata and Kumaon hills in Uttar Pradesh.

An essential oil obtained from the rhizomes is used in perfumery and pharmaceutical preparation.

29. *Homalomena aromatica* Schott. (Syn. *Calla aromatica* Roxb.); Araceae.

Na : Bengali, Hindi: Kuchugundubi.

Hab : It is an aromatic herb found mainly in Assam. .

The aromatic rhizomes are used for flavouring tobacco and snuff preparation.

H. rubescens Kunth

30. *H. rubescens* Kunth; Araceae.

Na : Hindi: Kuchugundibi.

Hab : The plant is found in Sikkim and Khasia hills.

Rhizomes yield an essential oil which is used in perfumery.

31. *Juniperus macropoda* Boiss.; Cupressaceae.

Na : Eng.: Indian Juniper, Himalayan pencil cedar; Hindi: Dhup, Padum; Punjab: Chalai, Lewar, Shukpa, Shur.

Hab : It is a small or medium sized tree found in the North Western Himalayas from Kashmir to Kumaon. Cedar oil is obtained by distilling the shavings and sawdust of this species which is used in perfumery.

Matricaria chamomilla **L.**

32. *Limnophylla rugosa* (Roth) Merr. (Syn. *L.roxburghii* G. Don); Scrophulariaceae.

Na : Bengali: Kalakarpur; Mundari: Losodba.

Hab : It is an herb found in the greater part of the country. Its aromatic leaves are used for perfuming the hair.

33. *Luvunga scandens* Buch.-Ham.; *Rutaceae.*

Na : Bangali: Lavangalata, Lavangaphal; Hindi: Sugandha Kokilas Sanskrit: Kakoli.

It is a large thorny climbing shrub found in Bengal, Assam and Khasia hills. Fruits yield an essential oil which is used in perfumery and pharmaceutical preparations.

Melaleuca leucadendron **L.**

34. *Matricaria chamomilla* L.; Asteraceae (Compositae.)

Na : Hindi: Babuna; Punjab & Hindi: Babuna, Suteigul.

Hab : and other details pl. see medicinal plants.

Leaves and twigs of this herb yield an essential oil which is used in perfumery.

35. *Melaleuca leucadendron* L.; Myrtaceae.

Na : Bengali: Cajaputte, Cajaputi; Engl: Cajeput tree Hindi: Kayaputi; Marathi: Cajuputa; Tamil: Kaiyapuppudai.

Hab : It is a tree native of Australia but is being grown in the country. It looks like bottle brush tree. Fresh leaves and twigs are aromatic and yield an essential oil on distillation. It is called cajaput oil in commerce. It contains 50 to 60 per cent cineole. It is used both in pharmaceutical preparations and perfumery.

Mentha arvensis **DC.**

36. *Mentha arvensis* DC. *var. piperascens* Holms; Labiatae.

Na : Hab : and other details pl. see medicinal plants.

Leaves and twigs on distillation yield an essential oil which on congealing yields menthol. It is used in pharmaceutical and culinary preparations as a flavouring agent.

M. piperita L.

Mesua ferrea L.

Michelia champaca L – Branch with flower

Mimusops elengi L.

37. *M. piperita* L. Labiatae.

(*Na :* *Hab :* and other details pl. see medicinal plants.)

Leaves and twigs on distillation yield an essential oil which is used in perfumery and as a flavouring agent in toothpaste and culinary preparations. It is also used medicinally.

38. Mesua ferrea L.; Guttiferae.

Na : Andaman: Gangane; Assam: Nahor, Diengngai, Ngai-ching; Bengali: Nagesar, Nagkesar; Eng.: Iron wood; Gujarathi & Marathi: Nagchampa; Hindi: Nagkesar, Nagchampa, Nageshwarchampa, Nagesar; Kanerese: Nagakesara, Nagasampige; Malayalam: Nanga, Peri, Veluthupala; Tamil: Nangu, Nangal, Irul, Nagachambagam; Telugu: Nagakesara, Kesaramu, Nagachampakamu, Trade: Mesua.

Hab : It is a tree found in the Eastern Himalayas, Assam, West Bengal, Western ghats, Kerala and Andaman Islands.

Flowers, flower buds and fragrant stamens are used in cosmetics soaps and perfumed ointments.

39. *Michelia champaca* L.;Magnoliaceae.

Na : Assam: Titasopa, Bolnabat, Shop; Bengali & Hindi: Champa, Champc; Gujrathi: Champo, Rae-champo, Pitochampo; Kanerese: Sampige, Kolasampige; Mal.: Champakam; Mar.: Pivaluchampa, Sonachampa, Kudchampa; Oriya: Chompa, Chompoko, Kanchana; Tamil: Shambuga, Chambugam; Telugu: Champakamu; Trade: Champac.

Hab : It is a large tree found in Assam, Sikkim, Bengal and South India. It is also grown in gardens for its aromatic yellow flowers. The oil obtained from the flowers is used in perfumery. The wood obtained from the species is also used for a variety of purposes.

40. *Microtoena insuavis* (Hance) Prain ex Dunn; (Syn. M. *cyrnosa* Prain); Labiatae.

Na : Eng.: Khasia patchouli

Hab : It is an under shrub found in Khasia hills.

The entire plant particularly the leaves produce an essential oil which is used in perfumery and also as a substitute for patchouli oil.

41. *Mimusops elengi* L.; Sapotaceae.

Na : Assam: Gokul; Bengali: Bakul; Eng.: Spanish cherry, Bullet wood; Guj.: Barsoi, Bolsari.; Hindi: Maulsari; Kanerese: Bakula, Pogade; Malayalam: Elengi, Ilanni; Oriya: Bokula, Baula; Tamil: Vagulam, Magodum, Ilanji; Telugu: Pogada.

Hab : It is a tree found in central and South India. It is also grown for its fruit in many parts of the country. The flowers yield an essential oil. An attar can be prepared with sandal wood oil. The oil has sweet scent.

42. *Mirabilis jalapa* L.; Nyctaginaceae.

Mirabilis jalapa **L.**

Na : Bengali: Krishnakeli, Sarpamani; Eng.: Four O' clock plant; Guj.: Gulbaji; Hindi: Gulabbas, Gulbash; Kanerese: Sanjemallige, Chandramallige; Malayalam: Antimalari, Antimanlaram; Mar.: Gulbas; Tamil: Andhimalligai; Telugu: Chandrakanta, Chandramalli.

Hab : It is an ornamental herb found in the North Western Himalayas, Bengal and Manipur and grown in other tropical parts in the country. Powdered seeds are used locally in cosmetics.

43. *Myristica fragrans* Houtt.; Myristicaceae.

***Myristica fragrans Houtt.* - branch**

Na : Bengali, Gujrathi, Eng.: Nutmeg; Hindi & Marathi: Jaiphal, Jaipatri; Kanerese, Malayalam, Tamil and Telugu:Jajikai, Japatri.

Hab : and other details pl. see medicinal plants.

Nutmeg oil is used in soap and perfumery industries. Essential oil obtained from the leaves is used in the preparations of chewing gums, flavouring essences and cosmetics.

Myristica fragrans **Houtt-fruit**

44. *Nardostachys jatamansi* DC.; Valerianaceae.

Na : Hindi: Jatamansi, Balchir.

Hab : and other details pl. see medicinal plants.

Essential oil obtained from the roots is used in perfumery.

Nardostachys jatamansi **DC.**

45. *Ocimum kilimandscharicum* Guerke; Labiatae.

Na : Eng.: Camphor basil; Hindi: Kapur tulsi.

Hab : It is an exotic perennial shrub growing to a height of about 1.2 to 1.5 m. It thrives best in open but can tolerate a certain amount of shade. It is being grown in teak plantations in South India. It is not fastidious about soil. It requires warm climate. In high rainfall areas growth is better. The plantation can be raised either by broadcast sowing directly in the field or by trans-planting of nursery raised seedlings. To economise seed and also to ensure the success of plantation it is better to sow in the nursery on raised beds. Since the seed is very fine it is mixed with 10 parts of sand and sown in lines 2.5 cm apart about 4 to 5 weeks before planting and covered with a thin layer of soil and dry leaf or grass. The beds should be irrigated by percolation or a fine spray should be used. The seeds germinate in a week's time. After germination cover should be removed. By the time of season of planting the seedlings would have attained 10 to 15 cm height when they are transplanted. This is usually done when the regular monsoon sets in. The plant can stand a good deal of rough handling and can be carried over long distances. The seedlings can be planted with naked roots in small holes made by wooden pegs. Earth round the roots is well packed

Ocimum kilimandscharicum **Guerke**

***Ocimum kilimandscharicum* Leaves**

so that no gap is left in the hole. About 75 gms of seeds can supply enough plants for planting one hectare at spacing of 60 × 30 cm. During the first monsoon one or two weeding may be necessary. When the plants are once established they can stand repeated cuttings for many years. In the first year one or two harvests can be had and from the second year onwards the crop can be harvested thrice in a year depending on climatic and edaphic conditions. Plants are cut with a scyth leaving 10 to 15 cm shoots from the ground level for future growth. The crop is then dried in the sun. After drying leaves are removed by threshing. The dried sticks do not contain any camphor or ocimum oil so they are either thrown away or used as fuel for distillation. The dried leaves do not deteriorate and also they do not lose any camphor content. The leaves are distilled to get camphor and ocimum oil by water-cum-steam distillation method. Camphor is congealed in a double jacket condenser. It gets deposited on the outer surface of inner jacket and on the outer surface of the outer jacket facing the inner jacket. About-2500 kg of dry raw material can be expected from hectare of plantation. The total distillate is 3 per cent of which 18 kg of camphor and 15 kg of ocimum oil per hectare of plantation can be obtained. Camphor is in great demand in celluloid and cinematography industry. It is also used in explosives and in many other industries

***Origanum vulgare* L.**

46. *Origanum vulgare* L.; Labiatae.

Na : Eng.: Potmorjoram, common or wild morjoram; Hindi: Satbra, Bantulsi; Kanerese: Maruga; Telugu: Mridumaruvamu.

Hab : It is a perennial herb found in the temperate Himalayas from Kashmir to Sikkim. Essential oil can be used in soap and perfumery industry.

***Pandanus tectorius* Soland *ex* Parkinson**

47. *Pandanus tectorius* Soland *ex* Parkinson (Syn. P. *odorotissimus* Roxb.); Pandanaceae.

Na -Bengali: Keya, Kedkikeya keori; Eng.: Screw-pine; Gujrathi: Kewoda; Hindi: Keora, Kewda, Ketki; Malayalam: Kaida, Thala; Marathi: Keora; Kanerese: Talemara, Kyadagegida; Sanskit: Ketaki; Tamil; Tazrai, Talay; Telugu: Mogali.

Hab : It is an ornamental shrub or small tree found along the sea coast or river banks and also in the Andaman Islands.

As essential oil obtained from the main inflorescence is used in perfumery.

***Pelargonium graveolens* L' Herit.**

48. *Pelargonium graveolens* L. Herit.; Geraniaceae.

Na : Eng. Rose geranium.

Hab : It is mostly cultivated in shevroy hills in Tamilnadu and in Karnataka states. The leafy twigs are the source of an essential oil which is used in soap and toilet preparation.

49. *Pinus roxburghii* Sar, (Syn.P. *longifolia* Roxb.); Pinaceae.

Na : Eng.: Chirpine, Long leaved pine; Hindi: Chil, Chir, Salla.

Hab : It is found in the Western and central sub temperate Himalayas.

Fresh needles and twigs yield an essential oil which is used in perfumery. The yield of oil is about 0.5 per cent. In commerce the oil obtained from the needles of fir and spruce is also called pine needle oil.

50. *Pogostemon perilloides* (L.) Mansf. (P. *heyneanus* Benth.; P. *patchouli* Hook. f. non pellet.); Labiatae.

Na : Bengali: Pachapat; Eng.: Patchouli; Gujarathi: Pachapandi; Hindi: Patchouli; Kanerese: Patchetene; Malayalam: Pachila Tamil: Kadirpachai.

Hab : It is a herbaceous perennial found in Karnatak. It is also cultivated in many other places.

The plant yields an essential oil which is a good substitute for patchouli oil of commerce which is obtained from P. cablin (Blanco) Benth.

It is a fairly hardy species and is capable of adapting itself to a wide range of soil and climatic conditions but thrives well in a warm and moist climate with an evenly distributed rainfall. It requires a fertile soil being quite soil exhausting. It is usually propagated from stem cuttings. Cuttings can be raised at any time of the year but the best time is during the rainy season. The cuttings should be 10 cm to 15 cm long with 3 to 4 nodes. They are planted in a slanting way in the nursery at spacing of 10 cm × 10 cm. Transplanting of nursery raised cuttings is done after 45 to 50 days when the root system is well developed. Planting is done at an espacement of 90 × 90 cm during the monsoon. It grows well under partial shade. Weeding, irrigation and manuring are carefully attended to throughout the growing period of the plant in order to get better yields. First harvesting can be had about 5 to 6 months after planting with another crop 6 months later and so on for 3 years. From the fourth year onwards, the vigour of the plantation declines. Harvesting in the morning or evening is more suitable. It is a practice that while harvesting, a few branches on each bush are left in order to allow the plant to grow again. The leaves are carefully dried in the shade when they develop the characteristic patchouli odour. Too much of drying of leaves has to be avoided. Sometimes the leaves are allowed to ferment slightly before they are distilled. A yield of 1345 kg per hectare of semidried leaves is regarded as the fair output per annum. The material is distilled either by steam or water-cum-steam distillation. The yield of oil from the semi-dried material is about 3.5 per cent. There is considerable demand for the oil in perfumery industry. It blends well with palmarosa, vetiver and clove oil. It is used as a fixative.

Pinus roxburghii Sar. **(chir pine)**

Santalum album L. – Tree

Santalum album L. branch

Sandal Heart Wood

Saussurea lappa C.B. Clarke

51. *Santalum album* L.; (Syn. *Sirium myrtifolium* L.); Santalaceae.

Na : Eng.: Sandal wood; Hindi.: Chandan.

Hab : It is a tree found in Karnataka, Andhra Pradesh, Tamilnadu and also in other states. There are as many as 20 spp. of *Santalum* but the most important species of this genus is *Santalum* album. Karnataka and Tamilnadu almost hold the monopoly in the production and export of sandal oil from S. *album*. Heart wood portion of the tree contains the essential oil, which is used as a fixative of floral essences and also in other perfumery industry.

The paste of sandal wood is used in religious functions by Hindus, Muslims etc. During Tippu Sulthan's times around 1792 AD, sandal wood had a flourishing trade and it was being treated as a royal tree and the owners of trees were entitled for 50% of the sale proceeds. With proper protection from fire and grazing sandal tree regenerates naturally. However artificial regeneration is also being attempted. Srimati R.A et.al., believes that the growth of sandal and formation of heart wood is not uniform under identical climate and edaphic conditions and that this phenomenon depends on genetic factors of individual trees. Therefore selection of proper genotypes is necessary in artificial regeneration programmes. The species is not fastidious about soil or climate. There is heavy mortality due to spike disease which is common in Karnataka and Tamilnadu. Normally spike effected trees are felled. The wood is chipped and distilled or converted into small billets. The yield of oil is highest from the root portion. Yield of oil from stem wood varies from 4.5 to 6.25 per cent. The wood is distilled under low pressure for 48 to 72 hours. The oil collected in the condenser is skimmed off. The crude oil is again purified. The chief constituent in the oil is santalal which is about 90 per cent. It blends well with other essential oils. There is a well established trade of this oil. During the year 2002-2003, 6.135 tonnes of sandal wood oil valued at Rs.11,39,51,783.00 was exported. Besides, 2.127 tonnes of sandal wood valued at Rs.3,14,58,977.00, and 726.780 tonnes of sandal wood chips and dust valued at Rs.31,79,30,280.00 were also exported.

52. *Saurauia roxbhurghii* Wall.; Sauraiaceae.

Na : Assam: Bon-posola, Hengunia, Paniposala, Porbotiasengunia; Cachar; Bon-loisu-phang;. Garo: Bon-jersinnig. Gengsning; Khasi: Dieng-soh-la-pied; Lushai: Terpui; Lepcha: Safarkung, Dangsipha; Manipur: Singkhrau; Nag.: Dia-ching, Tong-bahu; Nepal:.Aulegogun. Gogan.

Hab : It is a tree found in Assam, North-Eastern regions and Bengal. A mucilaginous substance obtained from the leaves is used as hair-pomade.

53. *Saussurea lappa* C.B. Clarke (Syn. *Aucklandia costus Falc.*); Asteraceae Syn. Compositae.

Na : Arabic: Kust: Bengali & Hindi: Kut. Kur. Pachak; Gujrathi: Upaleta, Kur; Kanerese: Koshta; Malayalam: Sepuddy; Marathi: Kushta; Sanskrit: Kushta; Tamil: Kostum, Putchuk, Goshtam; Telugu: Changala, Kustam; Trade: Costus, Kuth.

Hab : It is an herb found in Kashmir, Himachal Pradesh and Garhwal in Uttara Khand (Pl see medicinal plants). The roots are collected in September/ October. They contain an essential oil which is used in perfumery as a fixative and for blending high class perfumery for which it has a high reputation in countries like, U.K, France and U.S.A where it is sold at fabulous prices.

Valeriana officinalis L.

54. *Valeriana officinalis* L.; Valerianaceae

Na : Hab : and other details please see medicinal plants.

The dried rhizomes and roots are the source of an essential oil which is used in perfumery.

55. *V. wallichii* DC.; Valerianaceae.

Na : Hab : and other details please see medicinal plants.

The rhizomes and roots yield an essential oil which is used in perfumery.

56. *Vetiveria zizanoides* (L.) Nash; Poaceae Syn. Graminae.

(*Na : Hab :* and other details please see medicinal plants).

V. wallichii DC.

Vitever oil is obtained by steam distillation of Vitever (Khus)- Viteveria Zizanoides (L.) Nash; Poaceae, a perennial grass. The oil is one of the finest oriental perfumes with a persistent fragrance. The oil is used in the manufacture of soaps, cosmetics, perfumery, agarbathis, soft drinks, pan masala, sherbets etc. In blended perfumes, oil of vetiver acts as an excellent fixative for volatile components. Vetiver also has medicinal properties. It is known for its cooling properties. Hats, screens and hand fans made from Vetiver roots are sold in the markets.

Soil and Climate: Vetiver is a perennial grass found along the river banks and in marshy places in South India. It can grow in a wide variety of soils including poor, marginal alkaline soils (pH 9-10), waste lands and temporarily water logged soils. It is planted as a soil conservation measure. However, sandy loam soil is best suited for its cultivation.

Vetiver is a tropical grass and grows in areas having an annual rainfall of 100-200 cm with good amount of sub-soil moisture and temperature ranging from 21-43.5 °C with fair amount of humidity. It is not suited to temperate climates. Shady soils are avoided.

Vetiveria zizanoides (L.) Nash

Planting: There are two varieties of Vetiver namely, North Indian and South Indian. The N. Indian variety yields superior quality oil while the S. Indian variety gives higher yields of oil. CIMAP KS-1, CIMAP KS-2, Sugandha, Dharani, Gulabi and Kesari are improved varieties of Vetiver.

The land is cleared and deep ploughed. The best time for planting Vetiver is June/July. A basic dose of 5 tonnes/ha of farm yard manure along with 80 kg of superphos and 20 kg of potash will enhance the growth and yield of roots. Slips of Vetiver are prepared as follows: Leaves are cut off from old clumps of Vetiver and the clumps are uprooted. The slips are separated

Vetiveria zizanoides L. Nrash Roots

with 15-20 cm of long roots, intact. Planting pits are made at a distance of 30 cm between plants keeping 45 cm between the rows to facilitate easy movement. A slip is placed vertically in each pit, the soil around is pressed firmly and levelled. If irrigation is available, the crop is irrigated immediately and irrigation is given once in 10 days till the rains set in.

Irrigation and Weeding: During non-rainy periods, 2-3 irrigations are given per month. The seedling is kept weed free until the plant canopy develops and thereafter no weeding may be necessary.

Plant protection: Vetiver needs to be protected from grazing and trampling by animals during early growth. No conspicuous damage from pests and diseases is reported.

Harvesting: First harvesting is done after 15-18 months after planting, to get high yield of fruits and good quality oil. The above ground leafy part is first cut and the field is irrigated. The field is ploughed as deep as possible and all the roots including the small rootlets are collected. In heavy soils roots may have to be dug out.

Yield: Depending on the soil, climate and management, Vetiver gives 4-5 tonnes of roots per hectare. The fresh roots are chopped and steam distilled. The yield of oil from fresh roots is low. However, when it is intended to get oil from dried roots, they are soaked in water for 4-6 hours and then steam distilled. South Indian varieties yield 0.6-1.0 % oil where as improved varieties of North India yield 0.5-0.8 % of oil. Thus a minimum of 10-20 kg of oil is obtained per hectare.

Viola odorata L.

57. *Viola odorata* L.; Violaceae.

Na : & *Hab :* and other details Pl. see medicinal plants.

The flowers are used in perfumery.

6

Spices, Condiments and Savory Plants

From the earliest times the spices had considerable demand in all parts of the world and were eagerly sought after as gold. The craze for these has brought about many changes in the history and promoted international trade. It has resulted in the discovery of new lands and new sea routes. The majority of the spices originated in the Asiatic tropics and was among the first products of commerce between the East and the West. The Arabs were the first to establish trade links between the spice producing centres in Asia and the Europe and took spices such as pepper, cinnamon and cardamom from India and Sri Lanka to the European markets. Thereafter other countries followed and the trade remained in the hands of Venetians, the Dutch and the British for a long time. In the early days the spices were derived mainly from the forest but their ruthless exploitation in the past has resulted in the rarity or extermination of some of the species from the forest.

The produce, included under the name of "spices", consist of all aromatic vegetable products, which are used in flavouring or seasoning insipid food and drinks and are derived either from bark, flower bud, fruit, root or the underground stem – the rhizome. Condiments are such flavouring or seasoning substances which have a sharp taste. Fruits, seeds and leafy herbs whether fresh or dried, when used whole for flavouring food are called savory. Since the use of all these products is generally the same i.e., flavouring of food articles, no distinction is made in their usage and very often condiments and savory articles are called spices. Some of the spices, like clove, vanilla, cinnamon, pepper, cardamom, nutmegs, etc., in addition to their customary use as a flavouring agent of food stuffs are used for various other purposes such as perfumery, incense, soap making and medicine. Cinnamon and calamus are known to stimulate appetite and are used as carminative. In some cases they are used to disguise the unpleasant taste or odour of the drugs. Since the spices contain very little nutritive value they are not classed as food. Clove oil is used in histological studies or microscopy and also in dental prescriptions as medicine. Although different spices require different degrees of moisture and temperature the majority of them have their distribution in the tropical rain forest regions where the climate is hot and humid. The soil requirement also varies from species to species but in general they prefer deep well drained sandy loam soils. Although in recent years the use of spices as a flavouring agent by the western countries is diminished but still considerable demand in the international trade exists and offers good prospects for the cultivation of

Plant of *Alpinia galanga* (L.)Willd.

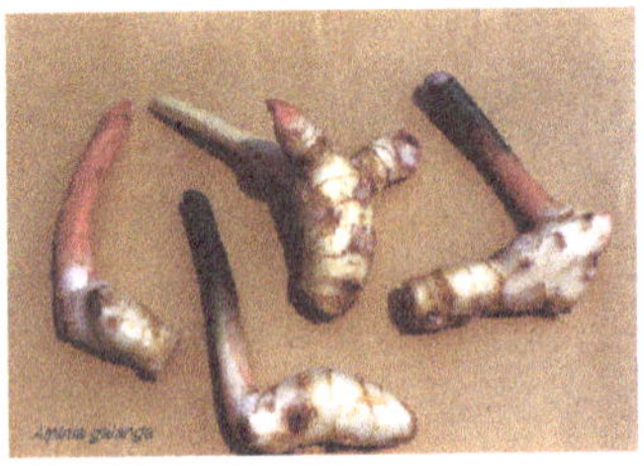

Rhizome of *Alpinia galanga* (L.) Willd

Plant of *Amomum subulatum* Roxb.

Large cardomon

large number of species. Cinnamon, cardamom, cloves, nutmegs and vanilla can be introduced in the forests as permanent crops lasting for a number of years while others like ginger, cumin, coriander can only be raised as agricultural or garden crops.

It is necessary that the spices produced should be in perfect condition. Inferior qualities and adulterants are often mixed with a motive of larger gains, which should be prevented. If by any chance any material is not up to the standard it should be sold separately. The material should be stored in a cool and dry place. If the place is excessively hot the volatile essential oil found in the material suffers. If the material is kept in a damp place it is likely to be affected by fungus. Care should also be taken to keep the produce free from attack of insects. It should be packed carefully so as to present a good appearance at the sales and also to avoid damage by insects or water during transportation.

1. *Alpinia galanga* (L.) Willd. (Syn. *Languas galanga* (L.) Stenz.; Zingiberaceae.

(*Na : Hab :* and other details pl. see essential oils)

The rhizome is used as a condiment and the seeds are used as spice.

2. *Amomum subulatum* Roxb.; Zingiberaceae.

Na. - Eng.: Large or Nepal cardamom; Hindi: Badi elaichi, Murrung elaichi.

Description of the plant: Large Cardamom (*Amomum subulatum Roxb.*) belongs to the family Zingiberaceae which the true cardamon (*Elettaria cardamomum*) and turmeric belong. It is commonly called large Cardamom, (Bari elachi) or Perivavelam while the true cardamom is commonly known as Chhoti elachi or velam. Large cardamom is a perennial plant and propagates both from the seeds as well as from the rhizomes.

Left uncared for, it thrives well for over 25 years. The long lanceolate leaves are borne on aerial shoots and the flowers are borne on long flexuous and prostrate panicles at the base. The aerial shoots die off after the fruits are mature but fresh leafy shoots and subsequently flower bearing shoots arise from the underground rhizomes at the base of the old ones. The flower bearing shoots arise from the base near the ground level and flowers are borne on short peduncles alternatively on the axis. The flowers have large white central strap-like lip and open from the base of the axis towards the tip. The fruit is more or less tri-angular in shape broad towards the base, green in colour when unripe and purple to brown when ripe. Each capsule contains 40 to 50 hard, round, greenish black to dark brown seeds embedded in soft slimy and jelly like pulp, which is sweet to taste. The spicy aromatic seeds contain a volatile oil known as cineole. There are three varieties of large cardamom, Ramshai, Sawney, Golshai.

Climatic and soil requirements: Large Cardamom is a native of Eastern Himalayan region where it is found in wild state by the side of springs and rivulets in the forests. It has been localised in a restricted area in Darjeeling district and in East-Sikkim, Nepal and Arunachal Pradesh. It grows well between 830 to 1830 meters above sea level in shady and damp places under evergreen forest canopy where enough of organic matter is naturally deposited due to the falling of leaves of the trees and light is available in restricted way through the canopy. The crop thrives well in places having an annual rainfall of 2000 mm to 2500 mm maximum temperature of 14 °C to 33 °C and minimum temperature of 6 °C to 22 °C. Frost is injurious to the crop as is the hail storm which cause irreparable damage to the growing leaves. Although the plant loves damp places but waterlogged areas are not suitable for its growth. Light well-drained soil having a gradual slope by the side of perennial springs where there is enough moisture but water flows down constantly without any patch of water logged area, is ideal for the cultivation of large cardamom.

Propagation: Large cardamom is generally propagated from the rhizomes like banana. More commonly a plantation is kept for 20 to 25 years without any replacement. When the aerial shoot dies in December-January after fruits are mature, the soil is lightly worked by means of forks and spades and the rhizomes are covered with soil. Fresh shoots appear during the rainy season in May-June. As all the plants do not mature at the same time, the plantation remains productive throughout its existence of 20 to 25 years but the vigour of the crop is gradually decreased and yield is reduced considerably. Pests and diseases and serious virus diseases like 'Foorkey' and 'Chirkey' also reduce the vitality of the plant.

(a) *Planting of rhizomes:* Planting of rhizomes, which method is gradually loosing popularity due to apprehension of diseases, is done with a view to avoid the labour and expenditure to raise seedlings and to get production about 2 years earlier than that from the seedlings. Rhizome pieces, collected from healthy and high yielding plants are planted in holes 12 to 15 cm deep and 15 cm in diameter at a distance of 2 meters both ways. Holes are dug and are filled with fine soil; wood ash, compost or well rotted leaves and powdered gravel along with a pinch of B.H.C. 50% powder captan and lime. The rhizome pieces are put in holes one piece in each, in such a way that the top portion is just at the ground level. The holes are covered with a fine layer of soil over which a layer of leaves and straw are put to serve as mulch. The holes are arranged in a horizontal manner to form a contour and serve as bench terracing for soil conservation purpose. Proper planting time is May-June at the onset of the monsoon rains. The growers at large, however, do not allow so much spacing as a result of which the plantation becomes so crowded that no intercultural operation becomes possible.

(b) *Planting from seeds:* When seedlings are planted care has to be taken to select the seeds from healthy and disease free plants. For this purpose a large number of healthy and disease free plants are selected in the month of September from large good plantations. The plantations are revisited at the time of harvesting in November-December and yield of those selected individual plants are taken. Seeds from those selected plants, which give high yield, are finally chosen to serve as seeds. The green fruits are brought to the nursery and dried from 7 to 10 days in open sun. The seeds are removed from the capsules and washed clean free of any mucilaginous substances and sown in the nursery beds. It has been observed that over-drying of the capsules or presence of any mucilaginous coating on the seeds hamper germination considerably. Pitting of seed coat, scarification by means of sulphuric or nitric acid or by rubbing on rough surface has been found to have no effect on improvement of seed germination. Attempts are being made to treat the seeds with different chemicals to improve their germination. It appears that presence of the volatile oil cineole and some alkaloid in the slimy substance may have some clastocholine activity to cause retardation in germination. Attempts are being made to isolate and identify the alkaloids, if any. On an average it has been observed that out of the total seeds sown in a seedbed, germination does not exceed 45%.

(i) *Primary nursery:* Two nurseries are required for raising seedlings of large cardamon, i.e., the primary nursery for sowing the seeds and raising one year old seedlings and the secondary nursery for care of the seedlings from one year to two year stage. The primary nursery is prepared in a clean place where a few trees are present to provide a little shade. It is better to avoid the slope facing south to protect the seedlings from strong wind and direct sunshine. Each nursery should not be longer than 3.5 meters and broader than 1.5 meters. It should have a shade of straw at a height of 2 meters from the beds. The shade should be slightly inclined so that the rainwater may be drained off and cannot drop on the seedbed. The nursery should be protected from stray animals by fencing all around. A few twigs may also be placed against the fencing so that strong wind or direct sunrays are prevented while allowing adequate ventilation. The nursery bed should be prepared by ploughing the soil at least 10 to 15 cm. deep, incorporating enough of compost, well rotted cow-dung and leaf manure and a little wood-ash and by raising the beds 10 cm above ground level. Lime at 5 kg per bed (3.5 × 1.5 metres) may also be incorporated at least 3 months prior to sowing of seeds in case the soils are acidic. Before sowing of seeds B.H.C. 50% powder and captan may be dusted on the beds. Aldrex may also be dusted where white-ant is present. There should be a small path between the beds for supervision and irrigation.

(ii) *Sowing:* Seeds are sown at the rate of 600 seeds per bed. Well-formed mature seeds are sown in lines 9 cm apart just below the soil (Seed bed) level and covered with a fine layer of dry soil over which a 3 cm thick layer of dry straw or leaves cut into small pieces should be scattered to form a mulch. Light irrigation by means of rose-cans is given daily if necessary in the morning and in the evening, to keep the seed bed moist but not damp. Too much water should always be avoided. During the dry season; the straw of the nursery shade should also be sprinkled with water to keep the nursery humid. The seeds begin to germinate from the 2nd month after sowing and continue to do so till about 10 to 12 months. Thus great care should be taken in the nursery stage to keep the seed beds moist and clean and weeding should be done without disturbing the seedlings. Sometimes, algal growth develops, particularly at the later period. Light scratching by means of stick removes the scum.

(iii) *Secondary nursery and care of seedlings:* As the seeds, sown at the same time, germinate at different periods, the age of different seedlings that appear in the 'primary seedbed is not the same. It has been observed that only 5 to 10 per cent of seeds sown germinate within 2 months i.e. July and attain an age of 1 year by next July but those which germinate after 10 months i.e. in March next year, become only 4 months old at that time. Hence, during replanting of one-year-old seedlings in the secondary nursery bed, care has to be taken to take out only those seedlings that are one year old keeping the younger seedlings undisturbed. The one year old seedlings are carefully taken out of seed bed by digging up the soil and are replanted in the secondary nursery bed which is made .in a separate place just like the primary nursery bed excepting that there is no shade on top and the distance between individual seedlings both ways is kept at 0.5 metre. Where rainfall is heavy, a temporary shed is provided for this secondary nursery also. Holes 15 cm deep and 15 cm in diameter are dug at 0.5 metre distance both ways and filled up with a mixture of dry soil, sand or powdered gravel, compost, cow dung or dried well rotted leaves, wood ash and a pinch of B.H.C. 50% powder and captan. The seedlings are planted in these holes so that the roots and the rootstalk are just completely covered with soil but do not go deep into it. The base of each seedling is covered with a thick mulch of dry leaves. As transplanting is done during rainy season, there is no need of any irrigation but during dry periods, frequent irrigation has to be provided to keep the soil moist. It is necessary to spray Blitox at 15 days interval during rainy season and at 30 days interval during the winter season in the secondary nursery to protect the tender seedlings from disease. Dusting of B.H.C. 50% powder has also to be done occasionally as a protection against insect attack. Drenching of the soil by captan at least once at the onset of dry period is also recommended.

(iv) *Planting of seedlings in field:* The 2 year old seedlings, i.e., seedlings that have spent a year in the secondary nursery bed are fit for transplantation in the field. The soil where final transplantation will be done has to be worked out at least 3 months prior to transplantation. Holes 22 cm deep and 30 cm in diameter are dug at a distance of 2 metres both ways horizontally to form a contour to make a sort of bench terracing. The seedlings are carefully dug out of the secondary nursery bed and transplanted in the field in a similar manner as in the case of their transplantation in the secondary nursery bed. As the transplantation is always done during the rainy season no irrigation is required. It is to be seen that the shade trees do not provide much shade to make the plantation dark. In such a case the shade trees are lopped off so that sunrays may be permitted to come through the canopy of the trees, which serves as a filter screen.

Cultural operations: The plants get established in the field within a month and vigorous growth starts. From the third year flowers are produced but it requires 5 years to attain the stage of full production potential. During this period, control of shade by lopping the shade trees as and when required, inter-cultural operations to prevent crust formation on the soil due to constant deposition of leaves and their decomposition, making of ridges to support the plants and cover the rhizomes with soil are all what are needed to be done. Occasional spraying of Blitox and dusting of B.H.C. 50%, application of compost or farm-yard manure every year before the onset of monsoon are necessary to keep the plants healthy. It has been observed that application of 30 Kg nitrogen along with 30 tonnes of cow-dung or compost per hectare every year increases the yield of large cardamom by 0.5 quintals. Wood ash has also been found to be effective. One hectare contains about 2500 Plants. Thus the dose of nitrogen per plant comes to about 10 gm and 10 Kg of cow dung or compost, 25 gm of superphosphate per plant may also be required. In most plantations, however, fertilizers are not applied at all as the soil is very rich in organic manures.

Replantation: The large cardamom crop starts bearing from the 3rd year after final transplantation in the field and continues to do so till 25 or 30 years but the optimum production period is from the 5th year till the 10th year after which the yield decreases. Thus re-plantation has to be done after the 9th year by gradual steps, as it is difficult and uneconomic to destroy the entire plantation at a time and wait for 3 to 5 years after replanting to have the new produce. However removal of the entire plantation and replantation has to be done in case of any indication of 'Foorkey' infection. While keeping the plantation undisturbed planting of new rhizomes in place of deceased plants only was in vogue sometime back but this method is not followed now due to the awareness of the growers of the 'Foorkey' disease. Replanting in place of otherwise damaged plants due to landslide; destruction by animals or white ants is however done. The growers in general, however, do not follow regular spacing nor take much care of their plantation in respect of manuring and intercultural operation as a result of which they do not get the maximum benefit out of their plantations.

Harvesting, yield and marketing: Harvesting starts from September and continues till December-January depending on the locality. Ripe, purple to brown coloured capsules are picked out once in a month. The fruits are dried in open on clean floor or on bamboo platforms in the sun for 7 to 10 days and sold in the markets or to the wholesale dealers. For keeping seeds for raising seedlings the seeds are taken out of the capsules, washed thoroughly and dried in shade or directly planted in the nursery beds. The capsules dried for 7 to 10 days in sun turn deep purple and contain enough of moisture. These green cardamoms, as they are commercially called, fetch much lower price as they contain about 50 percent water and are to be dried again for export or for storage otherwise fungal growth makes these unfit for consumption within a month or so. Many growers dry the green cardamom artificially, 3 to 6 platforms made of bamboo strips are placed one above the other keeping a space of 15 cm between the platforms, the bottom platform being at a height of 1 meter from the ground level and a tin shed is provided above the top one. Green cardamoms are spread on these platforms very loosely 2 to 3 layers thick. 2 or 3 small ovens are made on the ground where slightly damp wood along with saw dust are lighted so that instead of direct fire, heated smoke is liberated and passes up through pipes through the bamboo platform. Too much of smoke should be avoided as otherwise the capsules become smoky. The positions of the bamboo platforms with the cardamom capsules are inter changed i.e., the top one is brought to the bottoms and vice versa every day so that all the cardamoms on different platforms are uniformly dried. This heat treatment is continued for 5 to 6 days when the colour of the capsules turns black or deep brown. This dried cardamom contains 15 to 20 percent water and is sold at a much higher price. The yield of green cardamom varies from 2 to 6 quintals and that of dry cardamom from 1 to 4 quintals per hectare. A yield as high as 10 quintals of green Cardamom per hectare may easily be obtained by using scientific methods of cultivation.

Uses: Large cardamom seeds are much in demand not only for masticatory use but also for medicinal purposes. The highly aromatic properties make it an essential-flavouring agent for sweet dishes, cakes and pastries. It is highly valued for chewing and as an agent for making betel leaves delicious for chewing. In Ayurvedic and Unani medicines it is used as a preventive and curative for throat troubles, congestion of lungs, inflammation of eyelids, digestive disorders and even in the treatment of pulmonary tuberculosis.

Pests and diseases: Large cardamom has no major pest problem. Thrips and hairy caterpillars cause some damage to the young crop but dusting of B.H.C. 50% powder at an interval of 15 days during the rainy season and once a month during the dry period is quite effective in controlling them. White-ants sometimes cause damage particularly in the nursery stage, which can be controlled by Aldrex.

Of the diseases, the virus disease 'Foorkey' is the most important one. It causes havoc and cardamom plantations were on the point of total extinction some years back. The disease appears at drying of the tips of the leaves, which sometimes is mistaken for water logging effect. Within a fortnight to a month the entire plantation looks like a burnt field. There is no effective control measure that can be adopted against the disease excepting to uproot the plantation, burning out the plants, and keeping the field fallow for a year and then replanting with healthy, disease-free seedlings. Another virus disease called 'Chirkay' makes the affected plants stunted in growth and no fruit is obtained from such plants. This disease, however, is not as virulent as the former one. Uprooting of the affected plants and burning them out is the only measure that can be adopted against this disease. A disease of minor importance also caused by virus is called leaf-streak disease. This however is not at all serious at present. Not much of fungus diseases are encountered within large cardamom. Leaf blight and clump rot, caused by Alternaria sp. and Pythium sp. are found to occur in small patches particularly in very damp places or after a period of continuous heavy rainfall for several days. Spraying Blitox at regular intervals and liming check the spread of these diseases.

Angelica glauca Edgew.

3. *Angelica glauca* Edgew.; Umbelliferae.

Na : Hindi: Chora.

Hab : The plant is found in Kashmir.

The roots are used as spice and condiment.

Leaves of Artemisia dracunculus L.

4. *Artemisia dracunculus* L; Asteraceae (Compositae.)

Na : Eng.: Terragon, Estragon oil.

Hab : and other details pl. see Essential oils

The leaves and twigs yield an essential oil, which is used for flavouring foodstuffs.

5. *Bunium persicum* Boiss. Fedts. (*Carum bulbocastanum* Clarke non-Koch); Umbelliferae.

Na : Eng.: Earthnut, Black caraway; Hindi: Kalazira.

Hab : It is a perennial herb found in the temperate regions in Kashmir.

The seeds are used as a spice.

Plant of Carum carvi L.

6. *Carum carvi* L.; Umbelliferae.

Na : Eng.: Caraway; Hindi: Shiazira.

Hab : and other details pl. see medicinal plants.

The fruits are used as a condiment.

Seeds of Carum Carvi L.

7. *Cinnamomum tamala* (Buch.-Ham.) T. Nees and Eberm.; Lauraceae.

Na : Bengali & Hindi: Tejpat; Guj.: Tamalapatra; Sanskrit: Tamalaka, Tegpatra; Tamil: Talishappattiri; Telugu: Talisapatri; Trade: Indian cassia.

Tejpat (Cinnamomum tamala (Buch.-Ham.) T. Ness and Eberm) is a native of India. It is a moderate sized evergreen tree of the Himalayas, common eastwards in east Bengal, the Khasia Hills and Burma.

In the district of Sylhet the most extensive plantations of Tejpat are come across. In the Jaintia parganas the concentration of Tejpat cultivation is probably due to the fact that the plant was growing wild in this locality. These wild plants supplied the seed and seedlings for its extensive cultivation in the tract. In this tract a large number of the plantations are self-sown while others are planted.

Planting of seedlings: Seedlings are raised in beds during the rains. Seeds are sown in seedbeds in March-April and they sprout after a month or month and a half. The seedlings are planted out permanently in the field when they are four or five years of age. The seedlings are planted 3 to 4.5 m apart. The tree takes five or six years to grow, and comes into bearing at ten and may continue to give annual crops for 50 to 100 years. After an orchard is established it is not necessary to make any further plantations; the ripe seeds fall from the trees in the soil and germinate. These self-sown seedlings when about a foot high are transplanted. Great care is bestowed upon them when they are young and tender. As constant exposure to the sun would kill the shoots, they are planted behind bushes or trees for protection. The undergrowth is kept down twice a year in the plantations for the first eight or nine years; after that the jungle is cleared once a year in April. In some plantations the soil is dressed but in most cases it is never manured or cared for after planting. It is during the harvesting of the leaves that the plants are pruned.

***Cinnamomum tamala* (Buch.-Ham.) T.Nees and Eberm**

Harvesting: Tejpat is plucked in dry and mild weather from October to December and in some places the collection is continued to the month of March. The leaves are collected once a year from young trees, and every other year from old and weak ones. On an average 13 kg of dry leaves may be obtained from one tree, but the quantity depends upon local factors; a tree yields from 8 to 20 kg of dry leaves in a year. In harvesting the Tejpat, the small branches are cut down with the leaves and dried in the sun for three or four days. The leafy branches are then tied up into convenient bundles ready for the market. In the other case the leaves are separated from the branches and packed in bamboo nets of a cylindrical shape called bore or jungra. The growers usually sell the produce of the garden to the paikara who pluck the leaves and send them to the different markets of the district and also outside to the neighbouring districts. The cost of cultivation being very little it is a profitable crop. Harvesting also costs negligible. The leaves are used as spice but can be employed with myrobalans in dyeing and in the manufacture of vinegar. The essential oil of the leaves is also used as a flavouring agent in curries. The leaf oil is a rich source of eugenol.

Plant of *C.zeylanicum* Bl.

8. *C.zeylanicum* Bl.; Lauraceae.

Na : Eng.: Cinnamon; Hindi: Dalchini.

Hab : and other details pl. see medicinal plants.

Stem bark is used as a flavouring agent in curries. The bark oil is also used as a flavouring agent in culinary preparations and is very much priced.

9. *Coleus amboinicus* Lour.; (Syn. *C.aromaticus* Benth.); Labiatae.

Na : Bengali: Patherchur; Eng.: Indian borage; Hindi: Pathorchur, Amrod; Marathi; Patherchur; Sanskrit: Pashanabhedi; Tamil: Karpuravalli.

Hab : It is a common succulent herb with aromatic leaves and small pale purple flowers. Leaves are used for flavouring food products.

***Coleus amboinicus* Lour.**

10. *Crocus sativus* L.; Irideae

Na : Bengali, Eng. and Hindi: Kesar, Saffron; Guj., Kanerese, Marathi, Sanskrit and Telugu: Kesar; Tamil: Kungumapu.

Flower of Crocus sativus L.

Saffron is a product obtained from the stigma of the flowers of Crocus sativus, which is a native of South Europe, but also has long been cultivated at Panpur in Kashmir, Iran, Asia minor, Greece, France, Spain and Italy. It is grown only in Kashmir in India at Panpur 11 Kms from Srinagar. It has an altitude of about 1500 m above sea level. It requires cool climate. The bulbs are transferred to a freshly prepared land. They are not irrigated. Excessive moisture is injurious to the crop. The soil must be thoroughly pulverised by continuous ploughing. Transplanting is done in September and the plant flowers in autumn when the crop of yellow stigma is collected and dried in the sun or by means of artificial heat.

Plant of Curcuma aromatica **Salisb**

11. *Curcuma aromatica* Salisb. (*C.zedoaria* Roxb. - non Rosc.); Zingiberaceae.

Na : Bengali, Gujarathi, Eng.: Zedoari; Hindi, Kanerese, Marathi: Kachura; Hindi: Jangli haldi; Malayalam: Pulla kizhanna; Tamil: Kichilikizhangu; Telugu: Kachoram, Advipaspu.

Hab : It is a small herbaceous plant like turmeric.

The rhizomes are used as a flavouring agent and condiment. During 2002-2003, 24 tonnes of dry rhizomes valued at Rs. 5,71,595.00 were exported.

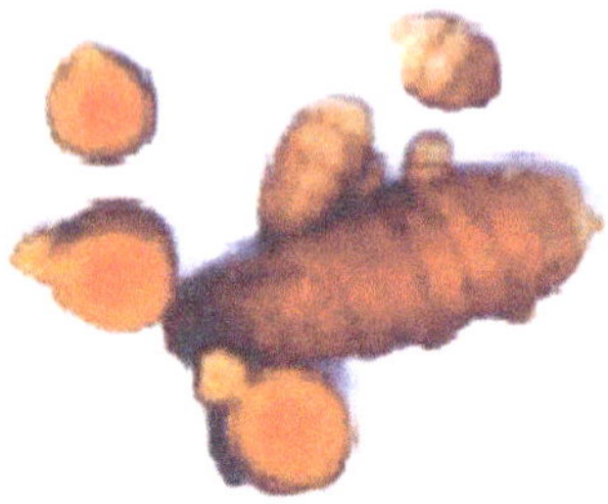

Rhizome of *Curcuma aromatica* Salisb

12. *Decalepis hamiltonii* Wight and Arn.; Asclepiadaceae.

Na : Kanerese: Makaliberu; Tamil: Mahali kizhangu.

Hab : It is a climbing shrub found in the sub-tropical regions in the Himalayas, Tamilnadu and Andhra Pradesh.

Aromatic roots are used as a spice and condiment.

Plant of *Decalepis hamiltonii* Wight and Arn.

13. *Elettaria cardamomum* (L.) Maton var. minor Watt; Zingiberaceae.

Na : Eng.: Cardamon; Hindi: Choti elaichi.

Hab : and other details pl. see medicinal plants.

Seeds are used as a spice, condiment and masticatory.

14. *Evodia lunarankenda* (Gaertn.) Merr.; Rutaceae.

Hab : It is a tree found in Assam, the Andaman Islands and Andhra Pradesh.

Leaves are used as a condiment.

Plant of *Elettaria cardamomum* (L.) Maton var. minor Watt

15. Fagaria budrunga Roxb. (Syn. Zanthoxylum rhetsa (Roxb.) DC.; Rutaceae.

Na : Hindi: Badrang.

Hab : and other details pl. see Essential oils.

Fruits are used as a spice.

16. ***Farsetia jacquemontii*** **Hook. f. and Thoms.; Cruciferae;**

Na : Hinai: Farid buti.

Hab : It is a herb found in Rajasthan and Northern India.

Tender twigs of this plant are used as a condiment.

Cardamon

17. ***Ferula assafoetida*** **L; Umbelliferae.**

Na : Eng.: Asafoetida; Hindi: Hing.

Hab : and other details pl. see medicinal plants and essential oils.

The gum-oleo-resin obtained from the roots is used as a flavouring agent in food products.

Fruits of Fagaria budrunga **Roxb.**

18. ***F.narthex*** **Boiss.; Umbelliferae.**

Na : Eng.: Narthex asafoetida; Hindi: Hing.

Hab : The plant is found in Kashmir in temperate regions.

Gum-oleo-resin obtained from this plant is used as a flavouring agent in food products.

Ferula assafoetida **L. Resin**

19. Gastrochilus pandurata Ridley; Zingiberaceae.

Hab : It is a perennial herb found in the Andaman Islands and West Coast.

Rhizomes are used as spice and condiments.

20. ***Gaultheria fragrantissima*** **Wall.; Ericaceae.**

Na : Eng.: Fragrant wintergreen: Hindi: Gandapuro.

Hab : and other details pl. see Essential oils.

The volatile oil obtained from the leaves is used as a flavouring agent in culinary preparations.

Ferula assafoetida **L. Plant**

21. ***Hedychium spicatum Sm.*****; Zingiberaceae.**

Na : Eng.: Spiked ginger lily; Hindi: Sitruti, Kapur Kachri.

Hab : This herb is found in the sub-tropical to temperate Himalayas particularly in Chakrata and Kumaon hills. (pl. see essential oils).

The rhizome as well as essential oil obtained from it is used as a spice in culinary preparations.

Plant of ***F.narthex*** **Boiss.**

22. ***Hibiscus ficulneus*** **L (Syn.** ***Abelmoschus ficulneus*** **Wight. and Arn.); Malvaceae.**

Na : Hindi: Ranbhindi, Kasturibhendi, Muskdana.

Gaultheria fragrantissima Wall.Plant of wintergreen oil

Hab : It is an annual aromatic herb growing to a height of 1 m found in the Punjab, Bengal and South India. Its leaves, flowers and fruits are similar to ladies fingers. Its seeds are small and dark in colour. They are aromatic and hence are called Muskdana.

Plant of *Hedychium spicatum* Sm.

Propagation: Climate and soil: It grows in tropical to sub-tropical regions where climate is not too severe. It requires good fertile neutral soils. For raising a hectare of plantation about 3 kg of seeds is required. After a good rainfall ploughing is done. 2-3 seeds are dibbled at each place in furrows ploughed 60 cm apart. Germination commences in about 4-5 days depending on the sub-soil moisture and temperature. If more than one seedling is noticed at each place vigorously growing seedling should be retained and rest removed during dry spells. Periodic watering may be necessary. Weedings may be done as and when required.

Hibiscus ficulneus L.Shrub

Pests and diseases: Like ladies finger the crop also gets damaged from pests and diseases. Periodic spraying of pesticides and fungicides would be beneficial. The crop is ready in 4-5 months when fruits show signs of ripening i.e. when the fruits turn whitish. After ripening they should be collected and dried. If the fruits are allowed to stay for a longer time after ripening they open and disperse the seeds. When the fruits are dried after collection they should be beaten up with a mallet and seeds separated. Fibre can also be collected from the stems of plants. A hectare of well maintained plantation may yield about 1 metric tonne of seed per hectare. With good and regular irrigation and manuring 2 harvests can be obtained.

Flowers of Hyssopus officinalis L.

Uses: Seed and its oil are used as a flavouring agent in *Pan masala*, chewing tobacco, in culinary preparations, in *agarbattis*, pomades and in medicines to mask off unpleasant odours.

23. *Hyssopus officinalis* L.; Labiatae.

Na : Eng.: Hyssop; Hindi: Zufah.

Hab : It is an aromatic shrub native of Europe and temperate Asia now found in the temperate Himalayas from Kashmir to Kumaon.

Leaves and flowering tops are used as flavouring agent in foodstuffs.

Leaves and Flowers of Hyssopus

24. *Juniperus communis* L.; Cupressaceae.

Na : Bengali: Havusha; Eng.: Juniper; Hindi: Aaraar, Abhal, Haubera; Marathi: Hosha.

Hab : It is a small tree found, at higher altitudes in the Western Himalayas from Kumaon westwards.

Fruits are used for flavouring gin and food products.

Plant of *Juniperus communis* L. along with berries

25. *Kedrostis rostrata* Cogn.; Cucurbitaceae

Na : Bombay: Narekvee; Tam.: Appakovay; Telugu: Kukumadanda.

Hab : It is a herb found in South India and Gujarat. Stem is angled and hairy.

Leaves are used for flavouring purposes.

Plant of *Murraya koenigii* (L.) Spreng. with leaves

26. *Limnophila rugosa* (Roth) Merr. (Syn. *L.roxburghii* G. Don); Scrophulariaceae.

Na : Bengali and Hindi: Kalakarpur, Mundari, Losodha.

Hab : It is an herb found throughout India. (pl. see essential oils).

Aromatic leaves are used in cooking.

27. *Murraya koenigii* (L.) Spreng. (Syn. *Burgera koenigii* L.); Rutaceae.

Na : Assamese: Bishahari, Narsingha; Bengali: Barsangu, Kariaphulli; Eng.: Curry leaf tree; Guj.: Gorunimb, Kadhilimbdo; Hindi: Kathnim, Mithanim, Kurrypata, Gandhela, Barsanga; Kanerese: Karibevu; Mal.: Kariveppilei; Marathi: Karhinimb, Poospala, Gandla, Jhirang; Tamil: Karivempu, Karuveppilei, Kattuveppilei; Telugu: Karepaku. .

Hab : It is a small tree found both wild and cultivated in Southern, Central and Northern India.

Leaves are used for flavouring curries.

28. *Myristica fragrans* Houtt.; Myristicaceae.

Na : Bengali, Hindi: Jaiphal (Nutmeg), Javitri (Mace); Tamil: Jadikaya; Telugu: Jaji-kaya. (Names, Habit and other details please see medicinal and essential oil plants).

Introduction: Nutmeg and mace, two of the most important of the spices, are derived from the fruit of this tree. The essential oils obtained from these spices, are also used as a flavouring agent and in medicine.

Myristica fragrans Houtt.

Description and Distribution: This is a handsome, evergreen tree, with dark leaves, reaching a height of 9 to 18 metres. It is normally unisexual, a tree bearing either male or female flowers, which are pale yellow and aromatic. The orange-or golden-yellow fruit, when ripe, is one of the most beautiful in nature, and on the whole resembles an apricot or a small peach. The size is variable but averages about 6 cm. When fully ripe, the husk splits along the groove, which runs down on side of the fruit, revealing shiny-brown seed covered almost completely with a brightened branching aril. Inside the seed is the aromatic Kernel which, when dried, is the nutmeg of commerce. The aril, when dried, is the mace of commerce.

The nutmeg tree is indigenous to the Moluccas or Spice Islands (Indonesia) and is now cultivated in the tropics of hemispheres, particularly in Indonesia and the West Indies. The tree was introduced into India nearly two centuries ago, but its cultivation even today remains confined to gardens, such as the Burliar Fruit Station (in the Nilgiris) and in Kerala.

Cultivation and Exploitation: The nutmeg tree requires a hot and moist climate and flourishes best near the sea, so, islands are very favourable for its growth. It flourishes on a wide variety of soils but thrives best in deep, loamy, well-drained soils and at elevations up to about 600 metres. Annual rainfall should be 2,030 to 2,540 mm and it should be well distributed.

The tree is usually propagated from seedlings, which are raised in nursery beds and later transplanted. It flourishes better in partial shade than in the open. It comes into full bearing at about 20 to 30 years of age and continues to bear for 30 to 40 years or even much longer. Fruits are produced all the year round and are collected when ripe. They may be collected before they split or afterwards. In any case, the pericarp is removed and the aril is separated from the shell of the seed and is carefully dried when it turns yellowish brown to reddish brown in colour. It is the mace of commerce. After removal of the aril (mace), the nuts are dried for several weeks in the sun. The shell is then cracked open and the kernel (nutmeg) is removed. The nutmegs are sorted into different grades according to size, smoothness and colour and often treated with lime or fumigated with methyl bromide to prevent insect attack.

Nutmegs are mostly oval in shape though some are nearly globular. Their size is also variable, the larger ones being about 3 cm long and about 2 cm in thickness. In appearance they are wrinkled but are quite smooth to the touch. They have a characteristic and strong aromatic odour and slightly bitter in taste.

Mace consists of yellowish to reddish-brown branched, flat pieces of about 3 cm in length and of no particular shape. It has a fragrant nutmeg like odour and an aromatic taste.

The two spices own their characteristic aroma chiefly to the presence of an essential oil, which can be isolated by steam distillation. Commercially, the oil is produced mostly from somewhat poorer grades of nutmegs since these are usually cheaper than mace. The yield of the oil from nutmeg is usually 6 to 16 per cent while that from mace is about 4 to 15 per cent.

Oil of nutmeg is almost colourless or pale-yellow liquid, possessing an odour and flavour characteristic of the spice, especially on dilution. Oil of mace is similar to the oil of nutmeg in odour and flavour.

The main constituents of the oil of nutmeg are d-pinene and d-camphene. Other constituents present are B-pinene, dipentin, p-cymene d-linalool, geraniol, eugenol etc. Mace oil is similar to nutmeg oil in composition and properties.

Uses: Nutmegs have been used medicinally and as a culinary spice for a long time. Their culinary uses include the flavouring of sweet dishes, pies, puddings, etc. They are also included in the recipes for many ground spices and seasonings. Medicinally, they are aromatic and carminative. Mace is one of the most delicately flavoured of spices and is extensively used for flavouring curries, meat and savoury dishes, cakes, preserves, sauces, etc., and in the commercial manufacture of certain foodstuffs such as tomato ketchup, sauces and pickles. Oils of nutmeg and mace are used for the same purposes as the spices themselves. They are also used to an extent in pharmaceutical and cosmetic preparations. It may, however, be noted that the oils are somewhat toxic owing to the presence of myristicin and the use of large quantities should be avoided.

***Nepeta cataria* L. – herb**

29. *Nepeta cataria* L; Labiatae.

Na : Eng.: Catnip, Catmint.

Hab : and other details pl. see medicinal plants.

The aromatic herb is used for flavouring purposes.

***Oenanthe javanica* (Bl.) DC. - Shoots**

30. *Oenanthe javanica* (Bl.) DC.; Umbelliferae.

Na : Bengali: Panturasi; Eng.: Waterdrop wort; Mundari: Ependung.

Hab : It is a perennial herb commonly found in Kashmir, the Punjab and Assam.

Young shoots are used as condiment.

***Origanum vulgare* L. – herb**

31. *Origanum vulgare* L.; Labiatae.

Na : Eng.: Pot marjoram; Hindi: Sathra, Bantulsi; Kanerese: Maruga; Telugu: Mridumaruvamu.

Hab : and other details pl. see essential oils.

The herb is used as condiment.

***Osmanthus fragrans* Lour. - herb**

32. *Osmanthus fragrans* Lour. (Syn. *Olea fragrans* Thunb.); Oleaceae.

Na : Bengal and Lepcha: Tungrung; U.P. hills: Silang.

Hab : It is an ornamental evergreen shrub of China and Japan. It is found in Kumaon, Garhwal and Sikkim. It bears fragrant flowers. Leaves are used to perfume tea.

***Perilla frutescens* (L.) Britt. - herb**

33. *Perilla frutescens* (L.) Britt.; (Syn. *P.ocimoides* L.); Labiatae.

Na : Assamese; Arim, Angami, Kenia; Bengali: Ban-tulsi; Eng.: Perilla; Hindi: Bhanjira, Bhasinda; Kumaon. Jhutela.

Hab : It is found in Kashmir, Bhutan and the Khasia hills.

Leaves are used for flavouring foodstuffs.

34. *Piper nigrum* L.; Piperaceae.

Na : Bengali and Hindi: Kalimirch, Kalamorich, Golmirch; Eng: Pepper; Gujrathi: Kalamari, Kalomirlch; Malayalam: Kurumulaku, Nallamulaku; Marathi: Kalimirich, Mire: Tamil: Milagu; Telugu: Miryalatige.

Pepper is a tropical large woody climbing plant. It requires heavy and well-distributed rainfall. The plant is not likely to grow very satisfactorily with either a long, hot dry season or a period of very low temperature. Gently sloping to level ground lying along the riverbanks, not too swampy and free from flooding is the best situation for Pepper.

Among the South Eastern Asian countries Indonesia is a rich source of pepper. It is indigenous to Sri Lanka, arid South India. The plant grows well on the rich decaying leaf mould or humus. There are three varieties of this species; some bear male flowers, others only female ones and while the third type bear bisexual flowers. Therefore material from the plant bearing bisexual flowers should be selected for propagation. The plant can be raised both by seeds and from cuttings. Cuttings 30 to 60 cm long with lateral branches taken from the upper part of the bearing vines are planted in shaded nursery beds. Watering in Nursery beds is necessary. The cuttings take root in about three month's time and are lifted during the rains for transplanting. The nursery beds require partial shade. Seeds give about 90 per cent germination. In about three months the seedlings are about 20 cm. But there are two objections for propagating through seeds – the possibility of plants not coming true to the required type and the possibility of longer time being taken by the time vines commence fruiting. It is therefore desirable to propagate through cuttings in preference to seeds. Rooted cuttings are planted at the foot of the tree with the beginning of monsoon. These supports or trees should have rough surface to afford better support to the vines. About 500 to 750 vines can be planted per hectare. Cow dung manure is added. Fish guano at the rate of about ½ kg and ½ kg of lime mixed with 10 kg of leaf mould per vine is given. The vines begin to bear in 3 to 5 years and continue to fruit at least for 20 years. The yield of pepper from each vine may be 3 to 7 kilograms. The improved variety "penury" has found favour in Kerala and is reported to be yielding as high as 7 kg of pepper per vine at maturity.

Piper nigrum L.

Area and production: India is the largest producer of Pepper with an annual production of 66,000 tonnes from 2.38 lakhs hectares, followed by Indonesia, Brazil, and Malaysia. In India, Kerala accounts for 97.4% of the area (2.30 lakhs ha) followed by Karnataka and Tamilnadu. Of the total production in the country 70% is exported annually. India, with more than 50% share of the world area under pepper, contributes hardly about 30% of the total world pepper production. Pepper productivity in India is one of the lowest in the world (315 kg/Ha). The average productivity of pepper/vine is highest around 1 kg in Karnataka, and lowest of 0.6 kg in Kerala.

In Andhra Pradesh, pepper is grown mainly in coffee plantations of APFDC Ltd. with annual production of 50 MTs and productivity of 0.70 kg per standard, which is comparable to the national average of 0.75 kg per standard. Apart from APFDC, coffee plantations owned by tribals, Coffee Board Research and Demonstration farms are producing about 15 to 20 MTs of black pepper annually in Andhra Pradesh.

The fruits are extensively used as spice and condiment.

***Satureja hortensis* L**

35. *Rumex hastatus* D. Don; polygonaceae.

Na : Hindi: Bhilmora; Kumaon: Amlora, Chulmora; Punjab: Khattimal, Katan bal.

Hab : It is a small shrub growing up to 90 cm in height. It is found in Western Himalayas at altitudes from 300 to 2400 m.

The leaves have a pleasant acid taste and are used as a condiment.

36. *Satureja hortensis* L; (Syn. *Calamintha hortensis* L); Labiatae.

Na : Eng.: Summer savory.

Hab : It is an erect pubescent annual herb 25 to 35 cm high with pinkish branches. It is found in Kashmir.

The stem and leaves are used for flavouring food products.

37. *Syzygium aromaticum* (L) Merr. and Perry; (Syn. *Eugenia aromatica* O. Kuntze; *E.caryophyllata* Thunb.); Myrtaceae.

Na : Bengali, Gujrathi, Marathi, Hindi: Laung; Eng.: Clove; Kanerese: Lavanga; Malayalam: Karayampu, Krambu; Tamil: Kirambu; Telugu: Lavangamu.

Hab : It is native of Malaya, Zanzibar, Tanzania, Indonesia, Mauritius and Sri Lanka. In India it is cultivated in the Nilgiris and Kerala.

The species normally grows from sea level to 900 m above sea level and requires humid conditions with an optimum of 2000 to 2500 mm well distributed rainfall and a mean temperature range of 16 to 38 °C. The tree is not likely to grow well with either a long hot dry season or a period of low temperature. It requires coastal climate. Soil should be dark loam. Gravely soil is also good but it should not be too hard or stony and should have a good mixture of clay in it. The seeds are soaked in water for three days and when germination has set in they are planted at a spacing of 7.5 to 15 cm apart in the shaded nursery beds or in alkathine bags with the bud end above ground. The young plants are watered by a fine-rase-can. In case of dry season watering may be done constantly so as to keep the soil moist and cool. Excessive water is fatal to the plant. The plants are allowed to remain in the nursery beds or bags for 9 months to one year. When they are about 15 cm tall they are gradually hardened off by partially removing the shade and are then left exposed to the sun for a month or two before they are planted out in the field. However care is taken that the plants do not die of prolonged dry weather. The seedlings are transplanted in pits with the beginning of monsoon at the rate of 250 trees per hectare. Planting is done

***Syzygium aromaticum* (L) Merr &Perry Plant**

Vanilla planifolia **Plant**

in 0.9 cube metre pits. Manure is added before planting. Grafts can also be used for planting. Partial shade of trees may be necessary in the early stages of growth. Cropping commences by about 7th year. Highest yields are obtained from 30 to 60 year old trees. The yield of clove per tree is 3 kg but it may go up to 15 kg depending on soil and climatic factor and also cultural practices. Sixty years is regarded as the economic life of the tree. One of the greatest impediments in the expansion of cultivation of the species is the severe seedling losses even up to 90 per cent, experienced during the first two years of planting. This is due to a fungus what is, known as collar rot of the young plant. The dried flower buds are used as spice and are also used medicinally as stimulant, carminative and in flatulence.

38. *Vanilla Planifolia Andrews*; Fam.: Orchidaceae Orchidaceae

Hab : There are three species of Vanilla, V.planifolia, Andrews; V.pompona, Schiede; V.tahitensis J.W.Moore.

Vanilla Planifolia **Flower**

Vanilla originated from Mexico and spread all over tropical and sub-tropical regions in the world. It is a low volume and high value species. Favoured species for cultivation is V.planifolia. In India it was grown over an area of 1600 ha during 2001 and the first yield obtained was 60 tonnes. In A.P., it has been grown in Krishna, West Godavari, East Godavari, Vizianagaram and Visakhapatnam Districts over an area of 80 ha during 2003-04. The quality of vanilla produced in India was much above the average of International standards.

It is a perennial climber and can be grown as an inter crop under the shade of trees. As such, it is ideally suited to be introduced in forest areas.

Climatic and edaphic requirement: Vanilla thrives well in equable climate of about 30 °C, with relative humidity of 60% and at an altitude of 1500 m. In tropical regions severity of summer temperature can be off-set by raising it as an inter crop under shade of trees and periodic irrigation. Fertile, neutral and well drained soils with well distributed annual rainfall of about 150 to 300 cm, good sub-soil moisture in summer are preferable. Periodic irrigation in summer will boost up the growth.

Vanilla Planifolia **Beans**

Planting: Plantation is raised by stem cuttings. Cuttings of about 15 cm in length with two-thirds of portion above the ground and one third below the ground are pressed into the soil. However, care should be taken to see that upside down planting is not done and the portion below the ground is firmly fixed so that there is no loose contact with the soil. They should be planted near the support if already available naturally. If natural supports are not available the cuttings should be planted at a distance of 2m × 2m fixing supports nearby to facilitate the vines to trail on them. Periodic weedings and cultural operations may be necessary. Periodic irrigation during dry season will boost up the growth and yield of vanilla beans is also higher. The climbers should be trained on the supports so that they remain at man's reach say about 1.4 m to facilitate pollination and collection of beans.

Pollination: Natural pollination is completely absent. Artificial pollination by hand is carried out to get commercial set (crop). The method of hand pollination is by holding a pointed bamboo splinter/stiff grass stick with right hand the rostellum is pushed back and the pollen sac is pressed against the stigma with the help of thumb smearing the pollen over it. This operation is labour intensive and one person may pollinate about 1200 flowers depending on his experience.

Yield: Plantations start flowering in 3rd year and fruiting in 4th year. The length of the bean is 18 to 25 cm. About 60 to 80 beans weigh one kg. The yield stabilizes at 5th or 6th year and plants remain economically productive till 13th or 14th year. Thereafter the yield declines when plantation may have to be replaced. Normally the yield of green beans is 600 gms per plant on an average. Beans should be harvested when they are just turning yellow. Immature beans, which are dark green in colour, should not be harvested, as they do not produce aroma.

Processing: When the tips of beans start yellowing they should be harvested. Even mature beans do not contain aroma but they are to be processed and cured for about two months when they develop aroma. Bourbon method of processing gives the best quality beans. The method consists of 4 stages:

1. *Killing:* The bean should be dipped in hot water at about 65 ^{0}C for 3 minutes to kill all living tissues in the bean and to initiate enzymatic reaction responsible for aroma.
2. *Sweating:* It is to be done for further rising of temperature of beans. The beans are exposed to sunlight for 1 to 1.5 hours and wrapped in woollen blankets and stored in wooden boxes. This process is repeated for 10 to 12 days. At this stage weight of beans is only half of its original weight noted at harvesting stage.
3. *Slow drying:* In the next stage, the beans are spread on wooden racks in a room at ambient temperature and agitated periodically for uniform drying. The process is repeated for 20 to 25 days when the beans are ready for conditioning. At this stage weight of the beans is one third to one fourth of its original weight. However precaution should be taken to see that constant weight is attained only then the process of conditioning is taken up.
4. *Conditioning:* Thereafter the beans are sorted and graded depending on size and general appearance and quality and bundled, each having 50 to 60 beans and tied at both the ends. They are then packed in material like butter paper, cellophane or polypropylene bags and stored in airtight containers to allow full development of aroma and flavour.

Zingiber cassumunar **Roxb. Rhizome**

Quality of processed beans: The colour of the bean should be dark brown. They should be soft and smooth to touch and unbreakable when bent. The beans should be free from insect damage or mould formation. The bean should contain 2.5% vanillin. Moisture content in the bean should not be more than 18% to 25%. Normally the yield of processed bean is one fifth of original bean.

Users and Marketing: Vanilla is worlds most famous and most sought after spice. Vanilla is used as a flavouring-agent in creams, chocolate, bakery and dairy products like ice-creams and puddings, pharmaceuticals, liquors, perfumes, tea, and soft drinks. Market price of fresh bean (soon after collection) is Rs.2,000/- per kg whereas processed bean is Rs.15,000/ - to Rs.20,000/- per kg.

***Zingiber cassumunar* Roxb. - Plant**

39. *Zingiber cassumunar* Roxb.; Zingiberaceae.

Na : Bengali: Banada; Eng.: Cassumunar ginger; Hindi: Jangli adrak; Kanerese: Kadushunti; Marathi: Nisan; Oriya: Banooda, Vanadraka; Sanskrit: Vanadraka; Telugu: Karrallamu.

It is a perennial herb found throughout India.

The rhizomes are used as a condiment.

7

Edible Wild Plants

All the familiar vegetables and fruits of our kitchen gardens as well as the cereals of our fields were once wild plants. Before agriculture came into existence the primitive man was a nomad and must have been dependent on various food plants growing wild in nature for his existence. He gathered edible fruits, grains, seeds, tubers leaves etc., and ate them to appease his hunger. At some later period in history he began to make primitive attempts at cultivating these useful plants by sowing seeds in favourable localities. Whether these attempts at agriculture were accidental or purposeful they were of great importance in changing the pattern of his existence. He soon gave up his nomadic life and cleared forest lands for growing food crops. He chose only those, which had the best taste and gave the largest yield as well as those, which were easiest to grow. Thus by way of selection many of the present day agricultural or horticultural crops were started to be grown. Although selection and breeding programme of desirable cultivars today has made tremendous strides and more and more varieties with better yields are coming up, yet the agricultural production in the country for some reason or the other has not been able to cope with the requirements of the increasing population. Coupled with this, frequent droughts, floods, cyclones etc., bring about scarcity conditions. It is not with such agricultural crops, however, that we are primarily concerned in this book but with those edible plants, which grow wild and which in the context of the existing food, situation can be scarcely overlooked. Someone has rightly pointed out that one cannot starve in the jungles. The forests in India provide a score of edible plants whose fruits, seeds, roots, etc., still make an important contribution to the dietary habits of the poor particularly those living near the forest. *Annona squamosa* L. (seethaphal, custard apple) for e.g. which is a delicacy, besides giving large revenue, forms the main diet of the poorer classes during the fruiting season in the regions where it is found. Accordingly bamboo seeds are collected and stored for edible purposes. These free gifts of nature are a boon to the poor. Certain nuts and seeds growing wild in the country have secured so firm a place even in the civilized dietary of the people in the country that every one, knows them, for instance, *Pinus gerardiana* Wall. (Chilgoza). *Buchanania lanzan* Spreng. (Chironji). In some cases they form ingredients in much of our cookery especially in sweet-meats and command a good trade. Stinging nettle has long been used as a vegetable not only in our country but in other countries as well. Goldsmith describing the food habits of French people in one of his books has mentioned that a French cook could make seven different dishes out of a nettle top. The plants possessing such irritating or bitter constituents are first boiled in water and the water is thrown away. Boiling has the effect of dissipating bitterness and undesirable taste. The boiled shoots, leaves and yams become soft and palatable and are made into curries. During emergency when food provisions run short wild edible plants can come to their rescue. Many of these plants offer grateful refreshment to the tourists, hunters, travellers, campers and naturalists.

A large number of mushrooms make their appearance during summer monsoon in tropical parts while in cooler regions they come out with the commencement of spring. They appear particularly in places, which are rich in humus. Many of them are edible and are collected and eaten. In fact they are treated as a table-delicacy. Besides, they have good food value. Quite a few of them are also poisonous which require to be identified and their collection avoided. There is thus a necessity to deal mushrooms separately before elaborating on the individual edible plants.

7.1 Mushrooms

Fungi are organisms which grow either in the ground or on an organic substratum that is either living or dead and have nuclei, lack the pigment chlorophyll, originate from spores and reproduce both sexually and asexually and whose filamentous (threadlike) and ramified (branching) somatic structures are surrounded by cellular walls containing cellulose or chitin or both.

Mushroom is a fungus comprising of two parts. The portion above the ground consists of a cap and a stalk and is edible. Underside of the cap, there are numerous radiating gills, which on maturity produce small microscope bodies, which assume the function of seeds. The underground root like part also known as mycelium grows extensively before it makes its appearance above the ground. The mycelium permeates partly decomposed organic matter.

The seed that triggers off the birth of the fungus is microscopic in size and is called a spore. The spore germinates and sends out a small tube, which becomes enlarged at the tip, where very large number of cells is formed in quick succession. A filament is thus formed which becomes ramified and each new tip of each ramification is a new growth centre. The filament formed by successively arranged cells is called hyphae and together the hyphae constitute the mycelium.

The development cycle of fungus : A spore germinates, producing a primary mycelium, and when it comes into contact with the mycelium produced by another spore, there is union giving rise to the secondary mycelium, which will form what is technically known as the carpophore (fruit-or-fruiting body)

The carpophores are formed by means of the sexual process from the mycelium. Two hypae of the same mycelium or more commonly of different mycelia, join together to form a new mycelium. The fruit bodies are also called mushrooms.

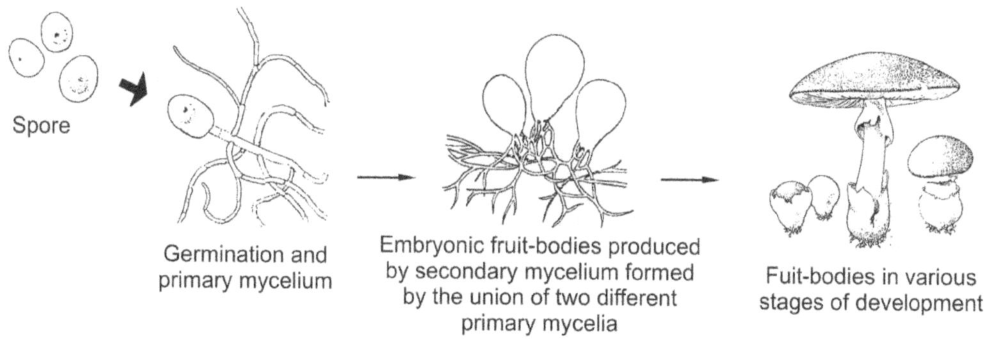

Fungi and Man

The presence of fungi has been known for thousands of years. The first hints of our mycological knowledge, relating to plant diseases are recorded as far back as 1200 B.C. in the ancient Vedas; and the effects of poisonous fungi were dealt with in an epigram written by Euripides in about 450 B.C.

There are about 2000 species of mushrooms. About 200 species are edible. Only 20 species are cultivated in the country. The oyster mushroom (Pleurotus sp.) and paddy straw mushroom (Volvariella sp.) are cultivated in tropical countries like India.

As far as ecological functions of fungi are concerned the picker should avoid picking mushrooms that are too large or specimen which are invaded by insect larvae or worms. Mushrooms with which he is well acquainted should only be picked up. When he is not sure of the authenticity of the fungi he should show the specimen to the expert and only when he confirms about its usefulness and edible nature, he should collect the material.

When to look for mushrooms

It is well known that periods of dry weather or intense cold are not favourable for the development of fruit-bodies of fungi. The best conditions for the occurrence of a massive production of carpophores have to do with the presence of humidity in the nutritional substrate and in the air and with a mild temperature.

7.2 Food Value

Mushroom is a valuable source of nourishment in as much as it contains protein, minerals and vitamins. Minerals such as calcium, phosphorous and potassium are present with fair quantity of iron and copper. Vitamins B, C & D are present in appreciable quantities. Young specimen is more nutritious than an old one. The cap is more nutritious than the stem. Mushrooms contain little or no starch. As such they are not harmful for diabetic patients.

A list of nutrients along with their percentages present in some of the mushrooms is given below: Percentages are on dry weight basis.

Mushroom	Moisture	Proteins	Fats	Sugars	Fibers	Energy (K. cal)
Agaricus bisporus	78-90	24-35	2-8	51-62	8-10	328-368
Agaricus campestris	89	33	2	57	8	354
Auricularia sp.	89	4	8	83	20	351
Flamulina sp.	89	17	2	73	3	378
Lentinus edodes	90	13-17	5-8	67-78	7-8	387-392
Pleurotus sajar – caju	90	26	2	51	13	300
Volvariella volvacea	89	26	2	-	9	276

The above nutrients may vary with the substrate and environmental conditions.

Nutrient contents in oyster species are as given below: (percentage on dry weight)

Nutrient	*Pleurotus florida*	*Pleurotus ostreatus*	*Pleurotus sajar - caju*
Proteins	27	10-30	26
Fats	1.6	1.6-2.2	2
Carbohydrates	58	47-82	51
Fibers	11.5	7.5-8.7	13
Amino acids	46	33	37

Nutritive values of mushrooms compared with various food items.

Essential Amino acid index	Amino acid score	Nutritional Indices
100 pork, chicken, beef	100 pork	59 chicken
99 milk	98 chicken, beef	43 beef
98 mushrooms	91 milk	35 soyabean
88 corn	89 mushrooms	28 mushroom (high among vegetables)
86 cucumbers	59 potatoes	-
79 peanuts	53 peanuts	26 spinach
76 spinach, soyabean	50 corn	25 milk
69 turnips	46 kidney beans	21 kidney beans
53 carrots	42 cucumbers	20 peanuts
44 tomatoes	33 turnips	17 cabbage
	27 spinach	14 cucumbers
	23 soyabean	11 corn
	18 Tomatoes	9 potatoes
		8 tomatoes
		6 carrots.

Compendium of lectures by NRCM Solan H.P. 1995

The Mushrooms which are commonly found and eaten are:

		Probable Botanical names
1.	(Bin ot) it is called snake mushroom because of its shape	*Lepiota mastoides*
2.	(Bunumot) White-ant hill mushroom as it is found on white ant hills. It is delicious	*Entoloma macrocarpum*
3.	(Bussup ot) straw mushroom. It grows on decayed straw	*Volvariella terastius*
4.	(Dak Mandi ot) Rice gravel mushroom. It is considered savoury.	*Entoloma microcarpum*
5.	(Gopha ot) It is a large kind of edible mushroom	*Variety of Collybia albtminosa*
6.	(Gundri gopha ot) it is an intoxicating mushroom	*Variety of Collybia albtminosa*
7.	(Masa ot) It is known as earth mushroom	*Agaricus campestris*
8.	(Hati ot) it is known as Elephant mushroom. It is large and bad smelling. It is eaten after boiling. It is rare.	*Boletus sp.*
9.	(Hurut ot) It is tree-stump mushroom. It is eaten after boiling	*There are two varieties of Lentinus subnudus one found during the rains and the other after the rains.*
10.	(Kat ot) It is called wood-mushroom	*Pleurotus ostreatus*
11.	(Karna ot) It is called bitter mushroom and (Karna patke)	*Puff-ball.*
12.	(Kot ot) It is black coloured	*Coprinus comatus*
13.	(Mat ot) Bamboo mushroom, very savoury	*Collybia sp.*
14.	(Matam ot)	*Variety of Collybia albuminosa*
15.	(Muci ot) Muchi's mushroom	*Entoloma microcarpum.*
16.	(Murum ot) Reddish mushroom	*Clavaria sp. C. dichotoma, C. fusiformis.*
17.	(Or tot ot)	*Collybia alnbuminosa*
18.	(Otec ot) Burst open mushroom white coloured commonly found in August.	*Variety of Collybia albuminosa.*
19.	(Bond tomar ot)	*Variety of Geaster*
20.	(Putka) the puff ball fungi of Lycoperdaceae family	*Lycoperdon giganteum, L.saccatum, L.gemmatum, L.piriforme. Calavatia sp.*
21.	Rote ot) frog mushroom and (Sim ot) dog mushroom. Reddish	*Small puffballs.*
22.	(Sisir ot) fowl mushroom. It is reddish	*Cantharellus cibarius & C. aurantiacus*
23.	(Sisir ot) dew mushroom found on stumps of Sal tree	*Lentinus subnudus*
24.	(Rormar ot) It is eaten but not very common	*Geaster sp.*
25.	(Tumba ot) gourd shell mushroom. It is large & round-shaped	*Bovista gigantean.*

Other common mushrooms are: *Volvariella diplasia, Morchella esculentum, Hydnum coralloidies* found on deodar trees, *Craterellius cornucopiode, Fistulina hepatica*, found in Darjeeling, *Polyporus sulphureus* found on *Eucalyptus* trees in Shillong, *Leocota procera* & *Collybia velutipes* (insipid).

Poisonous mushrooms: Following are some of the indications by which poisonous mushrooms can be recognized. Mushrooms, which have following characteristics, should be avoided. Amarita and false morels are deadly poisonous. Gyromitra takes the lead.

1. A volva or a sack. When young – "In the button" they are egg-shaped completely enclosed in a veil or sac (called volva) which ruptures, as the cap pushes upwards and remains attached at or near the bullbuous base as a cup with fragments or warts (appropriately known as the death cap). Digging of Anamita should be with care as the cup may be concealed below the ground.
2. Pink spores.
3. Growing in woods etc.
4. Which are no longer fresh and firm.
5. Which have a powerful peppery or nauseous taste.
6. Which have milky juice.
7. Anamita – have ring on the stem above the cup, just below the cap.
8. Anamita always grows on the ground never on trees.
9. Russula emetica is poisonous (muscarin is present)
10. In Gyromitra esculenta the cap is more or less convoluted, brain like in appearance. Its poisonous principle helvellic acid has been isolated.
11. Hypholoma fasciculave it is purple-brown spored.

Pleurotus florida

Following five types of mushroom poisoning are noticed:

1. Gastro-intestinal type.
2. Cholariform type.
3. Nerve-affecting type.
4. Blood dissolving type.
5. Cerebral type.

Among several edible mushrooms three species are commonly cultivated in India.

1. Button Mushroom (Agaricus Sp. European or temperate Mushroom)
2. Oyster or Dhingri Mushroom (Pleurotus spp).
3. Paddy straw mushroom (Volvariella sp)

Temperature and humidity conditions required for growing edible mushrooms are as follows:

Mushroom	**Temperature**	**Relative Humidity**
Button mushroom	15 – 18 °C	85 – 90%
Oyster mushroom	25 – 32 °C	75 – 85%
Paddy straw mushroom	28 – 38 °C	80 – 90%

Spawn Production

Spawn is a living ramified mycelium multiplied under highly hygienic conditions and used as seed material for cultivation of mushrooms under controlled climatic conditions. To prepare the spawn, chopped paddy or wheat straw is used as base material. In recent times, cereal grains are used instead of paddy straw. Normally the grains of wheat, sorghum, maize or Eleucine are used as base material. The cereal grains are cooked for thirty minutes and excess water is drained out. The cooked grains are dried on a cloth. 40 to 50 gms of calcium carbonate or chalk powder, 10 to 20 gms of calcium sulphate is mixed per one kg of cooked grains. The application of calcium carbonate prevents the aggregation of cooked grains and calcium sulphate is added to maintain the pH of 7.5-8.1. The mixture of cooked grains and chemicals is filled upto 3/4th of empty glucose bottles and the bottles are plugged tightly with sterilized cotton. The bottles are sterilized in an autoclave or pressure cooker for 90 minutes at 15 lb pressure. The sterilized bottles are cooled to room temperature. These sterilized bottles are inoculated with the mycelium from the mother or stock culture. The inoculation should be carried out under hygienic conditions in an inoculation chamber. The inoculated bottles are incubated at 25 + 1 °C for 10 to 15 days. The room can be kept cool by hanging gunny cloth and keeping it wet. The mycelium grows in the bottles. This mycelium is used as spawn for cultivating mushrooms on a large-scale.

Preparation of compost for commercial-scale cultivation of Button mushrooms:

In USA and Europe mainly Agaricus bisporus is grown. In India propagation of the species is gaining attention. In cold places it could be grown from March to October without additional heat. During winter in cooler localities extra heat will have to be supplied to maintain temp. of 15° to 19 °C during the cropping. For indoor cultivation mushrooms are commonly grown in wooden trays 10 to 15 cm deep since the trays can be moved without any inconvenience to any place or cellar maintaining a uniform temperature of 15° to 19 °C. Mushrooms can be grown outdoors but should be sheltered from direct sunlight because sunrays discolour them and dry the beds. Specially prepared compost is necessary for Button mushroom cultivation. Humidity of not less than 85 and not more than 90 with optimum of 85 may be necessary. Optimum temperature may vary between 15 to 20 °C. For production of spawn 25 ± 1 °C temperature is essential. The pH value of the soil should be 7.5. The trays of 90 × 60 × 10 cm are ideal for propagation. Following are the quantities of various ingredients for preparation of compost required for filling in 5 trays.

1. Saw dust	½ bag or say 10 to 15 kg
2. Wheat straw not more than 15 cm pieces	60 kg
3. Ammonium Sulphate	2 kg
4. Superphos	2 kg
5. Urea	1 kg
6. Wheat bran (choker)	3 kg
7. Gypsum (to bring down pH)	6 kg

The wheat straw is wetted for 48 hours, on a cemented platform. Thereafter, ammonium sulphate, superphos, urea, wheat bran, sawdust are mixed uniformly. The mixture is kept wet and heaped. During the process, the temperature rises to 60 to 71 °C on the 3rd day. First turn is given to the heap on 6th day. On the 10th day 2nd turning is given but the heap is kept wet. On the 13th day 3rd turning is given adding about 6 kg gypsum. In about 21 days, two or three more turnings are given adding sufficient quantity of water every time to keep the heap moist. In the last turning 250 gms of lindane is added to the compost against insect infestation. It is now ready for filling into trays. After filling the compost to 5 to 7 cm height in trays spawning is done. The spreading of mycelium under hygienic conditions and controlled climatic conditions is called the spawn running. Small pieces of spawn of marble to walnut size are pressed into the compost, 25 cm apart at a depth of 5 cm. A temperature of about 18 to 20 °C is maintained and the trays are kept one over the other. High humidity as indicated above, should also be maintained. In about 15 to 21 days, mycelium spreads to the entire compost. From this time onwards temperature should be lowered to less than 18 °C. Thereafter casing is done when the compost is covered with a 2.5 to 3 cm layer of soil. Before spreading, the soil should be mixed with sand in 50 : 50 ratio. Soil should be sterilized by drenching with 2% formalene or dry heat at about 65 °C for 12 to 15 hours. The beds are lightly sprayed with water twice a day to maintain high humidity. The mushrooms start appearing within 15-21 days after casing and the beds continue bearing for 2-3 months depending upon the quality and depth of the compost as well as other environmental conditions.

Cultivation of white button mushrooms in wooden trays

Oyster Mushrooms

For Oyster or Dhingri mushroom (Pleurotus species) cultivation, polythene bags of 30 × 45 cm or 12" × 18" with 100 guage thickness are selected and few holes are made at bottom and the sides so that there is good aeration and to facilitate the carpophores to come out. These polythene bags are filled with sterilized substrate prepared as above upto a thickness of 5 cm and pressed lightly handful of spawn is spread uniformly all along the periphery of substrate layer in the bag. The procedure of filling the polythene bag 4 to 5 times is repeated and once the bag is ¾th full the spawn is spread uniformly all over the top layer of substrate and the end of the bag is tied lightly by rubber band. Such bags should be kept in clean dark room on the racks or they can be hanged. One day before keeping the filled in bags for incubation in dark room, 2% formalin solution is sprayed in the room to disinfect the entire room, racks etc. Bags are kept for incubation for 18 to 25 days with temperature of 25 to 32 °C. After incubation period when the bag is fully covered with white cottony mycelium poly bag is cut open and the cover is shifted to cropping room. There should be 25-32 °C temperature and 80 to 90% relative humidity, sufficient light and ventilation should be provided in the cropping room. At the temperature of 25 + 1 °C the mycelium grows well and the humidity of the crop room is maintained beyond 80%. This can be achieved by spraying water on the gunny bags. The concentration of carbon dioxide should not exceed 0.1% in the cropping room. To maintain this concentration aeration is carried out 4 to 6 times in an hour. The operation is necessary when both oyster or button types of mushrooms are grown. The mushrooms start appearing within 10 to 15 days after opening the cover. The beds continue bearing for 2 to 3 crops depending upon the quality and depth of the substrate as well as other environmental conditions within a period of 2 months.

7.3 Picking of Mushrooms

Mushrooms are picked once a day just before they open out. Button mushrooms are usually plucked when pileus size is 1-1½" and oysters are plucked after opening of gills and ends become thin. The soil should not be disturbed. Holes left in the bed after picking are filled with fresh casing soil. A yield of ¾ kg to 1½" kg of fresh material per 30 × 30 sq.cm per season is expected. During cropping periods, the beds are usually sprayed with water lightly and frequently with a rose-can to prevent a casing soil becoming dry. The harvested mushrooms are collected in polythene bags. The yield of oyster is 300-400 gms/bag of 12 × 18" (3 crops put together)

Preservation of Mushrooms

Harvested mushrooms keep fresh for 24 hours at room temperature. To keep them fresh it is better to pack them in polythene bags with a few holes here and there. The bags are kept in refrigerator at a temperature of 4 °C. Button Mushrooms can be preserved in 18 to 20% brine solution and oysters are sun-dried or dehydrated in hot-air oven. Brine solution is prepared by adding 2% salt, 2% sugar and 2% citric acid in one liter of water.

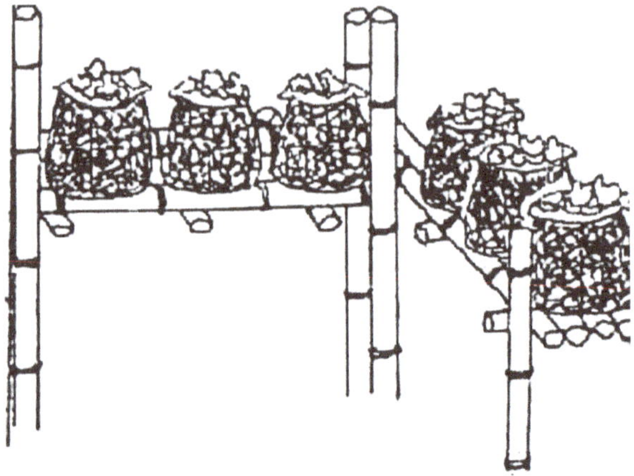

Polythene bags removed – keeping the growing mushrooms lengthwise on the shelf

Aegle marmelos (L.) Corr. Fruit

7.4 Other Edible Wild Plants

1. Aegle marmelos (L.) Corr. (Syn. Crataeva marmelos L.); Rutaceae.

Na : Eng.: Bael; Bengal: Quince; Hindi: Bel, Bilva.

Hab : The species is found all over the tropical regions in the country. (Pl. see medicinal plants).

Fruit pulp is eaten.

2. *Aeschynomene aspera* L.; Fabaceae. (Papilionaceae.)

Na : Bengali: Shola; Eng.: Solapith; Hindi: Sola; Kanerese: Bendu: Tamil: Attuneddi; Telugu: Nirjiluga.

Hab : It is a tall herb found in wet places in Bengal and South India.

The leaves are eaten.

3. *Agaricus campestris* L..; (Fungus).

Na : Eng.: Mushroom; Hindi-Dhingri.

Hab : It is found in cool places with rich humus content.

It is an edible mushroom. (Please see mushrooms)

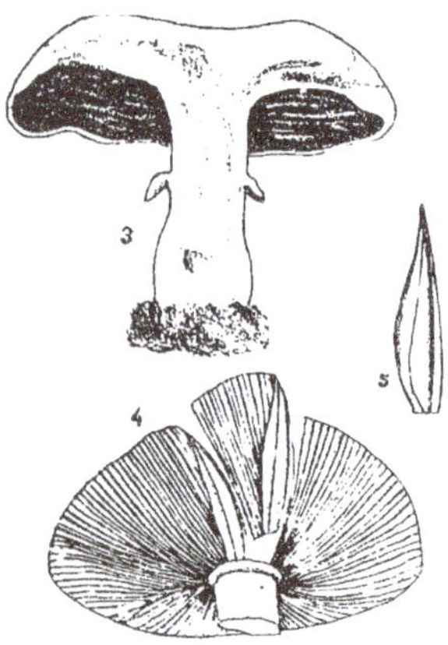

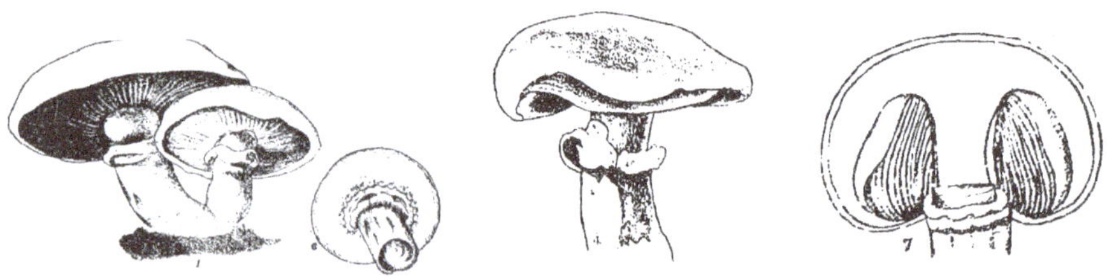

Agaricus campestris L.

***Anacardium occidentale* L.-Plant**

employment levels. Cashew plant was introduced into India from Brazil by the Portuguese in the 16th Century to prevent soil erosion and is grown over an area of about 7,70,000 ha as on 2001-02 in the country mostly by small land holders. Commercial and government sources estimate domestic consumption at about 1,60,000 tonnes and total raw nut production at 4,70,000 tonnes. Production fluctuates with weather conditions, pests and price of cashew. Pest attack particularly by Tea mosquito Helopeltis antonii is reported to reduce production by at least 30% in an average year and by more when infestation is severe.

Cashew products which are traded are cashew kernels mainly consumed as roasted and salted nuts (97% of total export value) and cashew nut shell liquid, a phenol used mainly in the manufacture of friction material (3% of export value). In certain states like Goa and Kerala "Cashew-Fenny" (alcoholic-beverage) is made from cashew apple. The export of other by-products of cashew is negligible. During the year 2002 cashew kernel export reached its peak of 97,550 tonnes of whole and broken cashew valued at Rs.1776.80 Crores. The decline is mainly due to dwindling of the import of raw cashew nuts into the country from African countries particularly Mozambiana and Tanzania for processing and the finished products viz. cashew-kernel was being exported. In the recent past the African countries have developed their own processing units and as such they have stopped sending out their consignments for processing. Since the demand of our processing-units for raw cashew nuts is high and production is less, forest departments, public sector corporations and small farmers in different states are taking up new plantations in a big way. Production of cashew can be increased by increasing the area under cashew, by intensive cultural practices and genetic improvement of the strains. The Government has also provided suitable subsidies for raising new plantations of cashew. Lending institutions like AR.D.C. (Agriculture Refinance Development Corporation) and commercial banks are funding cashew-plantation programmes. It is heartening to note that even, the International Monetary Organisations like World Bank have shown keen interest in cashew growing in India and a multi-state cashew project was implemented in the states of Andhra Pradesh, Orissa, Karnataka and Kerala since 1980.

Fruits

Cashew tree is hardy and can tolerate adverse climatic and edaphic conditions. In India it grows largely in coastal areas. For good yields deep and well-drained sandy loam soil with a rainfall exceeding 1000 mm is required.

Cultivation technique: The age old method adopted by villagers to raise cashew is by dibbling one or two seeds directly after scraping the soil but in the recent years basket or polythene bag plants of 3 to 4 months old are used for reliable and sure stocking as well as for beating up the casualties. When high yielding mother trees are to be propagated on a limited scale a sure method of propagating cashew is by air layering.

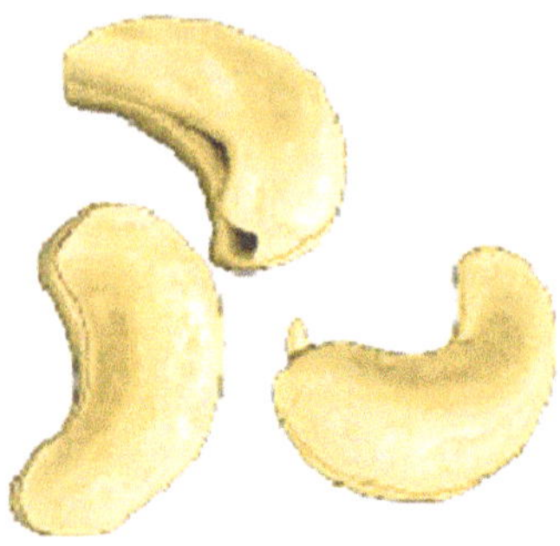

Kernel

Nursery technique: For raising seedlings in the nursery, polythene bags (150 to 200 gauge and size of 20 to 25 cm × 12 to 15 cm) with two holes punched at bottom are used. The bags are filled with soil mixture. Soil mixture varies according to the type and nature pf soil available at nursery site. In red loam to lateritic soils, the proportion of local soil to farm yard manure is 4:1 while in sandy and sandy loam it is 2:1 respectively. However in coastal sands and skeletal soils, local soil, tank silt and farmyard manure are mixed in equal proportion. Last mentioned tilth is the best mixture for raising nursery. Bag plants are watered and weeds removed periodically to obtain good growth. To have vigorous and healthy seedlings, they are sprayed twice a month with urea solution in water. They are shifted periodically to avoid the taproot penetrating the bag and striking into the ground. Gamaxine or aldrine in appropriate strengths are used to control insect pests.

Softwood grafting: The plants of seedling origin are showing tendency of variable yields, whereas the graft produced from high-yielding tree gives uniform and better yields. For this, cashew grafts are produced through softwood grafting and technique is as follows:

Select 60 days old seedlings as rootstocks for grafting when seedlings attain a height of 25-30 cm. On the selected rootstock seedling, 2 pairs of bottom leaves are retained and other leaves are removed using a sharp grafting knife at a height of 15-20 cm from ground level where softwood portion is available on the rootstock. A transverse cut is made and terminal shoot is removed and a cleft of 6-7 cm deep is made in the middle of decapitated stem by giving longitudinal cut. A little portion of wood is removed from the inner sides of the cleft at the top, so that after grafting, the joint will be perfect. A matching scion stick for the rootstock is selected and reduced to a length of 10 cm by cutting the excess portion at the bottom. The cut-end of scion is mended into a wedge of 6-7 cm length by chopping the bark and a little portion of wood from two opposite sides. The edge of scion is inserted into cleft of the rootstock and the graft joint is secured firmly with 2.5 cm wide and 30 cm length polythene sheet. A white polythene bag is inserted on the grafted plant, tied at the bottom with single knot. This will protect the scion stick from drying up and helps in building up of humidity. After two weeks the polythene caps are removed gently and the grafts are shifted to open nursery. The successful grafts show the sign of growth within 3-4 weeks.

Planting: In coastal areas the soil is sandy-to-sandy loam and loose with high porosity. Percolation in such soils is also high. In sandy loam soils pits of 30 × 30 × 30 cm are made and the bag plants are planted. In sandy soils earth is scooped in the form of saucer of 1m in diameter and planting is done in the centre. In red soils or red loams, trenches of 1m in length with a cross section of 30 × 30 cm are dug well in advance of planting. Central portion of the trench is filled back with excavated earth to form a mound 5 to 10 cm above the ground level. On these mounds planting is done in the centre. Generally 10 m espacement is adopted but sometimes 5 m espacement is also adopted. In the eastern coastal areas planting is done with the commencement of winter rains by end of September or beginning of October. In other places planting is done with the onset of summer-monsoon i.e., June/July. Three to four month old plants are planted in pits/

trenches/saucer pits as per soil working techniques already discussed, after cutting the bags but without disturbing the root system. Soon after planting if there is any gap in the rain, watering will have to be done. Periodic watering during non-rainy period will boost up the growth.

Application of fertilizer: In order to improve the quality and quantity of nut-yields artificial manuring with improved cultural practices is necessary. Dosage and nature of fertilizer depend on the type of soil and age of the plantation. In the 1st year of plantation ammonium sulphate and super phosphate is applied. Thereafter murate of potash is also added. With 160 plants per hectare the requirement of fertilizer in the first year is 40 kg of ammonium sulphate and 40 kg of super phosphate per hectare. In the 2nd year the quantity is doubled with an addition of 16 kg of murate of potash. In the 3rd and 4th year 160 kg of ammonium sulphate 120 kg of super phosphate and 40 kg of murate of potash is applied. Among the package practices in cashew orchards, ploughing interspaces in June/July with the onset of monsoon rains and raising wild indigo or Cassia tora or Calopagonium muconoides as cover crop or as a green manure crop, to improve soil tilth, increase moisture retention and fertility at the same time reducing soil erosion and surface run-off are recommended providing the plantations with a barbed wire fence to protect against straying cattle and pilferage of nuts during the fruiting season is also beneficial.

Injuries due to pests and diseases: Among the pests of cashew crop the most serious are the Tee mosquitoes, root and stem borer, leaf minor and the shoot caterpillar. Spraying of insecticides like B.H.C. or Folidol to control pests is necessary. Among the fungal diseases the pink or dieback disease is caused by *Corticium salmanicolor* and *Gloeosporium* fungus that can be controlled by prophyllactic spraying of copper based fungicides. Fungus *Colletotrichum gloeosporides* damages tender leaves, shoots, flower bunches, cashew apples and nuts causing dark shiny lesions and gum exudation. This can be controlled by spraying 1% Bordeaux mixture or other copper based fungicides such as Fytolan or Blitox at the time of new flushing and blossom.

Depending on growing conditions and care given to the plants cashew starts bearing in 3 to 6 years and reaches full production in 10 to 20 years. Tree flowers in November/December and nuts are harvested between March and May. Each nut forms at the end of a swollen fruit stalk, the apple, which drops to the ground on maturity. At this stage the nuts are detached and the apples are sold immediately for edible purpose locally or sent to breweries for making 'Fene'. Fresh nuts detached are called raw nuts. The raw nuts are properly dried and stored. Cashew nuts are processed when shell is removed from the kernel. Cashew shells are used as fuel in factories. Deshelled nuts are dried and graded and packed. About 260 factories employing 150000 people and having processing capacity of 450000 tonnes of raw nuts are located in Kerala alone mostly in Quilon. Preliminary data collected from processing industry indicates that the yield of whole is of the order of 65% and splits 30-35%. 100 kg of nuts give 30 kg of kernel giving about 30% yield in modern processing methods against 20-25% in the local country roasting method. The price of cashew kernel depending on size varies from Rs. 150 to Rs. 300 per kg.

Cashew fenny manufacture from the apple is indigenous to Goa and to some parts of Kerala. Cashew-fenny is manufactured locally as a cottage industry by crushing of cashew apples mannually and separating the juice. The juice is kept for 4 to 6 days for fermentation after which it is distilled. It is redistilled and cured and then bottling is done. One can of 18 litres juice can be obtained from 500 apples of medium size. Depending on the quality and blend the price of a bottle of fenny of 750 ml varies from Rs. 20 to Rs. 45.

***Annona squamosa* L**

10. *Annona squamosa* L.; Annonaceae.

Na : Eng.: Sweetsop, Sugar apple; Hindi: Sithaphal, Sharifa.

Hab : It grows wild in the drier parts in South India particularly Andhra Pradesh.

The fruit, besides being a delicacy and giving considerable revenue to the forest department, forms substantial part of food of the poor during the fruiting season in the regions where it is found. It can conveniently be grown in the forest as an under growth in South and Central India where its occurrence is either sparse or not present. Viability of seed is short. Germination is about 70%. Seeds are sown into Polybags in September or October after collection. Seedlings are transplanted in June – July after commencement of monsoon. Direct sowing is also done in the field with the beginning of monsoon. The plant is not browsed by cattle. In 4 to 5 years the plant starts yielding 15 to 20 fruits. Larger yields are obtainable from 10th year when about 40 to 50 fruits are collected annually.

***Anthocephalus indicus* A.Rich. Branch & Fruits**

11. *Anthocephalus indicus* A. Rich. (Syn. *A.cadamba* Miq.; *Nauclea cadamba* Roxb.); Rubiaceae.

Na : Eng.: Kadam; Hindi: Kadamba.

The fruits are edible.

12. *Antidesma acuminatum* Wall.; Euphorbiaceae.

Na : Hindi: Kalo bilauni.

Hab : It is a tree found in Assam, West Bengal and Bihar.

The fruits are edible.

13. *A.bunius* Spreng.; Euphorbiaceae.

Na : Eng.: Chinese laurel, Bignay; Hindi: Himalchari, Amati.

Hab : It is a small tree found in Assam, the Khasia hills and the Western Ghats.

The fruits are edible.

***A.bunius* Spreng.**

A.ghesaembilla **Gaertn**

14. *A. ghesaembilla* Gaertn. (Syn. *A.paniculata* Roxb.); Euphorbiaceae.

Na : Eng.: Black currant; Hindi: Umtao.

Hab : It is found in Sub-tropical Himalayas, Assam and Bengal.

The fruits are edible.

15. *Artocarpus heterophyllus* Lamk. (Syn. *A.integra* (Thunb.) Merr.; *A.integrifolia* L.; Moraceae.

Na : Bengali: Kanthal; Eng.; Jack tree, Jack fruit; Hindi: Kathal; Kanerese: Halasu; Mal.: Chakka; Sanskrit & Telugu: Panasa: Tamil: Pilapalam.

Hab : It is a moderate sized evergreen tree found in Western and Eastern Ghats but is also grown elsewhere. Fruits ripen from July to August. The viability of seeds is short. They are sown in polybags after washing pulp and drying. Germination is about 75%. Germination period is 15 days. About one year old seedlings are planted in the field. It starts yielding fruits in seventh year. In 15th year about 40 fruits can be collected. The average weight of the fruit may be 7 kg. The bigger fruit may be about 15 kg. The fruits and seeds are used for making curries. Fleshy perianths and carpels are also eaten. The seeds are also roasted and eaten.

Artocarpus heterophyllus **Lamk.**

16. *Asparagus adscendendens* Roxb.; Liliaceae. Syn. *A.integrifolia* L.Dholimusali

Na : Gujrathi: Ujlimusli; Hindi and Marathi: Salavar, Safed musli.

Hab : It is a branching sub-erect plant with densely crowded cladophylls. The plant is common upto 1600 m altitude in the Himalayas. The tubers are eaten as vegetable.

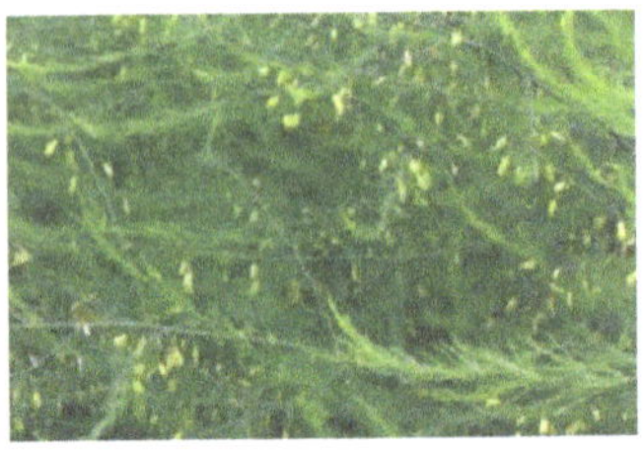

Asparagus officinalis **–L**

17. *A.officinalis* L.; Liliaceae.

Na : Bengali: Hillua; Hindi: Nagdown, Halyun.

Hab : It is a perennial branching plant common in temperate regions.

The new succulent shoots, which come up every year, are canned and also eaten green.

18. *A.racemosus* Willd.; Liliaceae.

Na : Bengali & Sanskrit: Satamuli; Guj.: Satawari; Hindi: Satawar; Kanerese: Majjige gadde; Mal.: Shatavali; Mar.: Satavarmul; Tamil: Shimai-shadavari; Telugu: Challa gadda, Chandamamgadda, Pillipichara, Pillitegalu.

Hab : It is a scandant climber common in tropical parts upto 1200 m in the Himalayas.

The roots are used in the preparation of curries.

A.racemosus **Willd**

40. *C.spinosa* L.; Capparidaceae.

C.spinosa L.

Na : Eng.: Caper bush; Hindi: Kabra; Punjabi: Bavrar, Kaur; Kanerese: Mullukatteri; Telugu: Kokilakshamu.

Hab : It is a small prostrate shrub found in rocky hilly localities in the Punjab, Rajasthan, Gujarat and the Western ghats.

The fruits and flower buds are eaten as vegetable.

41. *Carallia bracheata* (Lour) Merr.; Rhizo phoraceae.

Carallia bracheata (Lour) Merr. – Fruits

Na : Bengali & Hindi: Kirpa; Kanerese: Andipunaru; Mal.: Varanga; Marathi: Panasi, Punschi; Telugu: Karalli; Trade: Carallia-wood.

Hab : It is an evergreen tree found in Bengal, Assam, South India and Andaman Islands.

The fruits are fleshy and edible.

Carallia bracheata (Lour) Merr. Flowers

42. *Caralluma fimbriata* Wall.; Asclepiadaceae.

Na : Hindi: Maked singi.

Hab : It is a fleshy herb found in Andhra Pradesh.

The herb being fleshy is eaten as vegetable.

43. *Careya arborea* Roxb.; Barringtoniaceae.

Careya arborea Roxb. Flowering Branch

Na : Eng.: Patana oak; Guj.: Kumbi; Bengali & Hindi: Kalikatbhi, Kumbhi; Kan.: Doddala; Mal.: Alasoo, Pelu; Mar.: Kumbia; Tamil: Ayma; Telugu: Buddadarmi.

Hab : It is found practically throughout the tropical forests in the country. It is a medium sized tree.

The seeds are eaten.

44. *Carissa carandus* L.; Apocynaceae.

Carissa carandus L. Branch with fruits

Na : Eng.: Karanda; Bengali & Hindi: Karaunda, Karumcha; Guj.: Karamarda; Kanerese: Kare kagi; Marathi: Karavanda; Tamil: Kalakkay; Telugu: Vaka.

Hab : It is a spiny shrub found throughout the drier parts in the country.

The fruits are edible.

45. *Carya illinoensis* (Wang.) Koch (Syn. C.pecan Eng. & Craeb.; *C. olivaeformis* Nutt.); .Juglandaceae

Carya illinoensis (Wang.) Koch Tree

Na : Eng.: Pecan nut.

Hab : It is a tree of South Eastern United States and Mexico. It is being grown in the temperate regions of Western Himalayas.

Carya illinoensis (Wang.) Fruits

The fruits (Kernels) are a table delicacy and have the same taste as walnuts.

46. *Cassia auriculata* L.; Caesalpiniaceae.

Cassia auriculata L

Na : Eng.: Avaram; Guj.: Awal; Hindi: Tadwad, Avaram; Mal.: Avara; Marathi: Tarwad; Tamil: Avaram; Telugu: Tanged.

Hab : It is a shrub found in Western Central and Southern peninsula mostly in barren and poor soils.

It is a shrub with yellow flowers.

The flowers and flower buds are eaten as vegetable.

47. C.fistula L.; Caesalpiniaceae.

C.fistula L. Flowering Branch

Na : Bengali: Sundali. Amultas; Eng.: Indian labernum, Golden shower; Hindi: Amaltas; Guj.: Garmala; Kan.: Kakke; Marathi: Bahava; Sanskrit: Suvarnaka, Rajataru; Tamil: Konnei; Telugu: Rela.

Hab : It is a moderate sized tree with beautiful yellow flowers (pl. see medicinal plants).

Flowers and flower buds are eaten as vegetable.

48. *C.siamea Lamk*.; Caesalpiniaceae.

C.siamea Lamk.Branch

Na : Marathi & Trade: Kassod tree; Tamil: Manje konne; Telugu & Kanerese: Simatangedu.

Hab : It has been grown in afforestation and soil conservation areas.

Flowers and flower buds are eaten as vegetable.

49. *C.tora* L.; Caesalpiniaceae;

C.tora L.

Na : Eng.: Sickle senna; Guj.: Kariya; Hindi: Pamsar, Chakunda; Marathi: Taklu, Tankil; Tamil; Tagurai; Telugu: Tantemu.

Hab : It is a common weed with yellow flowers.

Young leaves are used as vegetable.

50. *Catunaregam spinosa* (Tuhnb.) Triveng. Syn. *Gardenia dumetorum* Roxb.; *Randia dumetorum* Lamk.; Rubiaceae

Catunaregam Spinosa (Tuhnb.) Triveng. fruit

Na : Eng.: Emetic nut; Hindi: Mainphal.

Hab : Pl. see Medicinal plants.

Fruits are eaten after roasting or cooking. The taste of the roasted fruits is said to be like that of almond.

***Celastrus paniculatus* Willd.**

51. *Celastrus paniculatus* Willd.; Celastraceae.

Na : Bengali & Hindi: Malkangni; Kanerese: Kariganne; Malayalam; Polulavam; Sanskrit: Jyotishmati; Tamil: Valuluvai.

Hab : It is a large climbing shrub with yellow fruits common all over the hilly parts upto 1200 m altitude. Unripe fruits are eaten.

52. *Ceropegia hirsuta* Wight & Arn.; Asclepiadaceae.

Na : Hindi: Khantodi.

Hab : It is a climbing herb commonly found in Andhra Pradesh, Tamilnadu & Kerala

The tubers are cooked and eaten.

***C.tuberosa* Roxb.**

53. *C.tuberosa* Roxb.; Asclepiadaceae.

Na : Hindi: Patala tumbi

Hab : It is found in Andhra Pradesh and Tamilnadu.

The tubers are cooked and eaten.

***Chenopodium album* L.**

54. *Chenopodium album* L.; Chenopodiaceae;

Na : Bengali: Chandanbethu, Betusag; Eng.: Pigweed, Lambs quarters; Hindi: Bhatua. Tamil: Parupukkerai; Telugu: Pappukura.

Hab : It is a common herb in wastelands.

Its leaves and tender twigs are used as vegetable.

***Chlorophytum tuberosum* Dalz.**

55. *Chlorophytum tuberosum* Dalz.; Liliaceae.

Na : Hindi: Kulai.

Hab : It is an under shrub found in Madhya Pradesh, Andhra Pradesh, Tamilnadu and Kerala.

The leaves are used as vegetable. Roots are also eaten.

***Citrus medica* L. Flower & Fruit**

56. *Citrus medica* L.; (Syn. *C.tuberosa* Mill.; *C.cetrata* Raf.; *C.aurantium* var. medica wight & Arn.; *C.medica* var. *medica proper* Hook; f.); *Rutaceae*.

Na : Bengali: Baranimbu, Begpura; Eng.: Citron; Guj.: Turanj, Bijoru; Hindi: Badanimbu, Madh-kankar, Madhkakree; Kenerese: Madala, Rueaka; Malayalam: Gilam, Rusakam; Marathi: Mahalunga, Mavalung: Tamil: Kadaranarathai; Telugu; Lungamu.

Hab : It is found in sub-mountainous Himalayan ranges of Khasia hills in Assam, Garhwal and Kumaon in U.P., in Sikkim, Pachmarhi in Madhya Pradesh and in Western Ghats.

The fruits are edible.

C.excavata **Burm. f**

57. *Clausena dentata* (Willd.) R. & S.; (Syn. *C.willdenowii* Wight & Arn.); Rutaceae.

Na : Mal.: Kariveppila; Tamil: Kattukkariveppilai.

Hab : It is a small tree foud in the Western Ghats and Assam.

Fruit is of the size of a cherry and is eaten.

58. *C.excavata* Burm. f.; (Syn. *Cookia punctata* Retz.) Rutaceae.

Na : Eng.: Wampee; Hindi: Agnijal.

Hab : It is a shrub found in Orissa and Bihar.

Fruits are edible.

59. *Commelina obliqua* Buch.-Ham.; Commelinaceae.

Na : Hindi: Kanjura.

Young shoots of the herb are eaten as vegetable.

Commelina obliqua **Buch.-Ham.**

60. *Commiphora caudata* (Wight & Arn.) Eng. (Syn. *Balsamodendron caudatum* Wight & Arn.); Burseraceae.

Na : Eng.: Hill mango; Kanerese: Kondamavu; Tamil: Kiluvai; Telugu: Kondamamidi.

Hab : It is a shrub or small tree found in Andhra Pradesh, Tamilnadu, Karanatak and Kerala. Pl. See medicinal plants.

Fruits are edible and are often pickled.

61. *Cordia dichotoma* Forst. F.; (Syn. *C.myxa* auct plur., non L.; *C.obliqua* Willd.); Ehretiaceae.

Na : Bengali: Bahubara; Eng.: Sebesten, Clammy cherry, Indian cherry; Guj.: Bargund; Hindi: Chokargond, Lassoora; Kanerese: Chikkachalle; Malayalam,: Viri, Cheruviri; Marathi: Shelvant; Tamil: Naruvili; Sanskrit: Bahuvaraka; Telugu: Chinnanakkeru.

Hab : It is a small tree found in the Punjab, Rajasthan, Madhya Pradesh, the western peninsula, and the Khasia hills.

Fruits are mucilaginous, sweetish and edible.

Cordia dichotoma **Forst. F**

62. *C.gharaf* (Forsk.) Ehrenb. & Asch.; (Syn. *C.rothii* Roem. & Schum.); Ehretiaceae.

Na : Hindi: Ghondi.

Hab : It is found in the Punjab, Rajasthan, Gujarat and Andhra Pradesh.

Fruits are edible.

4. *Agave americana* L..; Agavaceae.

Na : Eng.: Century plant; Hindi: Kantala.

Hab : It is a stout shrub native of South America.

The pulp of the flowering stock is eaten by the poor people.

Agave americana L.

5. *Aglaia roxburghiana* Miq.; Meliaceae.

Na : Hindi: Pupriyangu; Malayalam: Punyava; Tamil: Chokkalai; Telugu: Yerraaduga.

Hab : It is a large tree found in the Western Peninsula and Rajasthan.

The fruits are edible.

Aglaia roxburghiana Miq.

6. *Alangium salvifolium* (L.f.) Wang. (Syn. *A.lamarckii* Thw.); Alangiaceae.

Na : Eng.: Sage leaved Alangium; Hindi: Dhera, Akola.

Hab : It is a small tree found throughout the tropical forests in India particularly South India (Pl. see medicinal plants).

Pulp of the fruits is eaten.

Alangium salvifolium (L.f.) Wang. – Flower

7. *Alpinia galanga* (L.) Willd. (Syn. *Languas galanga* (L.) Stunz.); Zingiberaceae.

Na : Eng.: Greater galangal; Hindi: Kalinjan.

Hab : and other details pl. see essential oils.

The flowers are eaten.

Alpinia galanga (L.) Willd - Plant

Alpinia galanga (L.) Willd - Flower

8. *Amorphophallus campanulatus* (Roxb.) Blume ex. Dcne. (Syn. *Arum campanulalum* Roxb.); Araceae.

Na : Bengali: Ol; Eng.: Elephant food yam; Gujrathi & Marathi: Suran; Hindi: Suran, Zaminkand; Kanerese: Suvarnagadde; Malayalam: Chena; Marathi: Suran; Sanskrit: Arsaghna, Balukand; Tamil: Karnai-kilangu; Telugu: Kanda

Tuberous roots are eaten as vegetable.

Amorphophallus campanulatus (Roxb.) Blume ex Dcne Root

9. *Anacardium occidentale* L.; Anacardiaceae.

Na : Bengali: Hijlibadam; Eng.: Cashew; Marathi, Hindi: Kaju: Kanerese: Kerubija; Malayalam: Andiparuppu; Tamil: Mandiri: Telugu: Jidimamidi, Muntharnamidi.

Hab : It is a small tree.

Government of India is giving considerable importance to grow crops such as cashew nuts, which generate export earnings and require high

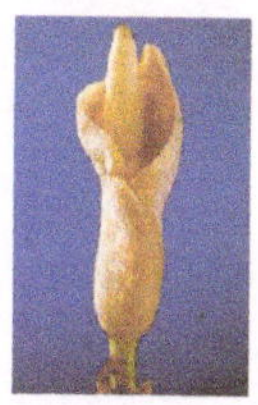

Amorphophallus campanulatus (Roxb.) Blume ex Dcne *Root*

63. *Corylus colurna* L.; Betulaceae.

Corylus colurna L. **Tree**

Na : Eng.: Turkish hazelnut; Hindi: Bhutia badam, Urni; Kashmir: Virin; Kumaon: Kapasi, Bhotiabadam; Punjab: Tungi, Urni.

Hab : It is commonly found in Kashmir and Kumaon in the temperate regions.

Nuts (seeds) are eaten.

64. *Corypha elata* Roxb.; Palmaceae.

Corypha elata **Roxb. Palm**

Na : Hindi: Bajoon.

Hab : It is a tall plam of Bengal.

Trunk of this palm is the source of starch. Buds are used as vegetable and the seeds are also eaten.

65. *Cudrania javanensis* Trecul.; Moraceae.

Na : Hindi: Manda: Mangei, Kangu.

Hab : It is a spiny shrub found in U.P., Orissa and Assam.

Ripe fruits and tender leaves are edible.

66. *Curculigo latifolia* Dry.; Amarillidaceae.

Curculiog latifolia **Dry**

Hab : It is an ornamental herb with broad curving graceful leaves and is commonly found in the Andaman Islands.

Fruits are sweet and edible.

67. *C.orchioides* Caertn.; Amarillidaceae.

C.orchioides **Caertn.**

Na : Bengali: Talamuli; Eng.: Black musale; Hindi: Kalimusli, Musalikand, Tallura; Kanerese: Nelatategadde; Sanskrit: Mushali; Telugu: Nelatatigadda.

Hab : It is a perennial herb found in the Western Ghats and the subtropical Himalayas.

Dried black root is ground and then eaten like flour.

68. *Curcuma leucorhiza* Roxb.; Zingiberaceae.

Curcuma leucorhiza **Roxb.Plant**

Na : Hindi: Tikar.

Hab : It is found in Bihar.

Rhizomes yield starch.

69. Cycas pectinata Griff.; Cycadaceae.

Na : Assam: Thaljimura; Nepal: Thakal.

It occurs in Nepal, Sikkim, Bengal and Assam. It yields coarse sago and edible fruits.

Cycas pectinata Griff. **Fruits**

Cyperus esculentus L. Herb

Cyperus esculentus L. Root Stocks

Dendrocalamus strictus (Roxb.) Nees Culms

Dillenia indica L. Fruits

D.pentagyna Roxb.

70. *Cyperus esculentus* L.; Cyperaceae.

Na : Eng.: Earth almond, Chufa; Hindi: Chichoda; Punjab: Koseru, Dela.

Hab : It is an erect grass like sedge and is commonly found in U.P., the Punjab and South India.

Tuberous rootstocks are eaten as food.

71. Debregeasia hypoleuca Wedd.; Urticaceae.

Na : Hindi: Sansaru; Punjab; Pincho, Sansaru.

Hab : It is a large shrub found in Kashmir and Kumaon hills of Uttar Pradesh.

Fruits are edible.

72. *Dendrocalamus strictus* (Roxb.) Nees; Poaceae (Gramineae.)

Na : Bengali: Karail; Eng.: Solid bamboo, Male bamboo; Guj.: Nakor vans; Hindi: Bans Narbans; Kanerese: Keribidur; Malayalam: Kalmungil; Marathi: Bhariyel; Oriya: Salia Bhanso; Sanskrit: Vansha; Tamil: Kalmungil: Telugu: Veduru.

Hab : It is found practically throughout the tropical forests in India.

Seeds are collected during the year of flowering and are eaten by the poor local people. Young and tender rhizomes are pickled and eaten as vegetable curry. (Pl. see Bambusa arundinacea)

73. *Desmos longiflorus* (Roxb.) Staff.; *(Unona longiflora Roxb.); Annonaceae.*

Hab : It is found in Orissa and Assam.

Fruits are eaten.

74. *Dillenia indica* L.; Dilleniaceae.

Na : Assamese: Chalita, Oulenga; Eng.: Elephant Apple; Bengali & Hindi: ChaIta; Guj. & Marathi: Karambel,KarmaI; Kanerese: Bettakanegala; Malayalam: Chatata, Punna; Tamil: Uva; Telugu: Pedda Kalinga, Ura; Trade; Dillenia.

Hab : It is a medium sized handsome evergreen tree with large white fragrant flowers and is found in the moist tropical forests of subHimalayan tracts and grows in Assam, Bengal, Madhya Pradesh and South India.

Flowers and fruits are edible. Dispersal of the seeds is helped by elephants, which are fond of fruits.

75. *D.pentagyna* Roxb.; *Dilleniaceae.*

Na : Bengali: Karkotta; Marathi & Hindi: Karmal; Kanerese: Kanegala, Kadukanigala: Malayalam: Punna, Kodupunnu; Tamil: Naytekku; Telugu: Chinnakaringa; Oriya: Rai.

Hab : It is a large deciduous tree with fragrant yellow flowers commonly found in the Western peninsula, Bihar and Assam.

Flowers are edible.

***Dioscorea bulbifera* L.Tubers**

76. *Dioscorea bulbifera* L.; *(Syn. D. sativa Thunb.); Dioscoreaceae.*

Na : Bengali: Banalu, Kukuralu, Gaicholu; Eng.: potato yam; Hindi: Gaithi, Suaralu, Pilaulu; Kanerese; Heggenasu; Malayalam: Kattukachil; Marathi: Manakund, Karukarinda, Gathalu; Tamil: Kodikilangu; Telugu: Chedupadudumpa. Malakakayapendalamu.

Hab : It is an unarmed climber twining to the left and is found throughout the country.

Tuber is eaten as vegetable. Wild forms are bitter and acrid but are rendered edible by processing with ash and steeping in cold water.

***D.oppositifolia* L.Climber with Tubers**

77. *D.oppositifolia* L.; *Dioscoreaceae*

Na : Tamil: Kavalu koti, Venilaivelli; Telugu: Yellagadda, Adairdumpa. .

Hab : It is a climber twining to the right.

Tubers are edible.

78. *D.pentaphylla* L.; *Dioscoreaceae.*

Na : Bengali: Suaralu; Hindi: Kanta alu; Kanerese: Nurai genusu; Mal.: Nurunnakilanga; Marathi: Chata vali, Mandi, Ulasi; Tamil: Kattukilangu; Telugu: Dukapendalamu.

Hab : It is a prickly climber twining to the left and is found throughout India.

Tubers are eaten after repeated boiling and washing.

***D.pentaphylla* L. Tubers**

79. *Diospyros ebenum* Koenig; *(Syn. D.assimilis* Bedd.; *D.sapota* Roxb.*); Ebenaceae.*

Na : Eng.: Mecasser ebony; Hindi: Abnus; Kanerese: Karemara; Mal.: Mushtumpi, Karu; Tamil: Tumbi, Karunkali; Telugu: Nalla valludu, Nalluti, Tunki.

Hab : It is a large sized ever green tree with a dense crown and is found in Andhra Pradesh, Tamilnadu and Kerala.

Fruits are edible.

80. *D.ferrea* (Willd.) Bakh.; *(Syn. Maha buxifolia* Pers.*); Ebenaceae.*

Na : Bengali & Hindi: Angaru; Kanerese: Karugana; Oriya: Goakuli, Guakuli; Tamil: Irumballi, Kuruvinei; Telugu: Chinnavullinji.

Hab : It is a large bushy shrub or small tree and is found in the Western Peninsula, Orissa and Tamilnadu.

Fruits are pulpy when ripe and edible.

***D.ferrea* (Willd.) Bakh.**

81. *D.lotus* L.; Ebenaceae.

Na : Eng.: Date plum; Hindi: Amlok.

Hab : It is a deciduous tree found in the Kashmir and other parts of western Himalayas.

Fruits are globose or ovoid and are sweetish and eaten fresh or dried.

82. *D.melanoxylon* Roxb. *(Syn. D.tupru* Buch.-Ham.); Ebenaceae.

Na : Eng.: Coromandel ebony; Gujrathi: Tamrug; Hindi: Abnus, Tendu; Kanerese: Abanasi, Bale, Tumari; Malayalam: Kari; Marathi: Tendu, Tumri; Sanskrit: Dirghapatraka; Tamil: Karai, Karundumbi, Tumbi; Oriya: Kendu.

Hab : It is found in Madhya Pradesh, Andhra Pradesh, Maharashtra, Mysore, Orissa and Bihar.

Pulp of the fruit is edible. Seeds are boiled and eaten (Also pl. see separate chapter on this species).

83. *D.peregrina* (Gaertn.) Geurke (Syn. *D.embryopteris* Pers.); Ebenaceae.

Na : Bengali: Makurkendi, Gab, Kalatendu; Eng.: Indian persimon; Hindi: Gab, Kalatendu; Kanerese: Holeterpare, Kusharta; Malayalam: Panachi, Vananji; Marathi: Timburi, Timburni; Sanskrit: Tinduko, Krishnasara, Virupaka; Tamil: Kattatti, Kavikatta, Tumbi; Telugu: Gabu, Tinduki.

Hab : It is a moderate sized evergreen tree with short fluted trunk and spreading branches and is found in Western peninsula and Madhya Pradesh.

Fruits are velvety in colour and are eaten.

***Diospyros lotus* L. Fruits**

84. *D.tomentosa* Roxb.; *Ebenaceae.*

Na : Bengali: Kend, Kyon; Eng.: Nepal ebony, Persimon; Hindi: Kendu, Tendu, Temru; Kanerese: Timburani, Tumari, Tindura; Tamil: Tumbi; Telugu: Chittatumiki, Tumiki; Oriya; Kendu.

Hab : It is large tree found in Madhya Pradesh, Bihar, Orissa and Rajasthan.

Fruits are eaten.

85. *Diploknema butyraceae* (Roxb.) H.J. Lam.; *(Syn. Bassia butyraceae* Roxb.; *Madhuca butyracea* (Roxb.) Macb.; *Sapotaceae.*

Na : Bengal: Gophal; Eng.: Bassia fat, India butter tree; Hindi: Phalwara, Chiura, Phulel.

Hab : It is a large deciduous tree found in the sub-Himalayan tract in U.P., Bihar, Bhutan upto 1500 m.

Flowers are eaten. Fruit is a berry blackish in colour with a sugary pericarp. Pericarp forms nearly 70% of the weight of the berry and is edible. The seed oil is used by the poor people for cooking purpose.

86. *Docynia indica* Decne; Rosaceae.

Na : Eng.: Indian crab apple; Lepcha: Likung; Khasi: Soh-phoh.

Hab : It is a moderate sized tree with handsome foliage and is found in the western Himalayas and the Khasia hills.

Fruits are fusiform to ellipsoid in shape and are edible.

Duchesnea indica Focke. Branch & Fruits

87. *Dracontomelum mangiferum* Blume; Anacardiaceae.

Hab : It is a tall and handsome tree of the Andaman Islands.

Fruit is yellowish when ripe and is edible.

88. *Duabanga sonneratioides* Buch.-Ham. (Syn. *D.grandiflora* Walp.*);* Sonneratiaceae.

Na : Assamese: Thora, Khukan; Bengali & Hindi: Bandorhulla; Nepali & Trade: Lampali.

Hab : It is a large tree found in Bengal, Assam, Manipur and Andaman and Nicobar Islands mainly along the rivers and streams.

Fruits are acidic and are eaten.

Echinochloa stagnina (Retz.) Beauv.

89. *Duchesnea indica* Focke (Syn. *Fragaria indica* Andre.*);* Rosaceae.

Na : Eng.: Indian strawberry; Hindi: Kiphaliya.

Hab : It is a perennial herb commonly found in the temperate Himalayas of Kashmir, the Punjab, Himachal Pradesh, Uttar Pradesh, Assam and in the Nilgiris.

Fruits are bright red, globose or oblong in shape. They are spongy and insipid and are eaten

Ehretia acuminate R.Br. Tree

90. *Echinochloa stagnina* (Retz.) Beauv.; Poaceae (Gramineae.)

Na : Bengali: Dul; Hindi: Banti; Kanerese: Kadu, Dabhaihullu; Telugu: Bontaoddu.

Hab : This grass is found throughout the greater part of the country.

The grains of this grass are sometimes eaten as an emergency food by the weaker sections of society.

91. *Ehretia acuminata* R. Br.; Ehretiaceae.

Na : Assam: Bual; Bengal: Kulaja; Eng.: Heliotrope tree; Hindi: Gual, Punyan, Punia, Lahichan, Koda; Punjabi: Punna, Punru, Kaulthaon.

Hab : It is a tree commonly found in the sub-Himalayan tract and also in Orissa.

Fruits are orange coloured and insipid to taste. However, they are eaten. Unripe fruit is pickled.

Ehretia acuminata R.Br. in flower

Elaegnus angustifolia **L.Fruits**

92. *Elaegnus angustifolia* L. (Syn. *E.hortensis* Bieb.); Elaeagnaceae.

Na : Eng.: Oleaster; Hindi: Shiulik.

Hab : It is a deciduous shrub or small tree found in the Western Himalayas, and Kashmir.

Fruit pulp is sweet when ripe and is edible.

E.latifolia **L.**

93. *E.latifolia* L.; *Elaeagnaceae.*

Na : Bengali: Guara; Eng.: Bastard oleaster; Hindi: Ghiwain; Kanerese: Hittale, Hunese, Balle; Malayalam: Kayalampuvalli; Marathi: Ambgod, Nildook, Nurgi; Tamil: Kolungai, Kalari.

Hab : It is a straggling shrub found in Andhra Pradesh, Khasia hills, Manipur and Western ghats.

Fruits are edible.

E.umbellata **Thunb.**

94. *E.umbellata* Thunb.; *Elaeagnaceae.*

Na : Hindi: Ghain; Punjabi: Bammerwa, Ghain, Kenkoli.

Hab : It is a much branched deciduous shrub with yellowish brown silvery branches and is found from Kashmir to Manipur in the temperate Himalayas.

The fruit is edible and is pickled or eaten in curries.

Elaeocarpus serratus **L.**

95. *Elaeocarpus serratus* L.; *Elaeocarpaceae.*

Na : Bengali: Jalpai; Eng.: Ceylon olive; Hindi: Jalpai; Kanerese: Perinkara; Mal.: Avil, Nallakara, Valiyakara, Karamava; Oriya: Jolopari; Tamil: Ulangkarei, Uttraccham.

Hab : It is a moderate sized tree found in North Kanara in Karnatak and in Kerala.

Fruit pulp is edible.

E.sphaericus **(Gaertn.) K. Schum.**

96. *E.sphaericus* (Gaertn.) K. Schum.; (Syn. *E.ganitrus* Roxb.); Elaeocarpaceae.

Na : Bengali: Rudrakya; Hindi: Rudrak; Rudraki; Kanerese, Malayalam, Marathi, Sanskrit, Tamil, Telugu: Rudraksha; Oriya: Rudrakyo.

Hab : It is a medium sized tree occuring in Bengal, Assam and Nepal.

Fruit is a bluish purple drupe. The pulp is edible.

97. ***Eleocharis dulcis*** **Burm.F.Trin.ex Henschel; (Syn.** ***E.plantaginea*** **R. Br.;** ***E.tuberosa*** **Schult.);Cyperaceae.**

Na : Eng.: Chinese water chestnut.

Hab : It is an herb found in U.P., Bihar, Bengal, Assam and Orissa.

Tubers are edible.

Eleocharis dulcis **Burm. f. Trin.** ***ex*** **Henschel**

98. ***Entada phaseoloides*** **(L.) Merr.; (Syn.** ***E.scandens*** **Benth.;** ***Lens phaseoides*** **L.); Mimosaceae.**

Na : Assamese: Gilalewa; Bengali: Gila; Eng.: Nicker bean; Elephant ereeper mackay bean, Garbee bean; Hindi: Badabi, Chian, Gilla; Kanerese: Dodda kampi, Halleykayiballe; Mal.: Kakkavalli, Malamanchadi; Marathi: Garamb, Garbe; Tamil: Vatta valli, Chilla; Oriya: Gredi.

Hab : It is a huge climbing shrub found in the tropical regions of central and eastern Himalayas, Bengal, Orissa, Bihar, U.P. and South India in high rainfall zones.

Kernel of the seed is eaten by the poor.

Entada phaseoloides **– (L).Merr. Climber**

99. ***Erioglossum rubinosum*** **(Roxb.) Blume; (Syn.** ***E.edule Blume);*** **Sapindaceae.**

Na : Hindi: Rilha; Oriya: Sonamahanga; Tamil: Korali, Manipangam. Telugu: Ishirashi, Undurugu.

Hab : It is a small tree with rusty pubescent on growing parts and is found in the Eastern and Western Peninsula and Madhya Pradesh.

Fruits are red to black and are astringent and edible. Tender leaves are also eaten.

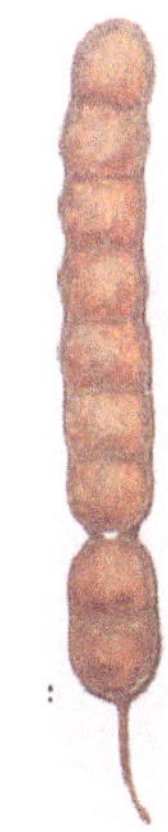

Entada phaseoloides **–(L). Merr. Fruit**

100. ***Erythrina variegata*** **L. var. orientalis (L.) Merr. (Syn.** ***E.indica*** **Lamk.); Fabaceae (Papilionaceae.)**

Na : Bengali: Palitamandar; Eng.: Coral tree, Bastard teak; Guj.: Bangaro, Panaraweo; Hindi: Pangra, Nasut, Dadap, Ferrud, Mandar; Kanerese: Vajipe, Harvana, Harvala; Malayalam: Kalyana, Murikku, Mandaram; Marathi: Mandar, Pangra; Sans.: Mandar; Tamil: Kalyanamurukku; Telugu: Badisa, Badita, Baridamu.

Hab : It is a small sized tree found in the Sundarbans, the Andaman and Nicober Islands, Assam, Tamilnadu and Orissa. Bark is smooth greenish grey peeling off in flakes. Branchlets are armed with prickles.

Tender pods and seeds are edible.

Erythrina variegata **L. Flowers**

101. ***Eugenia formosa*** **Wall.; (Syn.** ***Syzygium mappaceum*** **(Korth.) Mansf.; Myrtaceae.**

Na : Assamese: Lahajam; Bengali: Phuljamb; Hindi: Jamrul.

Hab : It is a tree found in Assam and other parts of Eastern Himalayas.

Fruits are edible.

Euryale ferox Salisb.

102. *Euryale ferox* Salisb.; Nymphaeaceae.

Na : Eng.: Fox-nut, Gorgan nut; Bengali & Hindi: Mokhana; Punjab: Jewar; Telugu: Mettunipadmamu.

Hab : It is a prickly aquatic herb found commonly in Kashmir, Bihar, Assam, Manipur, Tripura, Bengal and U.P. Leaves are floating. Fruits are round and prickly of the size of an orange containing 8-20 seeds.

Seeds of this aquatic herb are roasted and eaten and also used as a substitute for arrow root.

103. *Evodia fraxinifolia* Hook. f.; Rutaceae;

Na : Bengali: Kanukpa; Eng: American beach: Hindi: Kanukpa; Khasi: Diengsngiyat, Diengborapram.

Hab : It is a tree found in the sub-tropical Himalayas and the Khasia Hills. It also grows in Bengal. It bears pinnate leaves, greenish white flowers and red capsular fruits.

Fruits are edible.

Fagopyrum cymosum Moissn.

104. *Fagaria oxyphylla* (Edgew.) Engl.; (Syn. *Zanthoxylum oxyphylum* Edgew.); Rutaceae.

Na : Hindi: Timar, Marenga.

Hab : It is a shrub found in Kumaon and Khasia hills.

Tender shoots are eaten as vegetable.

105. *Fagopyrum cymosum* Moissn.; Polygonaccae.

Na : Eng.: Perennial buckwheat; Hindi & Punjabi: Banogal; Khasi: Ja-rain.

Hab : It is a tall herb found in the temperate Himalayas and the Khasia hills. Its leaves are long stalked and oval in shape.

Leaves are eaten as vegetable.

Fagus sylvatica L.

106. *Fagus sylvatica* L.; Fagaceae.

Na : Eng.: European beach, common beech,

Hab : It is a tall tree found in Kulu valley and in the Nilgiris.

Seeds (nuts) are sweet and are eaten only after baking or cooking as the raw nuts contain toxic substances.

Ficus auriculata Lour.

107. *Ficus auriculata* Lour.; (Syn. *F.roxburghii* Wall. *F.macrophylla* Roxb.); Moraceae.

Na : Eng.: Eve's apron; Hindi: Timla, Tirmal.

Hab : It is found in Bihar, Orissa, Khasia hills and Manipur.

Fruit, which is purplish in colour is edible and is used for making curries and jams.

108. *F.glomerata* Roxb.; (Syn. *F.goolereea* Roxb.) Moraceae.

Na : Bengali: Dumur, Jagyadumbar; Eng.: Cluster fig; Hindi: Gular, Udumbara, Domoor; Guj.: Umar, Gular; Kanerese, Malayalam, Tamil: Alhi; Marathi:. Umbar; Oriya: Dimri; Telugu: Athi, Bodda, Paidi, Udumbaramu.

Fruits when they mature have a pleasant odour and are edible but often they are full of maggots of the fertilising wasp and unfit for eating.

***F. glomerata* Roxb**

109. *F.hispida* L.F.; Moraceae.

Na : Hindi: Daduri, Dagurin, Gobla, Kagsha, Katgularia, Kathumbar, Konea-dumbar; Guj.: Dhedaumaro, J.anglianjir; Marathi: Bhokada, Bokria, Kalamber, Kharoti; Kanerese: Adavi Alhi, Kaddatti; Tamil, Malayalam: Peyatti; Telugu: Bodamamidi, Brammadi.

Hab : It is a shrub or small tree found in the greater parts of the country.

Fruits, which are greenish yellow when ripe, are eaten and made into jams.

***F.hispida* L.F.**

110. *F.palmata* Forsk.; (Syn. *F.virigata* Roxb.); Moraceae.

Na : Guj.: Pepri; Hindi: Anjiri, Bedu, Khemri; Telugu: Manjimedi; Punjabi: Phagwara.

Hab : It is found in North Western India and Rajasthan.

Fruits are edible. Young shoots are used as vegetable and the latex is used in the preparation of curd.

***F.palmata* Forsk**

111. *F.rumphii* Blume; *Moraceae.*

Na : Assam: Jakri, Porap; Bengali: Gaiaswat; Hindi: Gagjaira, Gajna, Kabar, Pakar, Pilkhan; Kanerese: Bettaarli; Marathi: Asht, Pair, Payar.

Hab : It is medium sized tree found in the Punjab, Madhya Pradesh and Assam. Fruits are sessile and black when ripe.

Ripe fruits are occasionally eaten.

***F.rumphii* Blume**

112. *Flacourtia indica* (Burm. f.) Merr. *(Syn. F.romontchi* L' Herit.); *Flacourtiaceae.*

Na : Bengali: Benchi, Baichi, Binja, Katai; Eng.; Romontchi, Governor's plum; Guj.: Kankod; Hindi: Kantai, Bilangra, Talispatri, Baichi, Kanju; Kanerese; Hattarimullu, Lunmunki; Marathi: Bhekal, Kaker, Paker; Tamil: Katukala, Sottaikala; Telugu: Kandregu.

Hab : It is a small tree found in the drier parts in the country. Leaves are variable in shape.

Fruits are dark purple or red with juicy pulp.

***Flacourtia indica* (Burm.f) Merr.**

F.jangomas (Lour.) Raeusch.

113. *F.jangomas* (Lour.) Raeusch. (Syn. *F.cataphracta* Roxb.*Ex* Willd.); Flacourtiaceae.

Na : Bengali: Paniala; Eng.: Puneala plum; Gujrathi: Talispatra; Hindi: Pani-amla, Talispatri; Kanerese: Chankali, Goraji; Malayalarn: Kanji, Talisam, Vayankatha; Marathi: Tambat; Oriya; Baincha; Tamil: Saralu, Vayangurai; Telugu: Gare.

Hab : It is a shrub or spreading tree growing in the drier parts in the country. Trunk has sharp spines.

Fruit is dark red or purple when ripe and is edible.

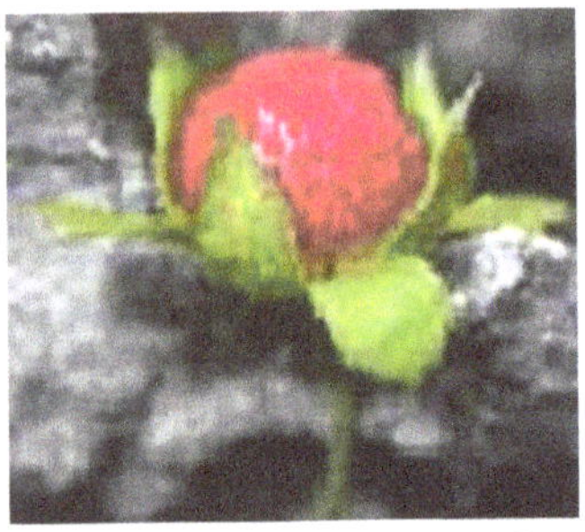

Fragaria nilgerrensis Schlecht. *ex* F. Gay

114. *Fragaria nilgerrensis* Schlecht. *ex* F. Gay; Rosaceae.

Na : Eng.: Houtbois strawberry; Niligiri strawberry.

Hab : It is a creeping perennial found in the Khasia hills, the Nilgiris and Pulney hills. .

Fruits are edible.

Garcinia atroviridis Griffi. – Tree

115. *Garcinia atroviridis* Griffi.; Guttiferae.

Hab : It is a moderate sized tree found mainly in Western Ghats, Assam.

Fruits are orange yellow, sub-globose. Fruit rind and pulp are eaten. Fruit rind is too sour to be eaten raw but tastes excellent when stewed with sugar.

116. *G. cambogia* Desr.; Guttiferae.

Na : Eng.: Gamboge tree; Hindi: Velaiti imli, Hila; Kanerese: Upagimara, Simehunese; Mal. & Tamil: Kadumpuli, Kodapuli; Marathi: Dharambe; Telugu: Simachinta.

Garcinia atroviridis Griffi. – Fruit

Hab : It is a small tree found in the Western Ghats in evergreen forests.

Fruits are ovoid yellow or red when ripe but are too sour to be eaten raw. Seeds yield 31 % of edible fat resembling kokam butter.

G.cambogia Desr.

117. *G.cowa* Roxb.; *(Syn. G.kydia* Roxb.*);* Guttiferae.

Na : Assam: Kujithekera, Kauthekera; Bangali: Kau; Hindi: Cowa; Nepal: Kaphal.

Hab : It is a medium sized dioecious tree occurring in West Bengal, Assam, the Nilgiris and the Andaman Islands.

Fruit is globose yellow or reddish with 6-8 grooves. It is edible though not very palatable because of its acid taste. Sun dried slices of the fruit are eaten in curries.

G. cowa **Roxb.**

118. *G. indica* (Thou.) Choisy; Guttiferae.

Na : Eng.: Kokam butter tree, Mangosteen oil tree, Brindonia tallow tree; Gujrathi: Kokan; Hindi: Vishambil, Kokam; Kanerese: Murgala; Malayalam: Punampuli; Marathi: Amsol, bhirand, Katambi, Kokam, Ratamba; Tamil: Murgal.

Hab : It is a slender evergreen tree found wild in Maharashtra, Karnatak and Kerala.

Fruits are dark purple when ripe and have an agreeable flavour and sweetish acid taste. It is used to give acid taste in curries. The seeds yield 23 to 26% edible fat on the weight of seed, which is known as kokam butter.

G. indica **(Thou.) Choisy**

119. *G. morella* Desr.; Guttiferae.

Na : Assamese: Kujithekera; Hindi, Bengali and Marathi: Tarnal; Eng.: Ceylon gamboge; Hindi: Gotaghanba; Kanerese: Hardala, Jarize; Mal.: Daramba, Karukkampali, Pinnarpuli; Chigeri; Tamil: Makki.

Hab : It is a medium sized evergreen tree found in the Khasia hills and in the Western Ghats from South Kanara in Karnataka and in Kerala. Fruits are globose of the size of a plum.

Seeds yield 30% fat similar to Kokam butter from G. indica. Fat is brownish yellow with a pleasant odour and is used in cooking and confectionery as a substitute for ghee.

120. *G. xanthochymus* Hook. f.; (Syn. *G.tinctoria* Dunn); Guttiferae.

Na : Bengali: Chalala, Tamal; Eng.: Egg tree; Gujrathi: Karamala, Ola; Hindi: Dampel, Tamala, Tumul; Kanerese: Devagarege; Malayalam: Anavaya; Marathi: Jharambi, Ota; Oriya: Cheoro, Sitambu; Tamil: Kulavi, Malaippachai, Mukki, Tamalam; Telugu: Ivarumidi, Tamalamu.

Hab : It is a medium sized evergreen tree found in Orissa, Maharashtra, Karnataka, the Nilgiris, Bengal and the Andaman Islands.

Fruit has a juicy pulp with a pleasant acid flavour. It is made into preserves and jams.

G. xanthochymus **Hook. f.**

121. *Gardenia campanulata.* Roxb.; *Rubiaceae.*

Na : Assamese: Bilmara, Dieng-chii, Bhimona; Hindi: Bitmara.

Hab : It is a spiny large shrub or small tree found in Assam and Bihar.

Fruits are ellipsoid and fleshy. Leaves and fruits are eaten after cooking.

G. turgida **Roxb.**

122. *G. turgida* Roxb.; Rubiaceae.

Na : Hindi: Thanella, Ghurgia; Kanerese: Bengeri, Bootbanguri; Malayalam: Malankara, Kharkar, karhar, Marathi: Pendra, Khurpendru; Oriya: Bomonia; Rajasthani: Karumba; Tamil: Malangarai; Telugu: Yerribikki.

Hab : It is a small deciduous tree found in drier regions in the country.

Fruits are eaten after cooking.

123. *Gaultheria fragrantissima* Wall.; Ericaceae.

Na : Eng.: Fragrant wintergreen; Hindi: Gandapuro; Khasihills: Jirhap, Soh-ling-thrait.

Hab : It is an aromatic shrub found in the Khasi hills, the Nilgiris and the Pulney hills in Tamilnadu and in Kerala. (Pl. see essential oils and spices and condiments).

Fruits are edible.

***Gaultheria fragrantissima* Wall. Flowers & Fruits**

124. *Girroniera cuspidata* (Bl.) Kurz; *(Syn. G. reticulata* Thw.*); Ulmaceae.*

Na : Hindi: Narakiya; Kanerese: Babbuchekke, Nyal, Naraka-bhutali; Khasihills: Dieng-charkhei; Lepcha: Shee-kung; Tamil: Kodilani, Khomanig.

Hab : It is a large evergreen-buttressed tree found in Sikkim, the Western Ghats, the Khasia hills and Karnataka.

Fruits are edible.

***Glycyrrhiza glabra* L.Plant**

125. *Gisekia pharmaceoides* L.; Phytolaccaceae.

Na : Hindi: Balukasag; Malayalam: Manalkeera; Marathi: Waluchi-Bhaji; Rajasthani: Morang, Sareli; Tamil: Manal keerai: Telugu: Isaka dosarikura, Isakadatukura.

Hab : It is a succulent herb found in the Punjab, Rajasthan, Gujarat and Tamilnadu.

Leaves are eaten as vegetable.

***Glycyrrhiza glabra* (L.) Roots**

126. *Glycyrrhiza glabra* L.; Fabaceae (Papillionaceae.)

Na : Eng.: Liquorice; Hindi: Mulhatti.

Hab : and other detail pl. see medicinal and essential oil yielding plants.

An extract from the roots is used in confectionery.

127. *Gmelina arborea* L.; (Syn. *Premna arborea* Roth); Verbenaceae.

Na : Assam: Gomari; Bengali: Gumbar, Gamari; Eng.: Malay bush beech; Hindi: Juganichukur, Gumbor Gamhari, Sewan; Marathi and Guj.: Shewan;

Kanerese: Shivani; Mal.: Kumbil; Oriya: Gambari, Bhodroporni; Punjab: Kumhar, Ban; Tamil: Kumadi, Umithekka, Perungumpil; Telugu: Gummad teak; Trade: Gumhar.

Hab : It is a large tree found in Maharashtra, Andhra Pradesh, Assam, Karnataka, Tamilnadu, Kerala and Bengal.

Fruit is fleshy ovoid and is edible.

***Gmelina arborea* L**

128. *Gnetum gnemon* L.; Gnetaceae.

Hab : It is an evergreen small or middle-sized tree found in Khasi hills, Western and Eastern Ghats. Trunk of the tree is marked with leaf scars. Fruit is ellipsoid orange or red when ripe containing single large starchy seed, which is eaten after roasting or cooking.

***Gnetum gnemon* L. *Branch &* Fruits**

129. *Grewia asiatica* L. (Syn. *G elastica* Royle) Tiliaceae

Na : Bengali: Phalsa. Shukri; Eng. and Hindi: Phalsa; Guj.: Phalsa, parush, shukri; Kanerese:, Tadasala; Marathi: Phalsi; Tamil: Palisa, Tadachi; Telugu: Nallajana, Phatiki.

Hab : It is a small tree found in the drier regions in Central and South India often cultivated.

Fruits which are purple in colour are eaten and a sherbet is also made.

***Grewia asiatica* L.**

130. *Gtenax* (Forsk.) Fiori (Syn. *Gpopulifolia* Vahl); Tiliaceae.

Na : Hindi: Chabeni, Gondni; Punjab: Gangu, Kanger; Rajasthan: Gangerun, Gangu: Tamil: Achchu; Telugu: Gundukadira, Kadadari, Kaladi.

Hab : It is a shrub found in Punjab, Rajasthan, Tamilnadu and other arid areas.

The orange-red acid drupe of the plant is edible.

***G. tenax* (Forsk.) Fiori**

131. *G tiliaefolia* Vahl; Tiliaceae.

Na : Bengali & Hindi: Dhamani, Pharasa; Gujrathi: Dalmon, Dhamana; Kanerese: Tadasal, Butale; Mal.: Chadicha; Oriya: Dhamandhamuro; Marathi: Daman, Damani; Tamil: Sadachi, Unnu; Telugu: Ettuladu; Trade: Dhaman.

Hab : It is a small tree found in Uttar Pradesh, Madhya Pradesh, Bihar, Orissa, Tamilnadu, Andhra Pradesh, Karnataka and Maharashtra.

The fruit is edible and has an agreeable acid flavour.

132. *G. villosa* Willd.; Tiliaceae.

Na : Gujrathi: Padekhado; Kanerese: Battigaragale, Garakela, Sunnudippe; Marathi: Kharmati: Punjabi: Jalidar; Rajasthan: Donkara; Tamil: Kullai; Telugu: Banta, Chenulu.

Hab : It is a shrub found in the Punjab, Rajasthan, Gujarat, Andhra Pradesh and Tamilnadu.

Fruit is a copper coloured drupe and is edible. The seed is also edible.

***G. villosa* Willd.**

***Heritiera fomes* Buch. Ham.**

133. *Heritiera fomes* Buch.-Ham.; (Syn. *H.minor* Roxb.); Sterculiaceae.

Na : Eng.: Sundri; Bengali, Oriya & Hindi: Sundri.

Hab : It is a small or medium sized evergreen mangrove tree with buttressed stem commonly found in the Sundarbans and the deltaic region of Mahanadi.

The leaves are edible. The seeds are starchy and are used as food in times of scarcity.

134. *Hibiscus furcatus* Roxb.; Malvaceae.

Na : Assamese, Eng.: Kiasi; Kanerese: Huligowri, Gumchi; Malayalam: Narnampupuli, Paccapuli; Oriya: Piri-pirika; Telugu: Adavigogu, kobdagogu, Danasonigogu.

Hab : It is a prickly herb found as an under-growth in forest in Assam, Orissa, Andhra Pradesh, Bengal and Karnataka.

Leaves are acidic and are eaten after cooking.

135. *H.tiliaceus* L.; (Syn. *Paritium tiliaceum* St. Hil.); Malvaceae.

Na : Eng.: Cuban bast, Mahoe; Bengali, Hindi: Bola, Chelwa; Mal. & Tamil: Nirparathi, Attuparathi; Marathi: Belapata; Oriya: Baniya; Telugu: Etagogu.

Hab : It is a much branched shrub found in South India and the Sundarbans.

Leaves are eaten.

136. *Hitchenia caulina* Baker; Zingiberaceae.

Na : Eng.: Indian arrow root; Bengali & Hindi: Tikhur; Marathi: Tavakhir.

Hab : It is a small tree found in Konkan and Mahableshwar.

Tubers are rich in starch. They are grated and washed repeatedly before eating. Starch also can be manufactured from them.

137. *Holostemma annulare* (Roxb.) Schum.; (Syn. *H.rheedii* Wall.); Asclepiadaceae.

Na : Guj.: Khamer, Khiravel, Khirdodi; Hindi: Chhirvel, Ranimarvi; Malayalam: Adakodien, Adamodien; Marathi: Dudurli, Shidodi, Tulatule; Sanskrit: Arkapushpi, Jivati: Tamil: Palaykirai; Telugu: Dudipalatige, Palagurugu, Palatura. .

Hab : It is a twining herb found in the tropical Himalayas and the Western Peninsula.

Leaves, flowers and the fruits are used as vegetable.

***H.tiliaceus* L.**

138. *Hygroryza aristata* (Retz.) Nees ex Wight & Arn.; Poaceae (Gramineae.)

Hygroryza aristata (Retz.) Nees *ex* Wight & Arn

Na : Assam: Phutidal, Phulki; Bengali: Uridhan; Guj.: Vanti; Hindi: Janglidal; Kanerese: Jyarahumedhe; Malayalam: Nirvallipulla; Marathi: Deobhata; Tamil: Vallipullu; Sanskrit: Aranyadhanya.

Hab : It forms floating mass in lakes and slow-moving streams. It is also cultivated for fodder, in cold weather in Assam.

Grains were being eaten in times of scarcity.

139. *Ipomoea aquatica* Forsk.; (Syn. *I. reptans* Poir. non; *Convolvulus reptans* L.); Convolvulaceae.

Ipomoea aquatica Forsk.Herb

Na : Bengali: Kalmisak; Eng.: Swamp cabbage; Hindi: Kalmisag, Ganthian, Karmi, Patuasag, Sarnali; Guj.: Nalanibhaji; Marathi: Nadishaka; Tamil: Vel1aikeerai; Telugu: Tutikura.

Hab : It is an aquatic floating herb with long hollow stem rooting at the nodes commonly found in Maharashtra, Bihar, Orissa and South India.

The leaves and young shoots are eaten as vegetable.

140. I.hispida (Vahl.) R & S; (*Syn. l.eriocarpa* R. Br.; *Convolvulus hispidus* Vahl.); *Colvolvulaceae.*

Na : Assam: Kalman; Hindi: Ghiabato, Boota; Malayalam: Pulichevidu; Oriya: Paninoi; Punjabi: Bhanwar; Telugu: Purititige.

Hab : It is a herbaceous twining annual found throughout greater parts of India. It is common on hedges and grasslands.

Leaves and stems are used as vegetable.

141. *1.muricata* (L.) Jacq. (Syn. *Calonyetion muricatum* G. Don); Convolvulaceae.

1.muricata (L.) Jacq.

Na : Hindi: Michai.

Hab : It is an herb found in Uttar Pradesh, Madhya Pradesh, Bengal and Bihar.

The flower pedicels are eaten as vegetable.

142. *Ischaemum rugosum* Salisb.; Poaceae (Gramineae.)

Ischaemum rugosum Salisb

Na : Assam: Joyajha; Bengali: Moraro; Hindi: Dhanka, Marainda, Maror, Murchi, Badaul, Bhador, Murdi, Tori, Jara, Bar, Bardi, Bher, Karkel, Log: Punjabi: Mehat, Munumuna; Tamil: Kudukken pillu.

Hab : It is a very variable species found in wet and dry habitats but it is common in wet places in the plains of India.

Grains were being eaten in times of scarcity.

Jatropha curcas L.

Juglans regia L. Tree

Juglans regia L. Fruit

Lactuca serriola –L. Branch

Lactuca serriola –L. Fruits

Lamium album L.Branch

143. *Jatropha curcas* L.: Euphorbiaceae.

Na : Eng.: Physic nut: Hindi: Safed arund, Bagbherenda, Jamalgutta.

Hab : It is a shrub or a small tree native of tropical America but now it is found growing wild in wastelands and along the fringes of forests in the tropical regions in the country (pl. see medicinal plants).

The plant is raised by seeds as well as by cuttings. Tender shoots are used in curries.

144. *Juglans regia* L.; Juglandaceae.

Na : Eng.: English walnut, Persian walnut; Hindi: Akrot. .

Hab : It is a large tree found in the temperate regions of the Himalayas; in Kashmir, Himachal Pradesh, Uttar Pradesh and Khasia hills.

Kernel (seed) is edible and a table delicacy.

145. *Lactuca serriola* L.; (*Syn. L.scariola* L.); *Asteraceae (Compositae.)*

Na : Eng.; Wild lettuce, Prickly lettuce; Bengali, Hindi: Kahu.

Hab : it is an erect leafy plant with sessile leaves and found in the Western temperate Himalayas.

Seeds are the source of semi drying oil with a pleasant odour, which can be used in food after refining.

146. *Lamium album* L.; Labiatae.

Na : Eng.: White dead nettle.

Hab : It is a perennial herb found in the Punjab, Kashmir and Kumaon.

Tops of the stems and leaves are eaten as vegetable.

147. *Lannea coromandelica* (Houtt) Merr. (Syn. *L.grandis* Dennst.) Engle.; *Odina wodier* Roxb.); Anacardiaceae.

Na : Assam: Jia; Bengali: Jiyal, Jeol, Bhadi; Gujrathi: Mavedi: Hindi: Jhingan, Kaimil, Mohin; Kanerese: Ajasringi, Kuratige, Gojal; Malayalam: Odiyamaram; Marathi: Moi, Shimti; Oriya: Indramai, Moi; Tamil: Wodier, Kalasan; Telugu: Gumpena.

Hab : It is a common tree found throughout tropical forests in the country.

Young leaves are eaten as vegetable by the people living in the vicinity of forests.

148. *Lasium anamalayamum* Bedd.; Meliaceae.

Na : Kanerese: Chigadmari; Malayalam: Vandakamin; Tamil: Sandana virai.

Hab : It is a medium sized tree found in the Western Ghats.

Fruit is of the size of grapes with one or two seeds enclosed in a juicy pulp and is edible.

Lannea coromandelica (Houtt) Merr.

149. *Lasia spinosa* (L.) Thw.; (Syn. L. *hetrophylla* Schott.); Araceae.

Na : Bengali, Hindi: Kanta-Katchu; Telugu: Mulasari, Kantakachoramu.

Hab : It is a prickly perennial herb found in Bengal, Assam and other parts of Sub-tropical Himalayas.

Tender leaves are eaten as vegetable. The fruit though fibrous is also eaten.

Lasia spinosa (L.) Thw.

150. *Leea edgeworthii* Santapau (Syn. *L.aspera* Edge. Non Wall.) Vitaceae.

Na : Hindi: Kumali, Kuram, Kuramchirpali, Kawaokhar; Khasi: Soh-phyrnou; Punjabi: Kumala, Holma.

Hab : It is a shrub found in the North Western Himalayas, Bihar and the Western Peninsula.

Fruits are black in colour and are edible. The leaves and roots are also eaten as vegetable.

L.macrophylla Roxb. ex Hornem

151. *L.macrophylla* Roxb. ex Hornem; Vitaceae.

Na : Bengali, Hindi: Dhola samudra, Hathikana, Hatkan; Lepcha: Dampamtom-Kung; Khasi: Pharun-bame; Marathi: Dinda.

Hab : It is a herb with tuberous roots.

Fruits are black in colour and are eaten. Leaves are eaten as vegetable.

Leptadenia pyrotechnica (Forsk.) Decne.

152. *Leptadenia pyrotechnica* (Forsk.) Decne.; (Syn. *L.spartium* Wight); Asclepiadaceae.

Na : Gujrathi, Hindi, and Punjabi: Kip, Khip; Rajasthani: Khimp.

Hab : It is a much branched, often leafless shrub found in the Punjab, Uttar Pradesh, Rajasthan and Gujarat.

Slimy fruits and young twigs are eaten.

153. *Limonia acidissima* L.Syn.*Feronia limonia* (L.) Swingle, *F.elephantum* Correa ; Rutaceae.

Na : Bengali: Kait, Katbel; Eng.: Elephant apple, wood apple; Guj.: Kavit, Kotha, Kothun; Hindi: Kavitha, Kaweet, Kutbel; Kanerese: Belavu; Malayalam: Vila, Vilatti; Marathi: Kavith, Kovit; Tamil: Vilanga; Telugu: Velaga.

Limonia acidissima L.

Macaranga peltata (Roxb.) Muell,Arg. Branch with Fruits

Hab : It is a deciduous large sized tree found in greater parts of tropical forests in the country. Branches are thorny. Leaves are pinnate. The fruits ripen in November and December. About 4 fruits make one kg. When fruit is ripe, seed is separated from pulp. Seed is mixed with ash and dried. Since viability is short, seed is sown soon after. Germination percent is 70 and period is 10-15 days. One to two year old seedlings are transplanted. It may start fruiting in seventh year when it may give 10-15 fruits. Well-grown trees may give about two quintal of fruits.

Fruit is large globose with hard woody pericarp. Sweetish aromatic pulp is edible. Pulp is also pickled.

154. *Macaranga peltata* (Roxb.) Muell. Arg.; (Syn. *M.raxburghii* Wight); Euphorbiaceae.

Na : Hindi: Chand Kal, Boddi; Kanerese: Chandakanne; Malayalam: Uppila, Vatta; Marathi: Chanda, Chandwar, Oriya: Piania, Gondaguria; Tamil: Vattakanne, Vattathamarai; Telugu: Boddi, Kondajaphara, Kondatamara.

Hab : It is found in South India and Orissa.

Fruits are eaten.

Madhuca indica J.F.Gmel.Tree

155. *Madhuca indica* J.F. Gmel. (Syn. *M.latifolia* (Roxb.) Macb.; *Bassia latifaia* Roxb.); Sapotaceae.

Na : Bengali: Mahwa, Maul, Mahula; Eng.: Mahua tree, Illipe butter; Gujrathi: Mabuda; Hindi: Mahua. Moha; Kanerese: Ippe; Malayalam. Poonam, Ilupa; Oriya: Mahula, Moha; Tamil: Illupei, Elupa; Telugu: Ippa.

Hab : It is a much branched lofty tree found in South and Central India.

Flowers are eaten. Fruits ripen during March-June. Seed oil is used for cooking by local people.

Madhuca indica J.F.Gmel Flowers

156. M. malabarica (Bedd.) Parker. (Syn. Bassia malabarica Bedd.); Sapotaceae.

Hab : It is a small to medium-sized tree found in the Western Ghats from kanara southwards.

Flowers are used as vegetable. Seed oil is used for cooking by local people.

Mangifera indica L. Tree

157. Mangifera indica L.; Anacardiaceae.

Na : Bengali: Am; Eng.: Mango; Guj.: Amri; Hindi: Am, Amra, Kanerese: Mavu; Malayalam: Amran, Cutam, Mavu; Marathi: Amba; Sanskrit: Amra, Chuta; Tamil: Manga, Mao; Telugu: Mamidi.

Hab : It is a large semi-evergreen to evergreen tree mostly cultivated and is also found naturally in the tropical forests along the streams and moister regions.

Fruits are edible. Raw mangoes are pickled. Fruits of naturally grown tree are small and less pulpy. Mango is a delicious fruit and is universally eaten. Grafted plants of good varieties are planted. Fruits ripen from April to June/July. Mango stones are separated from pulp and stored. Seeds are sown in poly bags. Germination percent is about 83. Germination period is about 15 days. When seedlings are of proper size they are grafted and planted with the commencement of monsoon or September/October depending on the availability of moisture in the soil. Nursery raised seedlings are also planted at that time. Spacing adopted may be 5 × 5 m. Grafted seedlings yield fruits in third year and five to ten year old plants may yield 20 to 100 fruits.

Manilkara hexandra **Roxb.Dub.Branch & Fruits**

158. *Manilkara hexandra* Roxb. Dub. (Syn. *Mimusops hexandra* Roxb,); Sapotaceae.

Na : Bengali: Khirkhejur; Guj.: Rayan, Khiri; Hindi: Khirni; Kanerese: Bakula; Mal.: Pala; Marathi: Ranjana, Rayan, Raini; Oriya: Khiri Khirakuli; Tamil: Palla, Palai; Telugu: Pala.

Hab : It is a small sized evergreen tree found naturally in tropical forests in Madhya Pradesh, Gujarat, Maharashtra and Andhra Pradesh.

Ripe fruits are eaten fresh or dried.

Melastoma malabathricum **L**

159. *Medinilla hasseltii* Blume; Malastomaceae.

Hab : It is a woody vine.

Leaves and fruits of this woody vine are edible.

160. *Melastoma malabathricum* L.; Melastomaceae.

Na : Assam: Phutuka, Eng.: Indian Rhododendron; Hindi: Phulki; Kanerese: Ankerki; Lepcha: Tungbram; Oriya: Gongoi, Koroti; Malayalam: Kalampalli; Marathi: Palore; Tamil: Nakkukaruppan; Telugu: Pattuda.

It is an ornamental large shrub in moist and humid places in the country.

Fruits are pulpy and sweet but slightly astringent and are eaten. Leaves are also used as vegetable.

Melochia corchorifolia **L**

161. *Melochia corchorifolia* L.; Sterculiaceae.

Na : Bengali: Tikiokra; Hindi: Bilpat, Tikiokra, Thuiak; Malayalam: Ceruvuram; Oriya: Chyeron, Dasakeroton; Santal: Thuiak; Tamil: Pinnak, Kuppundu; Telugu: Ganugapindikura.

Hab : It is an under shrub found in hotter parts of India, Gujarat and Tamilnadu.

Leaves are eaten as vegetable.

Memecylon edule Roxb.Branch with flowers

162. *Melothria heterophylla* Cogn.; Cucurbitaceae.

Na : Bengali: Klldari; Bombay: Gometta, Gometi; Hindi: Tarali; Khasi: Soh-Khia-phiang; Malayalam: Nierinjenpuli; Oriya: Karakia, Malka, Makirla; Punjab: Bankakra; Tamil: Pulivanji; Telugu: Thiyyadonda.

Hab : It is a prostrate or climbing herb with tuberous roots found through out India ascending upto 210 m in the hills.

Leaves, fruits and tuberous roots are edible.

163. *Memecylon edule* Roxb. (Syn. *M.tinctorium Koenig; M.umbellatum* Burn. f.); Melastomaceae.

Na : Assam: Lalidimabophang, Theihadam; Eng.: Iron wood tree, Hindi: Anjan; Kanerese: Archeti, Harchari, Lakhonde; Malayalam: Kashavu, Kanalei, Kannavu; Marathi: Anjani, Kurpa, Limba; Tamil: Alli; Anjani, Kaya; Telugu: Alli, Uddalalli.

Hab : It is a small ornamental tree found in Orissa, Assam and the Deccan Peninsula and the Andaman Islands.

Fruits are purplish or black, pulpy, astringent and edible.

164. *Mesua ferrea* L.; Guttiferae.

Na : Eng.: Ironwood; Hindi: Nagkesar, Nagchampa, Nageshwar champa.

Hab : Pl. see essential oils.

Fruits are sometimes eaten.

165. *Meyna laxiflora* Robyns (Syn. *Vangueria spinosa* Roxb. properte); Rubiaceae.

Na : Assam: Kalkormon; Bengali & Hindi: Pundrika, Muyuna, Maniphal; Guj.: Alu, Athu; Kanerese: Mullakare, Gundkare, Gobergally; Marathi: Alu, Huloo, Halawni; Oriya; Gurbeli, Mottakanta; Tamil: Manakkareu; Telugu: Cegagadda, Vellika, Visikilamu.

Hab : It is a large shrub or small tree found in Western Uttar Pradesh, Bengal, Assam and South India.

Fruit is green to brown succulent and eaten.

Mesua ferrea L. Branch with Flowers

166. *Miliusa velutina* (Dunal) Hk.f. & Thoms.; Annonaceae.

Na : Bihar: Ome; Garo: Borsamphol; Hindi: Domsal, Kari, Gidarrukh; Malayalam: Kanakaitha, Viluni; Oriya: Gandhapalas; Telugu: Peddachilkaduduga, Nalladuduga.

Ha : It is a deciduous tree found throughout the tropical forests in India.

Fruits are eaten.

167. *Mimusops elengi* L.; Sapotaceae.

Mimusops elengi L.

Na : Assam: Gokul; Bengal: Bakul; Gujrathi: Barsoli, Bolsari; Hindi: Maulsari; Kanerese: Bakula, Pagade; Malayalam: Elengi, Ilanni; Marathi: Ovalli; Oriya: Bokulobaula; Tamil: Vagulam, Magodom, Ilanji; Telugu: Pogada.

Hab : It is a small to medium sized evergreen tree found in South India and Andaman Islands.

The ripe fruit is edible.

168. *Mirabilis jalapa* L.; Nyctaginaceae;

Mirabilis jalapa L. Plant

Na : Eng.: Four O'clock plant; Hindi: Gulabbas.;Kanarese:Gulmasi

Hab : Pl. see essential oils.

Leaves are edible and are used as vegetable.

169. *Momordica balsamina* L.; Cucurbitaceae.

Momordica balsamina L.
Plant with fruits

Na : Eng.: Balsam apple; Hindi: Mokha.

Hab : It is an herbaceous vine found in the Punjab, Gujarat, Uttar Pradesh and Andhra Pradesh.

Tender fruits are eaten as vegetable.

170. M.dioica Roxb. ex Willd.; Cucurbitaceae;

M.dioica Roxb. ex Willd.

Na : Assam: Bhalkarela; Bengal: Bankarela; Eng.: Small bitter gourd; Hindi: Jangli karela, Murela, koksa; Kanerese: Karlekai, Adavihagalu; Marathi: Kartoli; Punjab: Kakawna, Kirara, Dharkarela; Tamil: Tholoopavai, Palupakai; Telugu: Agakara, Adavikakara.

Hab : It is a common climbing or spreading shrub found in the tropical forests in the country upto an altitude of 1500 m.

Fruits are used as vegetable.

171. *M.tuberosa* Cogn. (Syn. *M.cymbalaria* Fenzl ex Naud.*; Luffa tuberosa* Roxb.; Cucurbitaceae;

Na : Hindi: Kadavanchi; Malayalam: Kattupaval; Marathi: Kadavanchi; Tamil: Athalaikai.

Hab : It is a trailing herb found mainly in Maharashtra and Tamilnadu.

Fruits are edible.

172. *Morchella esculanta* L.; Fungi.

Morchella esculanta L.
Mushroom

Hab : It is found in damp temperate regions in the forests. It is a table delicacy and is in great demand. (pl. see mushrooms).

***Moringa oleifera* Lamk. Plant with fruits**

***Morus alba* L. with catkins**

***M.australis* Poir.**

***Mucuna gigantea* DC. Branch with fruits**

173. *Morinda pubescens*; Sm. Syn.M. tinctoria Roxb.; Rubiaceae.

Na : Eng.: Tagar wood of Madras; Hindi: Al.

Hab : It is a small to medium sized tree found in tropical regions in Bihar, Central and South India.

Green fruits are eaten as vegetable.

174. *Moringa oleifera* Lamk. Syn. *M.pteriogosperma* Gaertn.; *Guilandina moringa* L.; Moringaceae.

Na : Eng.: Drumstick tree, Horse radish tree; Hindi: Sahinjna, Soanjna.

Hab : Pl. see medicinal plants.

Green pods are used as vegetable. They are cut into slices and used in culinary preparations.

175. *Morus alba* L.; Moraceae.

Na : Bengal: Tut; Eng.: White mulberry; Gujrathi: Shetur; Hindi: Tut, Tul; Kanerese: Hiprerle; Kashmir: Tut; Marathi: Tut, Ambat; Oriya: Tuto, Tuticoli; Tamil: Musukette; Kambli chedi; Telugu: Tutturi chettu; Trade: Mulberry.

Hab : It is a small tree native of China. It is found as an escape in the forest. Its wood is valuable which is used in sports goods industry. Plants are raised by branch cuttings.

Fruits (catkins) are edible and are also used for making syrup and jam.

176. *M.australis* Poir. Syn. *M.acidosa* Griff.; *M. Indica* auct. non L.; Moraceae.

Na : Eng.: Wild mulberry; Hindi: Tut.

Hab : It is a shrub or small tree commonly found in the sub-Himalayan tracts in Assam and Khasi hills. Fruits (catkins) are edible and are also used for making syrup and jam.

177. *M.serrata* Roxb.; Moraceae.

Na : Hindi: Kimu, Khasi: Dieng Soh-tungkhar; Punjab: Karum, Kimu, Kartut.

Hab : It is found in the Western temperate Himalayas at altitudes from 1500 to 2700 m.

Fruits (catkins) are long, purple, sweet and edible.

178. *Mucuna gigantea* DC.; Fabaceae (Papilionaceae.)

Na : Eng.: Elephant cowitch; Kanerese: Turibilangi; Malayalam: Kakavalli; Tamil: Kalgaivalli; Telugu: Enuga dulagendi.

Hab : It is commonly found in the Sundarbans, the Andaman Islands and Malabar.

Seeds are said to be eaten by local people.

179. *Mutingia calabura* L.; Elaeocarpaceae.

Na : Eng.: Capulin, Chinese or Japanese cherry.

Hab : It is a native of South America cultivated as an ornamental plant.

Fruits are edible.

***Mutingia calabura* L.Branch with fruits**

180. *Myrica nagi* Thunb.; (Syn. *M.rubra* Sieb. & Zucc.); Myricaceae

Na : Assam; Nag-tenga; Bengali: Kaiphal, Satsarila; Eng.: Bayberry, Box myrtle; Gujrathi: Kariphal; Hindi: Kaiphal; Kanerese: Krishivani; Khasi: Sohphi; Kumaon: Kaphal; Punjab: Kaiphal, Kahela, Kahi; Marathi: Kayaphala; Tamil: Marudam; Telugu: Kaidaryamu.

Hab : It is a tree native of China and Japan but now it is found in the subtropical Himalayas upto 2100 m.

Fruits have a pleasant sour-sweet taste and are edible.

***Myrica nagi* Thunb.**

181. *Myrsine capitellata* Wall.; Myrsinaceae.

Na : Assam: Kachidria; Khasi; Dieng-ching.

Hab : It is a small tree found in the Khasia hills and other parts of Eastern Himalayas, in North Kanara and the Nilgiris.

Fruits are edible.

***Myrsine capitellata* Wall.**

182. *Nannaorrhops ritchiana* H. Wendl.; (Syn. *Chameerops ritchiena* Griff.); Palmaceae.

Na : Eng.: Mazari palm.

Hab : It is a small palm with large fan shaped leaves arising in tufts from underground stem found mainly in the Punjab and Haryana.

The inflorescens and the fruits are eaten as vegetables.

***Nannorrhops ritchiana* H. Wendl.**

183. *Nelumbo nucifera* Gaertn. (Syn. *Nymphaea nelumbo* L.; *Nelumbium speciosum* Willd.); Nymphaeaceae.

Na : Assam: Podum; Bengal: Padma; Eng.: East Indian Lotus; Gujrathi: Suryakamal; Hindi: Kamal, Kanwal; Kanerese: Kamala, Tavara, Tamara; Kashmir: Pamposh; Malayalam: Tamara Marathi: Kamal; Oriya: Padam; Punjab: Kanwal, Pamposh; Sanskrit: Ambuja, Padma, Pankaja, Kamala; Tamil: Ambal, Tamarai; Telugu: Tamaramu, Kamalamu.

Rhizomes and seeds of this aquatic herb are eaten as vegetable. The rhizomes are also pickled.

***Nelumbo nucifera* Gaertn.**

Nephelium mutabile **Blume. Branch with fruits**

184. *Nephelium mutabile* Blume; Sapindaceae.

Na : Eng.: Pulasan.

Fruits are edible.

185. *Neptunia oleracea* Lour. (Syn. *N.prostrata* Baill.); Mimosaceae.

Na : Bengali: Pani-najak; Bombay: Panilajak; Hindi: Lajalu; Malayalam: Nittetoddavaddi; Punjab: Lajalu, Panilajak; Tamil: Sadai, Sundaikkirai; Telugu: Neerutalavapu, Nidrayam.

Hab : It is a common floating aquatic herb and also found in rice fields, marshes etc.

This herb is used as vegetable.

Neptunia oleracea **Lour. Herb**

186. *Nipa fruiticans* Thunb.; Palmaceae.

Na : Bengali: Gulga, Gabna, Golphal; Eng.: Water coconut, Nipa palm; Hindi: Golphal, Gulga; Gujrathi: Pardeshi tadio.

Hab : It is a prostrate gregarious palm found in the Sundarbans and the Andaman Islands. It is also found in Saurashtra.

Inflorescens of this palm is the source of sugar and jaggery.

Nipa fruiticans **Thunb. Plant**

187. *Nymphaea alba* L.; Nymphaeaceae.

Na : Eng.: European white water lily; Hindi: Pandharenkamal: Kashmir: Brimposh, Nilofar, Kamud.

Hab : It is an aquatic herb found in Kashmir at altitude of 1800 m.

Root stocks of this aquatic herb are boiled before they are eaten and seeds are parched and eaten.

188. *N.esculenta* Roxb.; Nymphaeaceae.

Na : Eng.: Indian white water lily; Hindi: Chota sundi.

Hab : It is a floating aquatic herb.

Tubers and seeds of this aquatic herb are eaten.

Nymphaea alba **L. Aquatic Plant**

189. *N.nouchali* Burm. f. (Syn. *N.Pubescens* Willd.; *N.lotus* Hook.f. & Thoms.; *N.lotus* var. pubescens Hook.f. & Thoms.); Nymphaeaceae.

Na : Assam: Mokuva, Nal; Bengal: Shaluk, Rakta; Guj.: Kambal; Nilophal; Hindi: Kamal Kakri, Koka, Neelophal; Kanerese: Nyadale huvu; Malayalam: Periambal, Neerambal; Marathi: Lalkamal, Raktakamal; Oriya: Dhabalakain, Rangkain; Punjab: Chotakanval; Tamil: Allitamarai, Vellambal; Telugu: Tellakalava.

Hab : It is an aquatic ornamental herb commonly found along the fringes of lakes and ponds.

Rhizomes and flowers are eaten.

N.nouchali **Burm. f.**

190. *Opuntia dillenii* Haw.; Cactaceae.

Na : Eng.: Prickly pear; Bengali & Hindi: Nagphana; Gujrathi: Chorhatalo; Kanerese: Papaskalli; Mal.: Palakakkalli; Marathi: Chapal; Oriya: Nagphenia; Punjab: Chittartohar; Tamil: Nagathali; Telugu: Nagajamudu.

Hab : It is a thorny bush native of South America but introduced into India as a hedge plant.

Fruits are eaten.

Opuntia dillenii **Haw.**

191. Orchis latifolia L.; Orchidaceae.

Na : Eng. Orchis; Hindi: Salap.

Hab : It is a perennial herb found in damp places in Western Himalayas from 2500 to 5000 m attitude.

Tuberous roots are said to yield salep of commerce. Salep consists of washed scaled and dried tubers and they yield lot of mucilege with water and forms jelly which is nutritious.

Orchis latifolia **L.Plant**

192. *Oroxylum indicum* Vent.; (Syn. Calosanthes indica Blume); Bignoniaceae.

Na : Assam: Toguna, Bhatghila, Dingari; Bengali: Sona, Nasona; Guj.: Aralu; Hindi: sonapatha, Sora, Arlu; Kanerese: Tigdu; Mal.: Palagapaiyani; Marathi: Tetu; Oriya; Papni, Phonphonia; Punjab: Mulin, Tatmorang; Tamil: Achi, Peiarlanthei; Telugu: Dundilum, Pampini.

Hab : It is a small deciduous tree found practically throughout tropical and sub-tropical forests upto an altitude of 1200 m.

Tender and unripe fruits are eaten as vegetable.

Oroxylum indicum **Vent.**

193. *Osmanthus fragrans* Lour.; (Syn. *Olea fragrans* Thunb.); Oleaceae.

Na : N. Bengali & Lepcha; Tungrung; U.P. Hills; Silang

Hab : Pl. see spices.

Fruits are edible and the leaves are used to perfume tea.

Osmanthus fragrans **Lour.**

194. *Oxalis pes-caprae* L.; (Syn. *O.cernua* Thunb.); Oxalidaceae.

Na : Eng.: Bermuda Butter-cup; Hindi: Khatmitthi.

Hab : It is an herb native of South Africa but has spread wild along the moister places in wastelands.

Leaves are eaten as vegetable.

Oxalis pes-caprae **L.**

Parkia roxburghii G. Don

195. *Parkia roxburghii* G. Don; Mimosaceae.

Na : Assam: Aoelgap, Khorial, Unkampinching, Yongchak, Zongto; Hindi: Supota, Khorial.

Hab : It is a medium sized to tall handsome tree found in Assam.

Tender pods are eaten. Seeds though slightly bitter are eaten after roasting.

Pereskia aculeata Mill.

196. *Pedalium murex* L.; Pedaliaceae.

Na : Bengali: Baragokhru; Gujrathi: Mottogokharu, Kadvagokhru; Hindi: Bada gokhru, Kadvagokhru; Kanerese: Annegalugida; Malayalam: Kakamulla, Ananerinnil; Marathi: Hallicharate, Mothegokhru; Oriya: Gokara, Gokshura; Tamil: Anainerinji, Perunerunji; Telugu: Enugupalleru.

Hab : It is a succulent herb found in Gujrat, Rajasthan, Konkan and Andhra Pradesh.

The leaves are reported to be eaten as vegetable.

Phoenix acaulis Buch.-Ham. Ex Roxb.-Plant

197. Pereskia aculeata Mill.; Cactaceae.

Na : Eng.: West Indian gooseberry, Lemon-vine.

Hab : It is native of West Indies and in India it is stated to have run wild.

It is a woody vine with edible fruits.

198. *Phoenix acaulis* Buch.-Ham. ex Roxb.; Palmaceae.

Na : Hindi: Thakal, Khajeria; Marathi: Shevra, Shilind, Boichind; Oriya: Kojiri; Telugu: Yita.

Hab : It is a stemless palm found in Uttar Pradesh, Bihar and North Bengal, Orissa and parts of South India up to an altitude of 1500 m.

Fruits and pith of the stem are eaten.

P.pusilla Gaertn.

199. *P. pusilla* Gaertn. (Syn. *P.farinifera* Roxb.); Palmaceae.

Na : Eng.: Wild date; Hindi: Pala wat; Kanerese: Hallichala, Sannaichalu; Malayalam: Eentha, Chillintal; Telugu: Chiruta-ita, chitti-ita.

Hab : It is a bushy palm found along the Coromandel Coast. Pulp of fruits of this bushy palm is eaten.

The trunk contains sago.

Pheonix sylvestris (L) Roxb.-Tree

200. *P.sylvestris* (L.) Roxb.; (Syn. *Elate sylvestris* L.); Palmaceae.

Na : Bengal: Khajur. Kejur; Eng.: Wild date; Gujrathi: Kharak, Khakri; Hindi: Khajur, Sendhi, Kanerese: Ichalu: Marathi: Shindi, Boichand, Sendri; Oriya: Khajuri, Kojari; Telugu: Pedda Ita.

Hab : It is a tall palm found along the riverbanks in Karnataka, Andhra Pradesh, Maharashtra and Madhya Pradesh and also in other states upto 1500 m altitude.

Pulp of the fruits is eaten. The trunk contains sago, which is eaten during scarcity periods.

Phyllanthus amblica L.

201. *Phyllanthus emblica* L.; Syn. *Emblica officinalis* Gaertn ; Euphorbiaceae.

Na : Bengali: Amla, Amlaki; Eng.: Emblic myrobalan, Indian goosberry; Guj.: Amali, Ambala; Hindi: Amla, Amlika, Aonla; Kanerese, Tamil, Malayalam: Nelli; Sanskrit: Adiphala, Dhalri, Amaloko; Telugu: Usirikai..

Hab : It is a common medium sized tree found throughout the tropical forests in India.

Fruits ripen from October to February. Fruits are used in vegetables and curries and are pickled and made into jams.

Physalis minima L.

202. *Physalis minima* L.; Solanaceae.

Na : Bengali: Bantiparia; Eng.: Sunberry; Guj.: Purpoti, Popti; Hindi: Tulatipati, Papotan; Kanerese: Gadde hannu; Mal.: Ajodinjotta; Marathi: Dhanmori, Chirboti; Tamil: Thollakkali; Telugu: Kupanti, Buddabudama.

Hab : It is a pubescent herbaceous annual distributed nearly throughout India ascending upto 2300 m.

Fruits and leaves are reported to be edible.

Pinanga dicksonii Bl. Plant

203. *Pinanga dicksonii* Bl.; (Syn. *Areca dicksonii* Roxb.); Palmaceae.

Na : Kanerese: Kaduadike, Jandarige; Mal.: Kanakamuka; Telugu: Kondapoka.

Hab : It is a slender graceful palm found in North Kanara in Karnatak, in the Nilgiris in Tamilnadu and Kerala states.

Fruits of this palm are used as substitute for betel nut.

204. *P.hookeriana* Becc.; Palmaceae.

Hab : This palm is found in the Khasia hills.

Fruits of this palm are eaten.

Pinus gerardiana Wall.- Cone

205. *Pinus gerardiana* Wall. Pinaceae.

Na : Eng.: Neosia pine, chilgoza pine, Edible pine; Hindi: Chilgozah, Neoza; N.W. Himalayas; Cheri, Gumober, Prita.

Hab : It is a small to moderate sized tree found in North Western temperate to sub-alpine Himalayas. In patches it grows gregariously. Natural regeneration of the chilgoza pine suffers from extensive and ruthless collection of cones for procuring the edible seeds and also indiscriminate grazing and browsing. Chilgoza seeds are greatly valued as a dessert.

Pinus gerardiana Wall.- Seed

***Pistacia vera* L. Branch**

206. *Pistacia vera* L.; Anacardiaceae;

Na : Eng.: Pistachio, Green almond; Hindi: Pista.

Hab : It is a tree of West Asia, Afghanistan and Iran and is grown in Northern India. This could be raised in the forests for its edible kernels. Pista kernels have a delicious flavour and are used as ingredients in sweet meats, confectionery and ice-creams.

207. *Polyalthia cerasoides* Benth. & Hook. f.; Annonaceae.

Na : Gujrathi: Uma; Hindi: Kudumi; Kanerese: Nettalingamara, Sannabessare; Mal.: Narelai; Marathi: Hoom, Uma; Oriya: Potmossu; Santal: Panjon; Tamil: Nakulsi, Mullili; Telugu: Gutti.

Hab : It is a small tree found in Bihar, Orissa and the Western ghats.

Fruits are said to be eaten.

***Polyalthia cerasoides* Benth. & Hook. f.**

208. *Premna latifolia* Roxb.; Verbenaceae.

Na : Hindi: Bakar; Malayalam: Knappa; Oriya: Gondhona; Santal: Dandra sea; Tamil: Pachumullai; Telugu: Pedda nellikoora.

Hab : It is a small tree commonly found in the deciduous coastal forests in South India, Bengal, Orissa and Bihar.

Leaves are eaten in curries.

209. *Protium serratum* (Wall. ex Colebr.) Engl. (Syn. Bursera serrato Wall. ex Colebr.); Burseraceae.

Na : Assam: Mirtenga; Bengal: Gulgotya; Garo: Thikring; Hindi: Mirtenga; Khasi: Dieng-Soh-mir; Oriya: Nimburu, Nimburamoi, Sorupotrimoi; Telugu: Chitreka; Trade: Murtenga.

Hab : It is an evergreen tree found in Assam, Andhra Pradesh, Bihar and Orissa.

Fruits are pleasantly acidic and are eaten.

210. *Prunus cerasoides* D. Don. (Syn. *P.puddum* Roxb. ex wall.); Rosaceae.

***Prunus cerasoides* D. Don.**

Na : Assam: Dieng-Soh-iong-krem; Bengali: Padmak; Eng.: Wild Himalayan cherry; Guj.: Padmakathi; Hindi: Paddam, Phaya; Marathi: Padmakashta, Padmaka; Punjab: Paja.

Hab : It is a middle sized to large tree occurring in temperate Himalayas.

Fruits have scanty pulp and are eaten.

211. *P.jenkinsii* Hook. f. (Syn. *Cerasus jenkinsii* Hook. f. & Thoms.); Rosaceae.

Hab : It is found in Assam.

Fruits are eaten.

Pterocarupus marsupium Roxb.- Branch

212. *Pseudostreblus indica* Bur.; Moraceae.

Hab : It is a tree of the Khasia hills.

Fruits are eaten.

213. *Pterocarpus marsupium* Roxb.; Fabaceae (Papilionaceae.)

Na : Eng.: Indian Kino tree, Malabar Kino; Hindi: Bijasal, Pitsara.

Hab : Pl. see medicinal plants.

Flowers are eaten.

214. *Pueraria tuberosa* DC.; Fabaceae (Papilionaceae.)

Na : Bengali: Shimiabatraji; Gujrathi: Vidarikand, Phagvelo, Khakervel; HIndi: Siali, Bidari Kand, Sural, Bilaikand, Bharda, Tirra, Bankumra; Kanerese: Gumadigida; Marathi: Ghorbel; Telugu: Darigummadi.

Hab : It is a large perennial climber found in the Western Himalayas, Kumaon, Orissa, Punjab and Rajasthan.

Tubers and 1eaves are used as vegetable.

215. *Punica granatum* L.; Punicaceae.

Na : Bengal: Dalim; Eng: Pomegranate; Guj.: Dadam; Hindi: Anar; Kanerese: Dalimba; Malayalam: Matalam; Marathi: Dalimba; Tamil: Madulai; Telugu: Danimma.

Hab : It is a shrub or small tree mostly cultivated but grows wild in Kulu valley and other parts in the Himalayas at altitudes between 900 to 1800 m.

Seeds of the wild variety are sower and are eaten in curries.

216. *Pyrus vestita* Wall.; Rosaceae.

Na : Hindi: Mauli.

Hab : It is found in the Khasia hills and other parts of Eastern Himalayas.

Fruits are eaten.

217. *Randia uliginosa* DC. (Syn. *Gardenia uliginosa* Retz.; *Posoqueria uligillosa* Roxb.); Rubiaceae.

Na : Assam: Bon-bongana; Bengal: Piralo; Gujrathi: Gangeda, Gengdi; Hindi: Pindalu, Panar, Katul; Kanerese: Kare; Malayalam: Panikara; Marathi: Pendari, Telphetru, Phetra; Oriya: Pendra; Tamil: Wagatta. Perunkarai; Telugu: Nallakakisha, Nallaika, Goanka.

Punica granatum L

Hab : It is found in Central and South India.

Fruits are eaten boiled or roasted either alone or in curries.

Rheum nobile **Hook.f & Thoms.- Branch**

218. *Reptonia buxifolia* A. Dc.; (Syn. *Monotheca muscatensis* A. Dc.: *Edgeworthia buxifolia* Falc.); Sapotaceae,

Na : Hindi: Gurgura.

Hab : It is a small tree found in the Punjab.

Fruits are eaten.

219. *Rheum nobile* Hook. f. & Thoms.; Polygonaceae.

Na : Sikkim: Tchuka.

Hab : It is a pretty herb distributed in the Himalayas from Nepal to Bhutan at altitudes of 3900-4800 m.

Stems are pleasantly acidic in taste and are excellent for salads

Rhododendron arboreum **Smith**

220. *Rhododendron arboreum* Smith; Ericaceae.

Na : Bhutan and Lepcha: Elok; Bengali: Baras; Hindi: Burans; Kashmir: Kamri, Chhan, Chiu; Khasi; Tin-saw, Diengtin-thuin; Kumaon: Brus, Burans; Punjab: Andawal, Mandal, Chiu, Aru, Broa, Chacheon; Nepal: Bhorans, Gurans, Ghonas, Taggu, Dotial.

Hab : It is a small tree found in the temperate Himalayas from Kashmir to Bhutan; Khasia hills, in the Nilgiris and the Pulneys.

The large, beautiful scarlet flowers are used for preparing cold drinks and jams.

Rhodomyrtus tomentosa **Wight**

221. *Rhodomyrtus tomentosa* Wight: (Syn. *R.parviflora* Alston); Myrtaceae.

Na : Eng.: Hill guava, Hill gooseberry; Kanerese: Guddadenelli, Tavuti; Tamil & Malayalam: Thaontay, Koratta, Thavattu.

Hab : It is a handsome evergreen tomentose shrub found in the Nilgiris and the Pulney hills.

Fruits are aromatic, sweet and edible. Wild fruits are normally of poor quality.

Rhus chinensis **Mill.- Branch**

222. *Rhus chinensis* Mill; (Syn. R. *semialata* Murr. *R.javanica* Linn.); Anacardiaceae.

Na : Assam: Naga-tenga; Cachar; Gimbao; Garhwal: Darmila, Dharmil; Gao: Khilma; Hindi: Tatri; Jaunsar: Tibri, Arkhoi; Lepcha: Takhrit, Thanghaerkung; Punjab: Arkhar, Chechar, Dudla, Tetri, Thissa, Hulug, Rashlu, Wansh; Nepal: Bhaimlo, Bhagmili. .

Hab : It is a deciduous tree upto 11 m in height and is found in the temperate Himalayas.

The fruits have a sharp acidic taste and are eaten.

223. *R.mysorensis* Beyne ex Wight & Am.: Anacardiaceae,

Na : Hindi: Dansara, Dasni; Kanerese: Sabale; Rajasthan: Dasarni, Dasani, Davan; Marathi: Amboni; Tamil: Chippamaram, Sappala; Telugu: Sitha.

Hab : It is an aromatic, much branched thorny shrub or small tree found in the Punjab, Rajasthan and Andhra Pradesh.

Fruits are sour-bitter in taste and are eaten either cooked or roasted. Leaves are also sometimes eaten.

Ribes grossularia **L.- Branch & Fruits**

224. *R.punjabensis* J.L. Slew. ex Brand.: Anacardiaceae.

Na : Hindi: Tatri; Punjab: Dor, Kangar, Palai; Garhwal: Dasmila, Amlara.

Hab : It is an aromatic tree upto 15 m in height found in North Western Himalayas from Kashmir to Kumaon.

Fruit is acidic and eaten. It is also used in the preparation of sherbets.

R.nigrum **L.**

225. *Ribes glaciale* Wall.; Saxifragaceae.

Na : Hindi: Karu dhak; Punjab: Kukuliya, Manglia, Durbai; Nepal: Kembu.

Hab : It is a small shrub found in the Himalayas from Kashmir to Bhutan.

Berries are very sour and are sometimes eaten.

226. *R.grossularia* L.; Saxifragaceae.

Na : Bhotia: Chapehu, Lipchi; Eng.: European gooseberry; Hindi: Amlanch, Baikunti; Punjab: Amlanch, Kansi, Teila.

Hab : It is a small erect prickly shrub found in the Western Himalayas from Kashmir to Kuamon.

Fruits are eaten but not much relished.

Ribes rubrum **L.- Branch**

227. *R.nigrum* L.; Saxifragaceae.

Na : Eng.: European blackcurrant, Hindi: Nabar; Kumaon: Papear; Punjab: Muradh, Nabar, Beli, Shaktekas.

Hab : It is an erect shrub found in Kashmir.

Berries are black with a strong aromatic smell and taste. They are too acidic to be eaten raw. They are used for making jelly, jam or sauce.

228. *R.rubrum* L.; (Syn. *R.sylvestre* Syme); Saxifragaceae.

Na : Eng.: Red currant; Garhwal: Kinkolia; Hindi: Dak; Punjab: Dak, Phulanch.

Hab : It is a shrub found in the Western Himalayas from Kashmir to Kumaon.

Fruits are acidic and are used for making tarts, jellies and jams.

Ribes rubrum **L.- Fruits**

Rubus ellipticus **Smith Branch with fruits**

229. *Rubus ellipticus* Smith; Rosaceae.

Na : Assam: Jotelupoka; Eng.: Himalayan Yellow raspberry; Hindi: Anchu, Lalanchu, Hisalu; Kashmir: Gauriphal; Nepal: Tolu, Aselu; Punjab: Akhi.

Hab : It is a shrub found in the Western and Eastern Himalayas and South India.

Fruits are of very good flavour and taste and are eaten.

R.lanatus **Wall.**

230. *R.lanatus* Wall.; Rosaceae.

Na : Hindi: Hisalu: Kumaon: Hisalu.

Hab : It is found from Kumaon to Nepal in the temperate Himalayas.

Fruits are edible but are of inferior quality.

231. *R.lasiocarpus* Sm.; (Syn. *R.niveus* Thunb.); Rosaceae.

Na : Eng.: Ceylon raspberry, Mysore raspberry, Mahabaleshwar raspbeny; Hindi: Kala hisalu, Kalaanchu; Kashmir: Kandiari; Marathi: Gowriphal: Nepal: Kaloaselu; Punjab: Gunacha.

Hab : It is a large prickly shrub found in the Western Himalayas, Sikkim, the Khasia hills and the Western Ghats.

Fruit is black when ripe and has good taste. It is also used for making jam.

R.lasiocarpus **Sm.**

232. *R.moluccans* L.; (Syn. R.alcaefolius Poir.); Rosaceae.

Na : Eng.: Black cherry; Hindi: Katsol; Kumaon: Kalsoi; Lepcha: Safokji; Nepal: Bipenkanta.

Hab : It is found in Assam, the Nilgiris and Kerala.

Fruits are edible.

233. *Rumex acetosa* L.; Polygonaceae.

Na : Bengali: Chukapalam; Eng.: Garden sorrel; Hindi: Chuka, Khattapalak.

Hab : It is a perennial herb commonly found in the Western Himalayas from Kashmir to Kumaon in the temperate region

Leaves and tender twigs are eaten in salad or cooked like spinach.

Rumex acetosa **L.**

234. Sacciolepis interrupta (Willd.) Stapf; Poaceae (Gramineae.)

Na : Bengal: Nardula; Bombay: Pakalia; Kanerese: Modikehullu; Tamil: Tandanpillu; Telugu: Wolam.

Hab : This grass is usually found in moist places in tropical regions in Assam, Bengal, Orissa, Chota-Nagpur, Tamilnadu and Maharashtra.

The grain is eaten in times of scarcity.

235. *Sagitaria sagitifolia* L.; Alismaceae.

***Sagitaria sagitifolia* L.**

Na : Bengali: Chhotokut, Muyamuya; Eng.: Old-world arrowhead; Hindi: Chota kut.

Hab : This herb is found practically throughout India, in tanks and ponds.

The tubers are eaten as vegetable.

236. *Salsola kali* L.; Chenopodiaceae.

***Salsola kali* – l.-Twig**

Na : Eng.: Glass wort, Russian thistle; Hindi & Punjabi: Sajjibuti.

Hab : It is a rigid branched prickly herb or bush found in Kashmir and Himachal Pradesh in temperate regions.

Tender shoots and leaves are used as vegetable.

237. *Salvadora oleoides* Dcne.; Salvodoraceae.

Na : Bombay: Khakan, Kankhina: Gujrathi: Khakan, Mitijal, Mitijar, Pilava, Pilu; Hindi: Badapilu, Jhal, Pilu; Marathi: Diar, Godpilu, Khabbar kinkanela, Pilu: Rajasthan: Khabar, Pilu; Tamil: Kalawa, Karkol, Kohu, Ughai.

Hab : It is a shrub or small tree found in the Punjab, Rjasthan and Uttar Pradesh.

Fruits are sweet and edible.

238. S.persica L.; Salvadoraceae.

***S.persica* L.**

Na : Bengal: Jhal; Eng.: Mustard tree; Guj.: Kharijal, Kharijar, Pilu, Motijalya, Piludi; Hindi: Chota pilu, Jhal; Kanerese: Gonimara; Oriya: Kotungo, Toboto, Pilu; Marathi: Khakin, Kickni, Miraj, Mirajoli, Pelu, Pilwa, Rhakhan, Thorapilu; Tamil: Kalawa, Karkol, Perungoli, Ughaiputtai, Vivay; Telugu: Ghunia, Varagogu.

Hab : It is a much-branched evergreen shrub or small tree found in the Punjab, Rajasthan, Gujarat and Western ghats.

The leaves are eaten as vegetable and are also used for making sauce. Fruits are sweet and edible.

239. *Schleichera oleosa* (Lour.) Oken (Syn. *S.trijuga* Willd.; *Pistacia* oleosa Lour.); Sapindaceae.

***Schleichera oleosa* (Lour.) Oken-Fruits**

Na : Eng: Lac tree, Honey tree; Hindi: Kusum

Hab : and other details pl. see medicinal plants.

Raw fruits are pickled; the ripe ones possess whitish pulp and pleasant acidic taste and are often eaten. Seeds are also eaten raw or roasted. Young and tender shoots are eaten by the tribal people.

Semecarpus anacardium **L.f. (fruiting branch and leaf)**

Setaria glauca **(L.) Beauv.**

Setaria glauca **(L.) Beauv. Flowering stock**

Shorea robusta **Gaertn f.**

240. *Securinega leucopyrus* (Willd.) Muel.-Arg.; (Syn. *Phyllanthus leucopyrus* Koen. ex Roxb.; *Flueggea leucopyrus* (Koen.) Willd.); Euphorbiaceae.

Na : Guj.: Shinar; Humri; Hindi: Hartho, Ainta; Goa & Konkan: Parpo; Kanerese: Bilchuli, Gudahale, Hooli, Sooli; Marathi: Vorepuvan, Kandekuvana, Pamdharphali, Posheri; Nepal: Achal; Punjab: Karkun, Bhatbi, Rithei, Vanuthi, Girk: Rajastan: Salepan, Holepan: Tamil: Vellaippulanji, Maappulanathi, Kareioori; Telugu: Tellapulugudu, Tellapulisara, Challamunta.

Hab : It is a large shrub or small thorny woody small tree found in Uttar Pradesh, the Punjab, Maharashtra and Tamilnadu.

Fruits are sweet and edible. Leaves are also eaten.

241. *Semecarpus anacardium* L. f. (Syn. *Anacardium orientale* L.); Anacardiaceae.

Na : Assam: Bhala, Bholaguli; Bengal: Bhela, Bhelatuki; Eng.: Marking nut; Gujrathi: Bhilamu; Hindi: Bhela, Bhilawa; Kanerese: Karekeru; Malayalam: Chera; Nepal: Bhalai; Punjab: Bhela, Bhilawa; Marathi: Bibha, Bhilava; Tamil: Shenkollei, Erimugi; Telugu: Jidi.

Hab : It is a moderate sized deciduous tree found practically throughout the tropical forests in India. Seedlings of about 30-50 cm in height raised in the nursery by sowing seed can be transplanted in the field at an espacement of 5 × 5 m.

Kernel of the seed which has almond like taste is roasted and eaten and is also used in sweetmeats.

242. *Setaria glauca* (L.) Beauv. (Syn. *Punicum glaucum* L.); Poaceae (Graminae).

Na : Bengali: Pinginatchi; Eng.: Cat tail millet; Gujrathi: Kolaat, Kunchi, Ziptighass; Hindi: Bandra; Kanerese: Bilikorla hullu; Marathi: Bhadli; Punjab: Bandra, Dissi, Kotu, Bankangni; Santal: Kukra; Telugu: Nakkakona.

Grains of this grass are eaten in times of scarcity. They are either boiled and eaten or ground into flour to make bread.

243. *Shorea robusta*. Gaertn f.; Dipterocarpaceae.

Na : Eng.: Sal; Hindi: Sal, Sakhya, Karshya.

Hab : It grows gregariously along the foot hills of Himalayas, Uttar Pradesh, Bihar, Madhya Pradesh and South eastern parts of Orissa. (pl. see fatty oils)

Seeds are roasted and eaten.

244. *Sida veronicaefolia* Lamk.; Malvaceae.

Na : Bengali: Junka; Hindi: Bhiunli, Bananiyar; Guj.: Bhoybal; Kanerese; Bekkenalalegida; Marathi: Bhoybal, Bhuichikna; Santal: Bir, Tandi, Bariar, Jokhasakam, Rengla; Tamil: Palampasi; Telugu: Gayapaaku

Hab : It is more or less hairy herb and occurs practically throughout tropical parts of India

The leaves are eaten as vegetable.

Sida veronicaefolia **Lamk.**

245. *Smilax zeylanica* L. (Syn. *S.macrophylla* Roxb.); Smilacaceae.

Na : Hindi: Ramdataun, Jungli ushbah.

Hab : Pl. see medicinal plants.

The young branches and leaves are eaten as vegetable.

246. Solanum ferox L.; *Solanaceae.*

Na : Bengali & Hindi: Rambegan; Kanerese: Haladigulla Anesundegida; Malayalam: Velulhavazuthana; Tamil: Molakkai, Anachundai; Telugu: Mulakkayi.

Hab : It is a small prickly herb growing to about 2.5 m in tropical parts in Assam and South India and also in Andaman Islands.

The fruits are used in curries.

Solanum ferox **L.**

247. *S.torvum* Swartz; Solanaceae.

Na : Assam: Bhit-tita, Hathibhekuri: Bengal: Gotabegun; Bihar: Bengarbetahet, Kutunbi, Rangini, Marang; Hindi: Titbaigan; Kanerese: Kadusunde, Sundekkayi; Khasi: Dieng-Sohnonag; Malayalam: Kaatuchunla; Tamil: Sundaiai; Telugu: Chundaikai, Kottuvastu.

Hab : It is a spiny herb or shrub found in Orissa, Bihar, Bengal and Manipur in tropical parts.

The fruits are used as vegetable. They are also dried and preserved.

S.torvum **Swartz**

248. *S.violaceum* (L.) Ortega Syn *S.indicum* L.; Solanaceae.

Na : Assam: Tid bhagnri; Bengali: Byakun, Gurkamai; Eng.: Indian night shade; Hindi: Badikatai, Birhatta, Barhanla; Malayalam: Cheruchunda, Cheruvazhudena; Marathi: Ringani, Dorb, Inoti, Ringri; Punjab: Kalangkari; Tamil: Mulli, Papparamulli, Karimulli; Sanskrit: Virhati, Bhantaki; Telugu: Tellamulaka, Kakamunchi.

Hab : It is a common spiny herb found in the sub-Himalayan tract of Assam and Khasia hills.

Half ripe fruits are employed in the preparation of curries and chutnies. Leaves are also used as vegetable.

Solanum violaceum **(L.) Ortega- Branch**

Sorbus aucuparia L.Tree

Sorbus aucuparia L. fruiting branch

249. *Sorbus aucuparia* L. (Syn. *Pyrus aucuparia* (L.) Ehrh.; *Mespilus aucuparia* Web.); Rosaceae.

Na : Eng.: European mountain ash, Rowan; Hindi: Battal; Punjab: Baltal, Wampulitsi, Rangrek.

Hab : It is a small to moderate sized tree found in the temperate Himalayas from Kashmir to Kumaon.

The fruits are edible. They are used in the preparation of syrups and juices. In times of scarcity dried fruits are ground to a meal and made into bread.

Spondias pinnata (Koen.) Kurz

250. Spondias pinnata (Koen.) Kurz. (Syn. S.mangifera Willd.; Mangifera pinnata L. f.); Anacardiaceae.

Na : Andamans: Gue; Assam: Amora, Amratenga; Bengali: Amra, Ambra; Gujrathi: Ranamba; Hindi: Amara, Jangliam; Kanerese: Amatekaye, Ambattemara, Poondi, Kaadamate, Vrykshamla; Malayalam: Kataambolam, Ambazham; Marathi: Ranamba, Ambada; Oriya: Ambula, Amodo, Amaratoko; Punjab: Ambara: Tamil: Ambalam, Ambiram; Telugu: Adavimaamidi, Kondamaamidi.

Hab : It is a deciduous tree found wild throughout tropical forests in India upto 1500 m in the Himalayas and also in Andamans.

Fruits are eaten and also pickled.

Stellaria aquatica Scop.

251. *Stellaria aquatica* Scop.; Caryophyllaceae.

Na : Eng.: Chickweed; Hindi: Badeola.

Hab : It is often found trailing over bushes in the temperate Himalayas from Kashmir to Uttar Pradesh at altitudes of 1200-2700 m.

The leaves and tender stocks are eaten as vegetable.

Sterculia balanghas L.

252. *Sterculia balanghas* L.; Sterculiaceae.

Na : Malayalam & Tamil: Kavalam.

Hab : It is a small evergreen tree found in Kerala, Karnataka and Maharashtra.

The seeds are eaten after roasting.

253. *S.foetida* L.; Sterculiaceae.

Na : Eng.: Hill coconut, Wild almond; Bengali, Hindi Marathi: Jangli badam; Kanerese: Bhatala penari; Malayalam: Pottakavalam: Tamil: Pattaikavalam: Telugu: Adavibadam, Guttapubadam.

Hab : It is a moderate sized handsome tree found on the West Coast from Konkan Southwards.

The seeds are roasted and eaten.

Sterculia foetida – **L.Fruits**

254. *S.urens* Roxb.; Sterculiaceae.

Na : Eng.: Kateera gum, Gum karaya; Gujrathi: Karai, Kagdol; Hindi: Gulu, Katira, Kulu; Kanerese: Kempudal; Malayalam: Thondi; Marathi: Karai, Kandol; Oriya: Kavili; Tamil: Kavalam; Telugu: Kavili, Tapsi.

Hab : It is a decidous tree with a clean and smooth bole commonly found in Andhra Pradesh, Maharashtra, Madhya Pradesh, Rajasthan, Uttar Pradesh, Bihar and Orissa. Bark is grayish white or reddish, smooth peeling off in papery flakes.

Gum obtained from the stem is used for sweetmeats and in ice-cream preparations. The seeds are eaten after roasting. (Pl. see gums and resins).

***S.urens* Roxb.**

255. *S.villosa* Roxb.; Sterculiaceae.

Na : Assam: Udal, Odol; Bihar: Ganjher; Garo: Ubak; Hindi: Udal, Udar; Kanerese: Savaya, chawri; Khasi: Dieng-star; Punjab: Gulbodca, Gulkandar; Mal.: Vakka; Marathi: Sardal, Sarda; Oriya: Kodalo; Tamil: Murattam, Arnivakenar; Trade: Udal.

Hab : It is a large deciduous tree found in the Western Peninsula, Madhya Pradesh, Punjab, Uttar Pradesh, Khasia hills and other parts in Assam.

The gum is used for sweetmeats and in ice-cream preparations. The seeds are eaten cooked or roasted.

256. *Streblus asper* Lour.; Moraceae.

Na : Assam: Khorua; Bengali: Sehora; Cachar: Serphang; Garo: Kharanchibol; Hindi: Siora, Khorus, Dahia; Kanerese: Mitlemare, Mitligida; Mal.: Paruka, Parava; Marathi: Karera, Kharaali, Kharaota, Paraya, Poi: Tamil: Pira, Pasuna, Pukki, Vitti; Telugu: Barrenkachettu, Pakki.

Hab : It is an evergreen shrub found practically throughout the tropical forests in the country.

The fruits are sweet and eaten.

257. *Syzygium aquem* (Burm. f.) Alston; (Syn. *Eugenia aquea* Bum. f.; Myrtaceae.)

Na : Eng.: Watery rose apple; Hindi: Laljamrool; Telugu: Gulaabijammachettu.

Hab : It is a small tree found in the Khasia hills and Sikkim.

The ripe fruits are eaten as such or made into beverage or syrup.

***Syzigium cumini* (L.) Skeeels.**

Syzigium samarangense **(Bl.) Merr.& Perry.- Fruit**

***Tamarindus indica* L. Branch**

***Tamarindus indica* L. Fruits**

***Tephrosia purpurea* Pers.- Shrub**

258. *S.cumini* (L.) Skeels; (Syn. *Myrtus cumini* L.; *Eugenia jambolana* Lamk.); Myrtaceae.

Na : Andamans: Thabye. Jamun; Bengali: Kalajam; -Gujrathi: Jambu, Jamli; Hindi: Jamun, Jambu, Jambava; Kanerese: Nerale; Lepcha: Phoberkung; Malayalam: Naval, Perinnaral; Marathi: Jaman, Jambul; Oriya: Jamo; Punjab: Jammu; Tamil: Neredam, Naval, Sambal; Telugu: Neredu.

Hab : It is a large evergreen tree found wild as well as cultivated, throughout tropical to sub-tropical parts in the country.

Fruits are edible. They are generally sub acidic and astringent to taste. The juice of unripe fruits is used for preparing vinegar.

259. *S. samarangense* (Bl.) Merr. & Perry; (Syn. *Eugenia javanica* Lamk.); Myrtaceae.

Na : Eng.: Waxjambu; Bengali & Hindi: Jamml, Amrul; Malayalam: Paninirchampa; Telugu: Gulabijaamachettu.

Hab : It is commonly found in Andaman and Nicober Islands.

Fruits are spongy and fragrant with acidic flavour and are eaten.

260. *Tamarindus indica* L.; Caesalpiniaceae

Na : Assam: Teluli; Bengali: Tenlul, Ambli; Eng.: Tamarind, Tamarindo; Gujrathi: Ambli, Amli; Hindi: Imli; Kanerese: Hunesegida; Malayalam: Amlam; Marathi: Chincha; Tamil: Amilam; Telugu: Chintachettu.

Hab : It is a tree found in tropical regions in Central and South India. In the forest it is found near abandoned village sites. It is also grown as an avenue tree and in villages. Seed is collected during February/March. Its viability is long. Germination is 70%. Seed is given pretreatment by soaking in cold water and sown directly into bags. The seedling is ready for planting in the subsequent year. The plant starts yielding fruits in 8th year. The life of the tree is 60-80 years or even more. In the 10th year about 50 kg fruit may be available. Larger trees may yield about 2 qunitals of fruits annually.

The fruits and flowers and even the tender leaves are eaten. The unripe fruits are a rich source of tartaric acid. Tartaric acid and its salts are extensively used in various foods, chemical and pharmaceutical preparations. The polyose obtained from the seeds is a good substitute for fruit pectins. The wood makes a very good charcoal.

261. *Tephrosia purpurea* Pers.; Fabaceae (Papilionaceae.)

Na : Bengali: Ban-nilgachh; Eng.: Purple Tephrosia; Gujrathi: Ghodakan, Jhila, Sarphanko; Hindi: Sarphonka, Ban-nil: Kanerese: Empali, koggili; Malayalam: Kozhenjil, Kaatarniri; Marathi: Sirapakha, Udhadi, Unhali; Tamil: Kolingi, Paa-vali,. Kat-kolingi; Telugu: Neelavempali, Bonta vempali.

Hab : It is a common perennial herb.

The seeds are reported to be used as a substitute for coffee and are also eaten.

262. ***Themeda triandra*** **Forsk. (Syn.** ***Anthistiria imberbis*** **Retz.), Poaceae (Gramineae).**

Na : Eng.: Rooi grass; Guj.: Bhatolu, Bhatdi, Fuliu; Kanerese: Bhimanahanchi, Bellanchihullu, Gondaman chichullu; Tamil: Erigaithattu pullu; Telugu: Peddayerra kallakasuru.

Hab : It is a perennial grass found in the drier parts of the country.

The grains were eaten in times of scarcity.

Themeda triandra **Forsk.-Grass**

263. ***Thymus serpylum*** **L.: Labiatae.**

Na : Eng.: Mother-of-thyme, Creeping thyme; Hindi: Banajwain; Punjab: Kalandar zatar, Marizha, Masho, Rangsbur, Shakei.

Hab : This creeping herb is found in temperate regions in the Himalayas from Kashmir to Kumaon.

The leaves are eaten as vegetable.

Thymus serpylum **L.- Herb**

264. ***Toddalia asiatica*** **(L.) Larnk.; (Syn.** ***T. aculeata*** **Pers.); Rutaceae.**

Na : Bengali: Kada-todali; Eng.: Wild orange tree, Lopeztree, Forest pepper; Hindi: Kanj, Tindupara; Kanerese: Kaadumenagu, Mullumastikai; Kumaon: Chingatti; Khasi: Shia-sieng-ung, Sia-sohsat; Mal.: Kakatoddali, Kaaramullu; Marathi: Limri, Manger; Oriya: Tundpora; Sanskrit: Dahana, Kanchano: Tamil: Milagamai, Moolacamaymaram, Kattumilagu.

Hab : It is a shrub found in the sub-tropical Himalayas, the Khasia hills and Kumaon and also in Tamilnadu.

Leaves and ripe berries of this shrub are edible. Berries are pungent like black pepper and are pickled.

Toddalia asiatica **(L.) Larnk**

265. ***Trapa bispinosa*** **Roxb.; Trapaceae.**

Na : Bengali: Paniphal; Eng.: Water chestnut; Marathi, Gujrathi, Tamil, Kanerese & Hindi: Singhara; Malayalm: Karimpolam; Sanskrit: Shringala, Sringataka, Trikonaphalam, Jalaphala; Telugu: Kubyakama.

Hab : It is an aquatic herb mostly cultivated. Fruit is bony and 4-angled, 2-opposite angles each with a scabrous spine, 2 other spines sometimes obsolete, one seeded.

The raw seeds are edible and the dry ones are the source of starch.

Trapa bispinosa **Roxb.-Plant**

266. ***Trema orientalis*** **Blume (Syn. Celtis orientalis L.); Ulmaceae.**

Na : Bengali: Chikan, Jibon; Eng.: Charcoal tree, Indian nettle tree; Gujrathi: Gol; Hindi: Jiban, Gio; Kan.:Gorklu, Koruhale; Mal.: Ama, Malantotali; Marathi: Gol, Kapashi, Kargol; Tamil: Ambaratti, Chenkolam; Telugu: Budamuru, Chakamanu, Kakamushti.

Trema orientalis **Blume.-Plant**

Hab : It is a shrub or small tree found in tropical regions in Assam, Bengal, Uttar Pradesh, the Punjab and Gujarat and also South India.

The fruits are sweet and eaten.

267. *Trewia nudiflora* L.; Euphorbiaceae.

Na : Assam: Bhel-kol; Bengali: Pilai, Panigamhar; Hindi: Khamara, Bhillaura, Garnhar, Tumri, Pindara; Kanerese: Kaadukanji; Malayalam: Kattukumil, Pambarakkumbil; Marathi: Petari; Oriya: Monda, Pithaliapanijambhar; Tamil: Kanji, Annathuvarei; Telugu: Eruponuku; Trade: Gutel.

Hab : It is a large tree found in Kumaon Hills, Bihar, Bengal, Assam and other parts in the tropical regions.

The fruits are drupaceous and sweet and are eaten.

Trewia nudiflora **L.**

268. *Trigonella polycerata* L.; Fabaceae (Papilionaceae.)

Na : Eng.: Wild fenugreek; Hindi: Chani hari; Punjabi: Sainji, Chini, Khandarore.

Hab : It is a slender trailing annual found in the tropical parts in the Western Himalayas.

The leaves are used as vegetable.

269. *Tussilago farfara* L.; Asteraceae (Compositae.)

Na : Eng.: Colts foot, Cough wort; Hindi & Punjab: Watapana; Trade: Fanjuim.

Hab : This herb is found in the Western Himalayas from Kashmir to Kumaon at altitudes from 1500 to 3500 m.

The leaves are eaten as vegetable. Young leaves are used in soup.

Tussilago farfara **L.**

270. *Urtica parviflora* Roxb.; Urticaceae.

Na : Bengali: Pahara-bichuti; Punjabi & Hindi: Berain, Shishona, Bichu, Kanyali.

Hab : It is found in temperate Himalayas from Kashmir to Arunachal Pradesh and the Nilgiris in South India.

Tender leaves are boiled in water and eaten as vegetable.

271. *Uvaria cordata* Dunal Alston; Annonaceae.

Na : Bengali: Bagh-rungha.

Hab : It is a woody climbing shrub found mainly in Assam.

The fruits are fleshy, sweet and edible.

Urtica parviflora **Roxb.- Shrub**

272. *Vaccinium leschenaultii* Wight; (Syn. V. *symplocifolium* Alston); Ericaeeae.

Na : Tamil: Kilapalam.

Hab : It is a small handsome tree found in the Nilgiris and the Pulney hills.

Fruits are purple in colour when ripe. They are sweet and edible.

Veronica beccabunga **L-.Herb**

273. *Vangueria pubescens* Kurz; Rubiaceae.

Na : Hindi: Main phal, Mayena.

Hab : It is a large shrub or small tree found in Sikkim, Bengal and the Khasia hills.

The young leaves are eaten as vegetable and the fruits are also edible.

274. *Ventilago madraspatana* Gaertn.; Rhamnaceae.

Na : Assam: Bor-kalia; Bengali: Raktapita; Gujrathi: Ragatarohado; Hindi: Pitti; Kanerese: Pupli, Papudi; Khasi: Sher-kharung: Marathi: Kanvel, Lokhandi; Oriya: Roktopitta, Sajumalo, Toridi; Sanskrit: Raktavalli; Tamil: Papilli, Surali, Vembadam; Telugu: Ettasurugudu, Suralatige, Suratichekka.

Hab : It is a climbing shrub with branches hanging down and found in Maharashtra and Sosuth India. It is reported to be found in the North also.

The leaves and seeds are eaten.

Wrightia tomentosa **Roem. & Schult**

275. *Veronica beccabunga* L.; *Scrophulariaceae*.

Na : Eng.: Becca bunga.

Hab : It is a common herb found in the Western Himalayas in Kashmir.

The young shoots are used as vegetable. Young leaves are used as salad.

276. *Wrightia tomentosa* Roem. & Schult; (Syn.W. *mollissima Wall.: Nerium tomentosum* Roxb.); Apocynaceae.

Na : Assam: Alkuri, Garodudhkuri; Bengal: Dudhkoraiya; Garo: Bolmatra; Gujrathi: Dudhlo, Kala inderjau; Hindi: Dudhi, Dharauti, Daira; Kanerese: Bilikudegida, Bilikanagulugida; Malayalam: Mailampala, Nilampala; Marathi: Kalainderjau; Oriya: Palkurwan, Harido; Tamil: Pala; Telugu: Tellapaala, Peddapaala.

Hab : It is a small deciduous tree found in the Punjab, Rajasthan, Bihar and Assam in the warmer parts.

The leaves and tender fruits are edible.

277. *Ximenia americana* L.; Olacaceae.

Na : Eng.: Tallow wood; Wild plum, Wild olive; Hindi: Ingudi; Kanerese: Nagarigide, Nakkara, Kandanakkare; Tamil: Chiruillantai, Kadaranji, Siruilandai; Telugu: Kondanakkera, Uranakkara.

Hab : It is a large straggling shrub or a small tree with spiny branches. It is found in the drier parts in South India and the Andaman Islands.

The ripe fruits are aromatic and acidic sweet and are used as a substitute for lemon.

Ximenia americana **L. Shrub**

Ziziphus mauritiana **Lamk.**

278. *Ziziphus mauritiana* Lamk. (Syn. Z. jujuba Lam. Non Mill.); Rhamnaceae.

Na : Bengal: Kool, Ber, Boroi; Eng.: Chinese date, Jujube, Bor, Bordi; Hindi: Ber, Badara, Karkandhu, Kuvala, Pemdimber; Kanerese: Bhore; Malayalam: Elentha; Marathi: Bor, Bera; Sanskrit: Ajapriya, Badara, Karkandhu; Tamil: Elandai, Yellande, Elladu; Telugu: Regu, Gangaregu.

It is a small tree with a spreading crown and stipular spines growing wild all over the tropical forests in the country.

Ripe fruits are eaten and semi ripe ones are pickled

279. *Z. nummularia* Burm.f. Wight & Arn.; (Syn. *Z. rotundifolia* Lamk.; *Rhamnus nummularia* Burm. f.); Rhamnaceae.

Na : Gujrathi: Ghayabar adbauobordi; Hindi: Jhadber; Marathi: Gangar, Junglaber; Rajasthan: Bordi, Pala; Sanskrit: Bhukumtaka, Sukshma- phala; Tamil: Korgodi; Telugu: Neelaregu.

Hab : It is a thorny-shrub found in North West and South India ascending upto 900 m altitude.

The fruits are eaten.

Z. oenoplia **Mill.**

280. Z. oenoplia Mill.; Rhamnaceae.

Na : Bengal: Siakul, Makhua, Jhurkal; Hindi: Makoi, Makoh; Kanerese: Barige, Karisurimullu; Mal.: Kothavalli, Tutalimullu; Marathi: Kanerballi, Makor; Oriya: Barakoli, Kotakoli; Sanskrit: Srigalakoli, Bahukantaka, Karkhanda; Tamil: Ambulam, Suraimullu; Telugu: Paragi, Parki.

Hab : It is a straggling shrub with prickles, found in dry tropical forests of Northern, Central and South India.

The fruits are edible. The seeds are used as a substitute for Areca nut.

281. *Z. rugosa* Lamk.; Rhamnaceae.

Na : Hindi & Marathi: Suran, Churna; Malayalam: Malantutali, Todali; Oriya: Simukoli; Santal: Sekra; Tamil: Kattilandai; Telugu: Pinduparighamu.

Hab : It is an evergreen spiny shrub found in Bihar, Uttar Pradesh, Madhya Pradesh, Punjab, Rajasthan and Gujarat.

Fruits are eaten.

Z. sativa **Gaertn.**

282. *Z. sativa* Gaertn.; (Syn. Z. vulgaris Lamk.); Rhamnaceae.

Na : Bombay; Ranbor, Unab; Hindi: Pitniber, Kandika, Kandiari, Singli, Banber; Kashmir: Barikon-kamber, Phitni; Punjab: Amlai, Singli, Barari.

Hab : It is a deciduous spiny tree found in the Punjab and Bengal.

The fruits and seeds are eaten.

8

Gums, Resins and Oleoresins

Gums and resins are important forest products and their uses by man go back to the earliest times. The use of resins for torches and for caulking boats and canoes by the primitive people is well known. Lac, which is a secretion of minute insect *Laccifer lacca* Kerr. has long been used in India. The ancient Egyptians used resins on their mummy cases. Fragrant resins such as Frankincense and myrrh were used much before the Christian era. Similarly gums were also used for food as far back as history relates. The adhesive properties of these have also been utilized from early times particularly in preparing paints and white washing. Since gums and resins have distinctly different properties having different uses, they are dealt separately.

Section - I

8.1 Gums

8.1.1 The Occurrence and Origin of Gum in Plants

The vegetable gums are a group of plant products resembling carbohydrates. They are composed of carbon, hydrogen and oxygen with small quantities of mineral matter ash constituents and sometimes a little nitrogen. Small quantities of tannin may also be present in some gums. Gums are found in a large number of families. Notable among them are Leguminosae and Sterculiaceae. Other important gum yielding families are Anacardiaceae, Combretaceae, Meliaceae, Rosaceae and Rutaceae.

8.1.2 Function of Gum in the Life of the Plant

There are various views about the function of gum in the plant. Some believe that the gum is part and parcel of normal metabolism of the plant. In some cases the production of the gum has been attributed to fungi attacking the plant. This fungus is responsible for enzymes that penetrate the tissues and transform the cellulose and hemicellulose of the cell wall into gum. In some cases it is believed that certain bacteria cause exudation of gum while in *Astragalus* spp. (true tragacanth) the gum is produced inside the plant by the transformation of cell walls and adjacent layers of medullary rays and the pith into gum. This readily absorbs water, causing it to swell and exerts pressure on the surrounding tissue. As a result, gum oozes from the stem of its own accord or due to injury.

8.1.3 Different Kinds of Gums

The true gums are generally divided into two main classes or groups:

(a) ***Soluble gums:*** Such gums dissolve in water and form more or less transparent, viscous and adhesive solutions as in Indian gum arabic.

(b) ***Insoluble gums :*** They absorb water and swell into a thick jelly or jelly like mucilage or, with sufficient water, breakdown into a very thick translucent solution as in gum karaya obtained from *Sterculia urens* Roxb.

There is yet a third group of gums with properties intermediate between the above two classes. They may be classed as "semi-insoluble gums". These are considered inferior to the other groups of gums.

8.1.4 Properties of Gums

They emanate in a liquid form, from the body of the plant which on exposure soon nearly dry up into amorphous translucent tear shaped bodies or flakes. It has been seen that the gums from the same species when collected from the same plant at different seasons of the year exhibit noticeable differences. Length of time, it has remained attached to the tree after secretion, may also affect some of its physical properties. The gums which remain attached for long have impurities like dust and are dark coloured in nature. The rain water also has the effect of darkening the colour, perhaps due to substances washed on to the gum. Scorching from forest fires also darkens the exuded gum. Colour is of great importance in the commercial valuation of gums, a strong preference being always shown for those that are light coloured. Gums collected are irregularly globular or drop or pear shaped. Some gums however have the appearance of irregular red shaped fragments for e.g., *Anacardium occidentale* L. (cashew). The surface of the freshly collected gum is smooth but this may soon become rough or covered with minute cracks or striations. Gums are scentless and tasteless but some are sweet or bitter. Most gums break with a clear glassy fracture when properly dried and may be readily pulverized. Gums are hygroscopic and absorb moisture and become soft. On heating they decompose completely without melting usually showing charring.

8.1.5 Tapping Method

Different species have different size and shape and even the nature of flow of gum differs considerably. Therefore a general method of tapping cannot be applied uniformly *to* all the species. Usually the collector inflicts a large number of cuts with an axe on the tree and collects the gum during his spare time. However in the recent years, since some of the gums like gum karaya *Sterculia urens* Roxb. have assumed great commercial importance. Efforts are being made to tap the trees systematically. As it is difficult to cover tapping methods for all the species in the scope of this book, the method suggested for tapping *Sterculia urens* Roxb. is described.

Season of tapping and flow of gum: The gum exudes from blazes all the year round but the flow is more in hot weather. Often, diseased or unhealthy trees produce more gum than healthy trees. It has also been noticed that some trees which appear to be quite healthy, do not produce any gum at all on tapping. In some places tapping is done from September to June while in others most of the collection is done during 15th February to 15th June. No collection is made during rainy season because most of it is either washed away during this period or spoilt.

Collection: Collection of gum is done by handpicking. It is collected daily in summer and on alternate days in winter by labourers. Each labourer may collect about one kilogram of gum daily. Collection of gum before it is dry on the tree should be avoided. Trees get damaged if uncontrolled heavy blazing is resorted to.

8.1.6 Processing, Grading and Storage

After collection, the gum may be sun-bleached before it is sent to the market. Cleaning to remove particles of bark and sand or earth may be carried out as a preliminary to grading. The better grades of gum show uniformity in size of fragment and are pale in colour with an absence of dust or powder. Grading, however, depends on the kind of species and the use for which the gum is put to. The gum should be stored in dry and rat proof godown.

Acacia catechu (L.f.) Willd.

8.1.7 Commercial Valuation of Gums

Colour and viscosity are probably the two most important factors in assessing the value of the gum but good solubility is also important. The value of any particular gum however depends on the purpose for which it is required, for example in the case of gum used for edible or pharmaceutical purposes great stress is laid on taste or rather the absence of it. Usually paler coloured gums are preferred.

8.1.8 Industrial and Other Uses of Gums

Many natural gums are sweetish in taste and not at all unpleasant although some may be astringent and bitter. One of the main uses of gums by man is still as food, for the bulk of the gum is used in confectionery. In India it is used in the preparation of "Laddu" and various other kinds of sweetmeats syrups etc. Another common use for gums, especially the poor grade gums, is for mixing in whitewash for walls. Gum is used in the manufacture of certain fine water colours. It is used in medicine as a demulcent, as a binding agent in the preparation of face powders, lozenges and also in fumigating pastilles. It is an ingredient in glycerine soap creams, hair curling powders and eye cosmetics. Other uses are in textile and calico-printing and as a body builder in transparent finishes used for artificial silk or rayon and in the manufacture of felts. It is also used in the paint and polish industries, in clearing liquors, in lithographic work and in certain dyeing processes. It is used for imparting smoothness to ice-cream and in making syrup, jellies, pickles and sauces. Gums are also used in the preparation of envelopes, labels and postage stamps. It is used in the preparation of skin creams and as a basis for lubricants for surgical instruments and for catheters.

8.1.9 Gum Yielding Plants

1. *Acacia catechu* (L.f.) Willd.; Mimosaceae.

Na : Bengali & Marathi: Khair; Eng.: Black cutch, Catechu; Hindi: Khair, Khadira; Kanerese: Kaggeli; Sanskrit: Khadira; Tamil: Karangalli; Telugu: Sundra.

Hab : It is a tree found in the Punjab, U.P., Bihar, Madhya Pradesh, Andhra Pradesh, Karnataka and Maharashtra.

Gum obtained from the bark of the stem is more similar to the imported gum arabic, from *Acacia senegal* (L.) Willd. than the babul gum obtained from *A.nilotica* (L.) Del. (Syn. *A.arabica* Willd.). The tears are about 3 cm in diameter and are pale yellow to dark amber in colour.

A. farnesiana (L.) Willd.

2. *A.farnesiana* (L.) Willd.; Mimosaceae.

Na : Eng.: Sweet Acacia, Cassie; Hindi: Gandhbabul, Divanababul.

Hab : Pl. see essential oils.

The stem bark yields gum.

3. *A.ferruginea* DC.; Mimosaceae.

Na : Guj.: Kaigu; Hindi: Son-Khair, Kaiger; Kanerese: Banni; Marathi: Pandhrakho; Tamil: Velvelam; Telugu: Ansandra.

Hab : It is a medium sized tree found in Andhra Pradesh, Western Ghats, Gujarat and Maharashtra.

The stem bark yields gum similar to gum arabic.

A.leucophloea **(Roxb.) Willd.**

4. *A.leucophloea* (Roxb.)Willd. (Syn. *Mimosa leucophloea* Roxb.); Mimosaceae.

Na : Bengal: Safed babul; Gujarat: Haribanal; Hindi: Safed babul, Safed Kikar, Ronj, Arinj; Kanerese: Bilinjali; Mal.: Pattacimrayamaram; Mar.: Hewar; Tamil: Velvayalam; Telugu: Tellatumma.

Hab : It is a moderate sized tree found in the Western and Southern parts of India. The bark of the stem yields gum of lesser importance.

A.nilotica (L.) Del.

5. *A.modesta* Wall.; Mimosaceae.

Na : Punjab: Phulai.

Hab : It is a moderate sized thorny tree found in the sub-Himalayan tract and the outer Himalayas in Jammu & Kashmir. Stem-bark yields a translucent and pale yellowish gum but in small quantities.

6. *A.nilotica* (L.) Del.; (Syn. *A.arabica* Willd.); Mimosaceae.

Na : Bengali & Marathi: Babul; Guj.: Banel; Eng.: Babulgumtree; Hindi: Babul, Kikar; Kanerese: Jali; Babli; Malayalam: Karuvelam; Punjab: Kikar; Sanskrit: Barburaha; Tamil: Karuveli; Telugu: Nalla tumma.

Hab : It is a moderate sized spiny tree with dark-brown or black bark and is found in many parts of the country especially in drier localities. It is also grown both along the fringes of fields in the foreshores of tanks and in forest areas.

It forms the main source of indigenous gum Arabic, which is also known as Indian gum arabic in trade. The gum is emulsifying and suspending agent. It is almost entirely soluble in twice its weight of water yielding a very viscous slightly acid solution. It is insoluble in alcohol .Wounds are made on the bark of the stem for production of gum.

A.senegal **(L.) Willd.**

7. *A.senegal* (L.) Willd. (Syn. *A.verek* Guill. & Perr.); Mimosaceae.

Na : Eng.: Gum Arabic; Hindi & Rajasthan: Kumta.

Hab : This tree is native of North Africa but is found in Rajasthan and the Punjab. It is the main source of imported gum arabic. Yield of gum from the stem bark of the species in India is very low and hence it cannot be ranked as an important Indian product.

A.senegal **(L.) Gum**

Aegle marmelos **(L.) Corr.**

Albizia lebbeck **Benth.**

A.odoratissima **(L.f.) Benth.**

A.procera **Benth.**

8. *Aegle marmelos* (L.) Corr. (Syn. *Crataeva marmlos* L.); Rutaceae.

Na : Eng.: Bael, Bengal quince; Hindi: Bel, Bilva.

Hab : It is found in parts of tropical and sub-tropical forests in the country (Pl. also see medicinal plants).

The mucilagenous substance secreted round the seed is used as an adhesive and also in varnish and whitewashing. The stem also yields gum.

9. *Albizia lebbeck* Benth.; (Syn. *Mimosa lebbeck* L.; *Acacia lebbeck* Willd.); Mimosaceae.

Na : Bengali: Sirish; Eng.: Lebbeck tree; Guj.: Pilo Sarshio; Hindi: Siris; Kanerese: Bagemara; Mal.: Vago; Mar. & Sanskrit: Sirisha; Tamil: Vagei; Telugu: Dirsanam; Trade; Kokko.

Hab : It is a tall deciduous tree found in the drier and hotter parts in the country.

Gum obtained from the bark is used in furniture industry.

10. *A.odoratissima* (L.f.) Benth. (Syn. *Mimosa odorotissima* L.f.); Mimosaceae.

Na : Bengal: Kakur Siris; Eng.: Black Siris, Fragrant Albezia; Guj.: Kalo Sarasio; Hindi: Kala siris; Kanerese: Bilvara; Mal.: Pulivaga; Marathi: Chikunda; Tamil: Karuvagei; Telugu: Chinduga.

Hab : It is a large erect tree found in the tropical to sub-tropical regions in the country. Stem bark yields a dark brown insoluble gum in the form of rounded tears. It is similar to the gum of *A. lebbeck* (L.) Benth.

11. *A.procera* Benth.; Mimosaceae.

Na : Bengali: Koroi; Hindi: Safed Siris; Kanerese: Bellate: Mal.: Karun thagara; Marathi: Kinhai; Tamil: Konda vagei; Telugu: Tellachinduga; Trade: White siris.

Hab : It is a large deciduous tree found in tropical regions in the country in moister places.

Stem exudes gum.

12. *A.stipulata* Boiv.; Mimosaceae.

Na : Bengali: Chakua; Hindi: Siran; Kanerese: Hottabage, Kalbage; Mal.: Pottavaga; Marathi: Laeli; Tamil: Pilivagei; Telugu: Kondachiragu.

Hab : It is a tall deciduous tree occurring in the north eastern regions and also South India. It yields an insoluble gum which can be used in paper industry.

13. *Anogeissus latifolia* Wall.; Combretaceae..

Na : Guj.: Dhavdo; Hindi: Dhawa, Dhawra; Kanerese: Dinduga; Mal.: Marukinchiram; Marathi: Dhaura; Tamil: Vellainaga: Telugu: Tirumanu, Chiruman; Trade: Axle-wood.

Hab : It is a large or moderate sized gregarious tree found in the sub-Himalayan tract, Madhya Pradesh, Maharashtra and South India. It is an important associate of teak. Bark has whitish grey depressions caused by exfoliation of bark. Bark of the stem is the source of commercially important ghatti gum. The colour of the gum varies from whitish yellow to amber. It is partly soluble in water forming a colourless mucilege and possesses double the viscosity of acacia gum. It is a good substitute of Indian gum arabic. It is used for edible purposes, calico printing, sizing of paper and for pharmaceutical purposes.

***Anogeissus latifolia* Wall.**

14. *Astragalus heratensis* Bunge; Fabaceae (Papilionaceae.)

Na : Eng.: Indian tragacanth. The stem is the source of Katira or tragacanth gum.

15. *A.prolixus* Sieb.: *Fabaceae (Papilionaceae.)*

Hab : It is an under shrub found in the Punjab.

The stem is the source of gum tragacanth used in confectionery and in many other ways.

***Azadirachta indica* A.Juss.**

16. *A.strebileferus* Royle. Fabaceae (Papilionaceae.)

Na : Hindi: Garmezu; Persian: Jib.

Hab : It is found in Kashmir. The gum obtained from the stem is used in confectionary.

17. *Azadirachta indica* A. Juss. (Syn. *Melia azadirachta* L.); Meliaceae.

Na : Bengali Hindi: Nim; Guj. & Marathi: Limba; Kanerese: Bevu; Malayalam, Tamil and Telugu: Vepa; Persian: Nib, Azab darkht hind; Sanskrit: Nimba.

Hab : It is a medium sized deciduous tree found scattered all over the tropical regions in the country. It is also grown as a roadside tree and also for shade. The gum obtained from the stem bark is bitter in taste and is used as an adhesive.

***Bauhinia purpurea* L.**

18. *Bauhinia purpurea* L.; Caesalpiniaceae.

Na : Bengali & Marathi: Deva or rakiakanchan; Eng.: Camel's foot tree, Pink Bauhinia; Hindi: Kaliar, Lalkachnar, Khairwal; Tamil: Mandari; Telugu: Kanchanar.

Hab : It is a moderate sized ornamental evergreen tree with pink flowers and is found in sub-Himalayan tract, in Western India and the Khasia hills rising up to 1200 m.

Stem bark produces gum but it is not commercially exploited.

B.racemosa **Lamk.**

19. *B.racemosa* Lamk.; Caesalpiniaceae.

Na : Hindi: Kachnal, Gurial.

Hab : It is a small tree found all over the tropical parts of the country. Bark of the stem is the source of gum but it is not commercially exploited.

Bombax ceiba **L.**

20. *B.retusa* Roxb.; Caesalpiniaceae.

Na : Hindi: Kandla, Semla; Telugu: Nirpa.

Hab : It is a medium sized deciduous tree found in Bihar, Uttar Pradesh, foot hills of Himachal Pradesh and Madhya Pradesh.

This species is often tapped for gum. The gum obtained from bark is used for edible purposes and is also used for sizing cloth, paper and for white washing.

21. *Bombax ceiba* L. (Syn. *Bombax malabaricum* DC.; *Salmalia malabarica* (DC.) Schott & Endl.); Bombacaceae.

Na : Eng.: Red silk cotton; Hindi: Semul.

Hab : Pl. see medicinal plants.

Bark of the tree is the source of hog gum.

Buchanania lanzan **Spreng. - Branch**

22. *Buchanania lanzan* Spreng.; (Syn. *B.latifolia* Roxb.); Anacardiaceae.

Na : Eng.: Kuddapah almond; Hindi: Chironji.

Hab : Pl see edible plants.

The bark yields gum. It is similar to benzoin gum which is obtained from *Styrax benzoin* but is said to be inferior.

Butea monosperma **(Lamk.)Taubert**

23. *Butea monosperma* (Lamk.) Taubert; (Syn. *B.frondosa* Koenig ex Roxb.); Fabaceae (Papilionaceae.)

Na : Beng. & Marathi: Palas; Eng.: Flame of the forest, Bengal kino; Guj.: Khakra; Hindi: Dhak, Palas; Kanerese: Muthulu; Mal.: Palasin samatha; Sanskrit: Palasha; Tamil: Parasa; Telugu: Mothugu.

Hab : It is a small tree found practically throughout tropical forests in India.

It yields ruby-red gum from the bark which is known as Bengal kino in trade.

24. *Chukrassia tabularis* A. Juss.; Meliaceae.

***Cochlospermum religiosum* (L.)Alston.**

Na : Assam: Bogapoma; Bengali: Chikrassi, Pabba; Eng.: Bastard cedar; Hindi: Chikrasi; Kanerese: Delmara; Malayalam: Akil, Malavepu; Marathi: Pabha; Tamil: Agil, Malaivembu; Telugu: Errapogada.

Hab : It is a tall handsome tree found in the Western Ghats, Andhra Pradesh, Andaman Islands, Assam and Bengal

The bark yields reddish to amber coloured gum which is used as an adhesive.

25. *Chloroxylon swietenia* DC., Rutaceae.

Na : Eng.: East Indian satin wood; Gujarathi & Marathi: Halda, Bheria, Billu, Hardi; Hindi: Bhirra, Satin, Girya; Kanerese: Bittula, Hurihuli, Masula; Tamil: Porasu; Telugu: Billu, Billudu, Bittu.

Hab : It is a small tree common in dry deciduous forests in Central and South India.

It yields gum of lesser importance. As such it is not tapped.

26. *Cochlospermum religiosum* (L.) Alston; (Syn. *C.gossypium* DC.); Cochlospermaceae.

Na : Arabic: Katira; Eng.: White silk cotton; Hindi: Pilikapas, Kumbi, Galgal; Kanerese: Arasina burga; Malayalam: Appakutakka; Marathi: Ganeri, Ganglay; Oriya: Kontopalas; Tamil: Kongilam, Tanakku; Telugu: Kondagogu.

Hab : It is a small deciduous tree commonly found in Andhra Pradesh, Maharashtra, Madhya Pradesh, Uttar Pradesh and Bihar. It bears conspicuous yellowish flowers.

The bark on tapping yields gum which is quite similar to Karaya gum obtained from *Sterculia urens* Roxb. It is used for edible and various other purposes. It is hygroscopic and when water is added it swells like karaya gum. It has commercial importance.

27. *Lannea coromandelica* (Houtt.) Merr.(Syn. *L.grandis* (Dennst.) Engl. Odina wodier Roxb.; Anacardiaceae.

***Lannea coromandelica* (Houtt.) Merr.**

Na : Hindi: Jhingan.

Hab : Pl. see edible plants.

Bark of stem and branches is the source of Jhingan gum. This gum is used in confectionery and as mucilage for making ink. The yield of gum in the first year of tapping is about 4 kilograms per tree but at the end of 5 years the yield declines. Therefore, after tapping for five years, the plant should be given rest until wounds get healed up and the tree has regained vitality.

Limonia acidissima L.

M.peltata
(Roxb.) Muell.-Arg.

Mangifera indica L.

Moringa oleifera Lamk.

Ostodes paniculata Bl.

28. *Limonia acidissima* L Syn.*Feronia limonia* (L.) Swingle; *F.elephantum* Correa; Rutaceae.

Na : Eng.: Elephant apple, Wood apple; Hindi: Kavitha, Kaweet, Katbel.

Hab : Pl. see edible plants.

The gum obtained from the bark of stem and branches is used as a substitute for gum arabic.

29. *Macaranga indica* Wight; Euphorbiaceae.

Na : Hindi: Ramalo.

Hab : Pl. see edible plants.

Bark on wounding yields crimson coloured gum which is used for taking casts.

30. *M.peltata* (Roxb.) Muell.-Arg. (Syn. *M.roxburghii* Wight); Euphorbiaceae.

Na : Hindi: Chand-kal, Boddi; Kanerese: Chanda Kanne; Malayalam: Uppilavatta; Marathi: Chanda, Chandwar; Oriya: Piania, Gondaguria; Tamil: Vattakanni, Vattatamarei; Telugu: Kondataphara, kondatamara.

Hab : It is a small tree found in South India, Orissa and Bihar.

Leaves are peltate. Reddish gum exudes from the cut branches, bases of petioles, young shoots and fruits. It can be used for sizing paper.

31. Mangifera indica L.; *Anacardiaceae.*

Na : Eng.: Mango; Hindi: Am. Amra.

Hab : Pl. see edible plants.

The species exudes gum but it is not exploited for commercial purposes.

32. *Moringa oleifera* Lamk. (Syn. *M.pterygosperma* Gaertn. *Guilandina moringa* L.); Moringaceae.

Na : Eng.: Drumstick, Horse radish; Hindi: Sahinjna, Soanjna.

Hab : Pl. see medicinal plants.

It is the source of hog gum.

33. Ostodes paniculata Bl.; Euphorbiaceae.

Na : Assam: Dieng-ja-tung, Tasichange; Lepcha; Palok-kung; Nepal: Bepari.

Hab : It is a large handsome ever green tree found in Assam, Sikkim, Bhutan and the Khasia hills.

Stem bark yields gum which is useful for sizing paper.

34. *Pterocarpus marsupium* Roxb.; Fabaceae (Papilionaceae.)

Na : Eng.: Indian kino tree, Malabar kino; Hindi: Bijasal, Pitsara.

Hab : Pl. see medicinal plants.

It is the source of Malabar kino. This gum is officinal in Indian Pharmacopoeia. To tap a tree, fishbone type of blaze is made into its main trunk. One vertical incision with a number of oblique ones is cut just exposing the wood. The brick-red kino flows out quickly in a liquid form which is collected. Incisions are made usually in the evening and by the following morning the flow generally stops. It is strained to remove impurities and then dried. Good quality kino is ruby red in colour containing 70 per cent of tannic acid. Therefore it is used as astringent.

Pterocarpus marsupium **Roxb.**

35. *Spondias pinnata* (Koen.) Kurz (Syn. *S.mangifera* Willd.; *Mangifera pinnata* LF.) Anacardiaceae.

Na : Eng.: Hog plum; Hindi: Ambara, Amra.

Hab : Pl. see edible plants:

Stem yields gum which can be used in confectionery.

Spondias pinnata **(Koen.) Kurz.**

36. *Sterculia urens* Roxb.; Sterculiaceae

Na : Eng.: Indian tragacanth, Kateera gum, Gum Karaya; Hindi: Gulu, Kadaya, Karaya, Kateta, Katilo, Kullo; Telugu: Tapsi.

Hab : Pl: see edible plants.

Gum karaya which is obtained as an exudation from *Sterculia urens* Roxb. a species confined, mainly to India, is one of the most important forest products of our country. It has assumed greater importance in external trade during the last few years and the exports from India are gradually increasing. India exported about 1119.838 tonnes valued at about Rs. 131.845718 millions during 2002-03. Amongst gums it ranks second only to gum arabic in commercial importance. The gum is similar to gum tragacanth (*Astragalus gummifer* lab.) in properties as it swells in cold water to form an opaque-gel. In fact it was first sold as an adulterant of gum tragacanth in U.S.A. but since 1920 it has gained considerable importance and U.S.A. is now one of the important importers of this gum. It is used for a variety of purposes such as pharmaceuticals, paper and food and textile industries and in the preparation of dentures and cosmetics. It is used in U.S.A. for the preparation of finger-wave lotions for the Beauty trade. Its superior grades find use in ice-cream, ketchups and sauces, various food products and in the preparation of some medicines.

The lower grades are useful in many branches of textiles, and in the preparation of varnishes, inks, rubber compositions, linoleum, oil-cloth, paper coating, polishes and engraving process. It is considered to be an important industrial ingredient.

Sterculia urens **Roxb Tree**

Distribution: *Sterculia urens* is indigenous to India having a wide distribution. It is abundantly found in Madhya Pradesh contributing about 50% of gum production in the country and the remaining coming from Rajasthan, Gujarat, Orissa, Maharashtra, Andhra Pradesh, Tamilnadu and Karnataka. In addition to these states, the species is also found in Uttar Pradesh and Bihar.

Field Characteristics for identification: *Sterculia urens* is characteristic of tropical dry rocky hills and plateau in association with *Boswellia serrata*. It is always conspicuous, especially in the hot season, from its smooth greenish-grey bark or white bark peeling off in large papery exfoliation and the gaunt white stem with stiff spreading branches. The tree starts shedding its leaves from November depending on the locality. The panicles of flowers appear from December to March at the end of leafless branches. The red follicles covered with stinging bristles ripen by about April and the fruits are seen on the leafless tree. Leafing starts by about May.

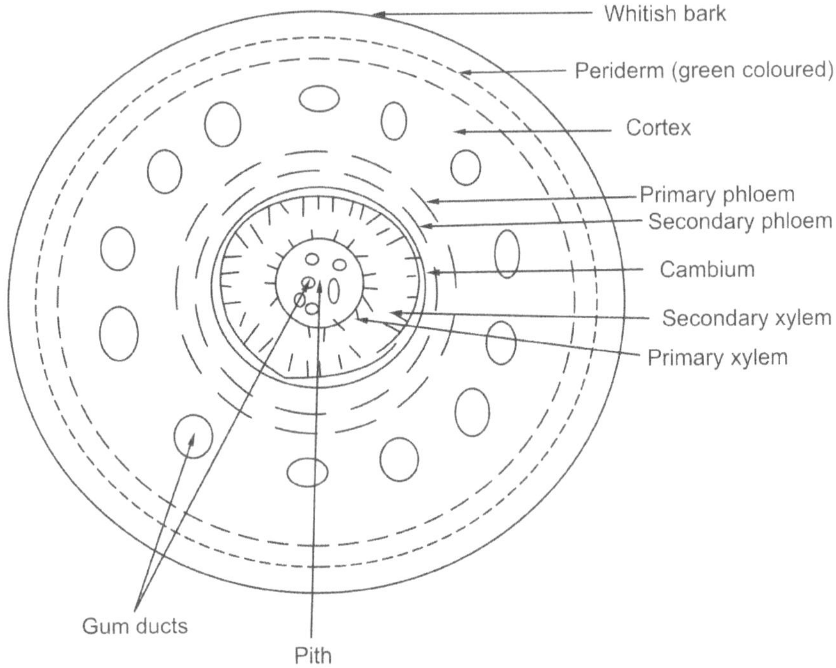

Fig. 8.1 T.S of *Sterculia urens* Roxb.

Gum exudation in relation, to wood structure-Structure of normal wood: (Fig. 8.1). Gum canals are normally absent in the wood (xylum) as well as in the bark (Phloem). They appear to be traumatic (gum ducts formed by wounding) and post cambial in origin and are lysigenous i.e., formed by disintegration of parenchymatous (thin walled) cells in the wood near the cambium into gummy substance. Figure shows the normal structure of a twig. Greenish coloured periderm is seen below the upper outermost whitish bark which is followed by creamy coloured parenchymatous tissue known as cortex wherein some mucilaginous (gum) ducts or canals are noticed. These ducts, however, disappear once the secondary growth starts as they are pushed out and peeled off along with the dead bark. Below the cortex lies the phloem which is followed by a thin layered cambium. Below the cambium, xylum is found. In the centre of the twig lies the pith region. Formation of mucilage or gum canals in this region starts early and is seen before secondary growth starts. Some of the cells in the pith start dividing and form small groups which are easily recognisable because of their denser protoplasm. The walls of these cells soon disintegrate and contents disorganise into mucilaginous or gummy mass. These canals are however, not available for tapping of gum firstly because of their interior location and secondly because that once secondary growth starts, the pith gets compressed owing to the formation of wood and looses its identity as gum producing tissue. During the process of secondary growth owing to meristematic nature of cambium, primary and secondary phloem (bark) on the outer side and the primary and secondary xylem (wood) on the inner side are formed.

As seen in the Fig. 8.2 the wood is characterised by a few large vessels, mostly in short radial multiples of 2-3, predominantly diffused parenchyma with occasional broad bands at irregular intervals broad to very broad rays and thick walled fibres. The phloem lies in a wedge shaped mass separated by wide rays and consists of alternating layers of thick walled fibrous tissue and thin cells, comprising of sieve tubes, companion cells and phloem parenchyma. Wounding the tree was found to result in the disintegration of the Parenchyma cells in the wood near the cambium into gummy substance. In fact this is the tissue which contributes much in making available the commercial quantities of gum. Before desolution, these parenchymatous cells are distinguished from others by denser protoplasm. The gum canals are often confluent; canal orifices 200-250 μ wide radially, 500-700 μ tangentially with pale yellowish-brown translucent deposits. Formation of abnormal wood and gum canals in it is continued for several years once the tree is wounded. The phloem region is not affected very much by wounding and no gum canals are usually formed. However, severe tapping may sometimes result in the disintegration of the ray cells of the phloem into gummy substance. But this does not probably contribute much to the gum exudation. Therefore since most of the gum exudes from the ducts in the wood formed after wounding, it is necessary that the blazes should be deep enough just to expose the newly formed wood for obtaining the gum.

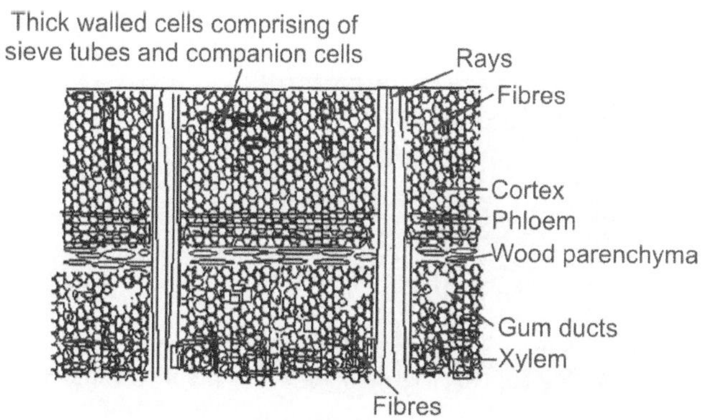

Fig. 8.2 T.S. of wood prior to blazing – Sterculia urens Roxb.

Changes in wood after blazing: (Fig. 8.3) In the vicinity of the wound, the wood formed after tapping becomes spongy, weak and coarse-textured with the development of rows of gum canals.Most of the tissues formed after wounding are parenchymatous in nature with relatively thin walls. The parenchyma cells differ also from the normal type in shape and size being generally larger and radially flattened. The fibres are usually very poorly developed. The vessels become smaller and more numerous and are arranged in tangential multiples. The rays, however, do not appear to be affected very much by wounding. The older tissues also undergo some change. The vessels continuous to the newly formed tissue are generally plugged with tyloses and reddish-brown gummy deposits. In the entire space between the rays, gum ducts are formed and usually these are the largest and most numerous in the wood formed immediately after wounding. After blazing, the wound responses have been observed to extend on all the four sides of the blaze but more on the vertical direction than laterally, maximum being on the upper side of the wound. The gum canals become fewer and smaller on the farther side of the wound and the structure remains more or less normal at a distance of about 15 cm from the wound.

In the initial stages of blazing no gum exudation takes place for several days. The formation of gum canals in the wood starts with in a week or so, but the exudation is noticeable only after 2-3 weeks. Immediate exudation in some cases is due to occurrence of previous injuries. The gum canals are continuously formed in the wood for several years once the tree is wounded.

Suggestion for making blazes: Since formation and exudation of gum is the result of wounding and disintegration of parenchyma cells in the wood near the cambium, it is necessary that blazes should be just deep enough to expose the newly formed wood and also as gum canals so formed after injury become fewer and smaller on all the four sides as the distance from the wound increases, scooping out wood or making of very deep blazes should be avoided. Proportionately more gum ducts are formed in the wood just below the cambium at a depth between 1.3 cm to 3.00 cm in trees from 30 cm to 120 cm girth and at a depth between 2.5 cm to 3.8 cm in trees with 135 cm girth or above. Therefore, parenchymatous tissue in the wood just below the cambium where gum ducts are formed is required to be exposed by the tappers.

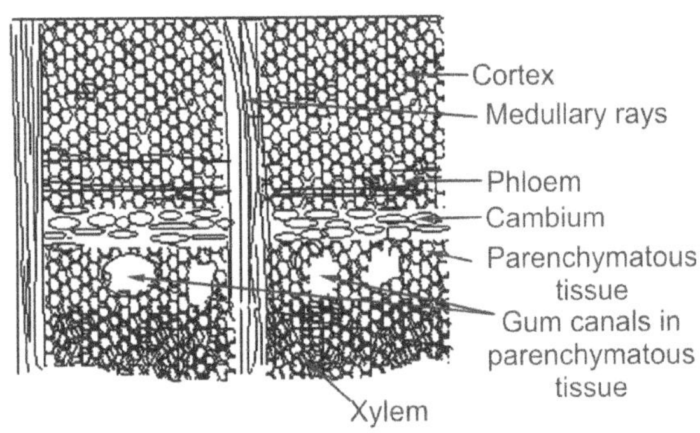

Fig. 8.3 T.S. of wood after Blazing – Sterculia urens.

***Gum tapping method:* (Fig. 8.4)**

(a) *Selection of trees for tapping:* Trees of and above 90 cm in girth at breast height are tapped. Trees of lower girth classes are not tapped since any damage to them may adversely affect their growth. Middle aged trees yield more gum. Selected trees are marked and numbered.

(b) *Season of tapping:* The gum exudes from the trees all the year round. The flow is more in hot weather. Tapping may be done from October to June. No collections should be made during the rainy season as most of the exuded gum is either washed away or becomes darker in colour.

(c) *Tapping Method:* Regular tapping of trees for obtaining gum karaya is reported to have been started round about 1930. So far no definite and uniform method of tapping of *Sterculia urens* basing on scientific data is being followed in the country. The method is varying from region to region. In most cases it is left to the discretion of the tappers with the result they, with a view to obtaining maximum yield of gum, are making deep and very large blazes of about 60 cm or so resulting almost in girdling of trees. In the subsequent years the same blazes are tapped vertically and laterally without giving any chance for the tree to heal up its wounds. This is resulting in the death of trees.

***Sterculia urens* – Defective method of Tapping**

(d) *Blazes: Initial work:* The initial blaze of 15 × 10 × 2.5 cm to 3 cm (6" long 4" wide and one to 1.2 inch deep) may be made on trees with sharp adze having 7.5 cm (3") wide blades. The blaze should be just deep enough to take out 0.5 cm (0.2") deep wood portion below the cambium. Deeper blazing may damage the tree since it will delay the process of healing.

The initial blazes may be made as in the figure at 10 cm or 4" above the ground level. Number of blazes will depend on the girth of the trees and initially a tree of 90 cm or 3 ft girth one blaze may be made and for every increase of 50 cm or 20 inches one more may be added. Blazes may be made horizontally leaving approximately equal space between the blazes. The blazes should not have any loose fibre. The lower surface should be made slightly sloping outwards to avoid rotting on account of water collection. The surface should also be smooth. This will avoid contamination of gum with bark.

Freshening of the blazes: The gum starts oozing out soon after the blazes are made, which is collected with an interval of about fortnight till the rain commences. Depending upon the availability of labour, weekly collections may be useful. To avoid clogging of gum in the gum canals and consequently for getting optimum yield of gum timely and repeated freshenings are essential which can be done at the time of collection of gum. In case, fortnightly freshenings are done in places where tapping is carried out from October to mid June, 16 freshening will be required. In each freshening about 2 cm (0.75") bark should be shaved off from the upper end. The sides and surface of the blaze may also be scraped off so that the clogged canals are opened up and at the end of year's freshening only 1.25 cm (0.5") more space is covered on either side of the blaze but in no case the depth in the blazed surface should be increased. Similarly subsequent freshenings are done. As the effect of wounding is greatest above the blaze, freshenings should be done only from the upper edge of the wound. Edges are freshened in order to expose gum ducts. The size of the blaze at the end of the tapping season should not be more than 47.5 cm (19") in length, 12.5 cm (5") in width and the depth should not be further increased.

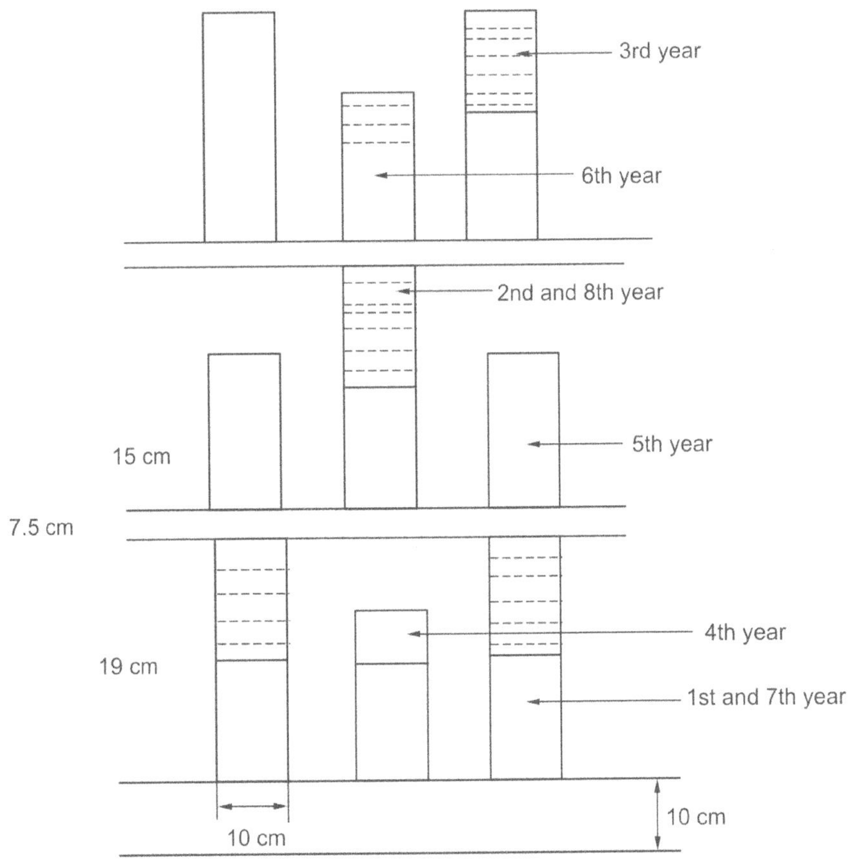

Fig. 8.4 Plan for light and continuous tapping method-sterculia urens.

Blazing in the subsequent years: For continuous tapping the bole is divided into three zones, each one being tapped for one year. Thus the three zones are covered in three years. For making another horizontal row of blazes in subsequent years 5 cm (2"), space may, be left above the blazed portion of the previous years. The blazes of the subsequent year's horizontal row should be alternating or staggering with the previous years blazes i.e. the old and fresh blazes will not be in the same longitudinal row as shown in the Fig. 8.4.

The size, shape etc., of the blaze before and after freshening may be the same. Similarly in the third year same method will be followed. In the fourth year blazing may be continued in the zone where it was done in the first year. But this time the blazes may be in between the previous ones. With this method the blazed region gets complete rest for two years, and one year's partial rest. This cycle could be repeated making use of all the trees available simultaneously. In the 7th year, the tapping can be reverted to the blazed area of initial exploitation i.e. the blazes may be made overlapping the first years' blaze. Thus the blazed portion will get complete rest for six years during which period the wound would get healed up and will resume normal activity. After blazing if the space between the two adjacent blazes is not sufficient, for further blazing, it should not be taken up but stopped.

Girth of the tree	No. of blazes	Total space to be covered horizontally by the number of blazes indicated in the previous column in the 1st cycle	Space left unblazed on the tree horizontally at the time of initial tapping	Space left between blazes in the first cycle	Space left between blazes after taping in the next cycle
1	2	3	4	5	6
Cm. (Inches)		Cm. (Inches)	Cm. (Inches)	Cm. (Inches)	Cm. (Inches)
90-140 (36"-56")	1	12.5(5") × 1 = 12.5(5")	77.5 (31")	77.5 (31")	65 (26")
140-190 (56"-76")	2	12.5(5") × 2 = 25.0(10")	115.0 (46")	57.7 (23")	45 (18")
190-240 (76"-96")	3	12.5(5") × 3 = 37.5(15")	152.5 (61")	50.8 (20.3")	37.5 (15")
240-290 (96"-116")	4	12.5(5") × 4=50.0(20")	190.0 (76")	47.5 (19")	35 (14")
290-340 (116"-136")	5	12.5(5") × 5 = 62.5(25")	227.5 (91")	45.5 (18.2")	32.5 (13")
340-390 (136"-156")	6	12.5(5") × 6 = 75.0(30")	265.0 (106")	44 (17.3")	31.6 (12.7")

Note : Space between the blazes has been calculated taking the minimum of girth class.

Collection: Collection of gum may be done by picking at fortnightly intervals when freshening is also done. Gum may be collected when it is completely dry. This will reduce processing cost and it will not contain lose bark adhering to it.

Yield: A tree of about 1.4 to 1.9 m girth with two blazes yields 2 to 5 kg of gum in a year and may produce up to 10 kg depending upon the locality, size and vigour of the growth of tree and method of tapping. The yield per 100 blazes is estimated to be 75 to 150 kg per year on an average. Gum exudes from wounds of the trees throughout the year but is at its maximum in summer, from mid April to mid June being about 25-30% and also immediately after the rains from September-October being 15-20% of the years collection. Exudation is less from November to February and starts increasing again from March onwards till rains. Gum also exudes in the form of coiled threads from small holes or cracks on the stem of the tree, made by borers healing wounds etc. However, the yield being low it has no commercial value.

Processing and Grading: (Fig. 8.5) Processing and grading of raw gum is done manually or mechanically and it is almost the same in principle in various gum grading centres. In Madhya Pradesh raw gum is cut with choppers into 2.5 to 5 cm pieces. These are dried in Sun (Sun drying also bleaches the colour to some extent) and graded further by breaking and sorting according to size, colour and impurities. Drying takes about 5 to 15 days depending on the initial moisture of the raw gum and also weather conditions. Depending on colour and bark and impurity, gum is roughly graded into I, II, III and last grade known as shiftings. As per Indian Standard Institution specification (I.S. 5025-1969) these are

Grade I white to amber colour

Grade II Reddish pale yellow.

Grade III Brown to Black.

Each grade is broken again to separate bark impurity if any and finally graded. Grading of size at each stage is done by passing the gum through the standard size sieves from No. 1-16 and for separation of powder fine sieve is used. The sieves are made up of a standard size. Wire mesh fitted in a circular wooden frame with 8 cm high walls and about 25 cm diameter. At Indore in Madhya Pradesh sieves of 4 different mesh sizes are used.

Sieve No. Mesh Size (No. of divisions per 6.25 Sq. Cm. (per Sq. inch)

1..	4	A. grade
2..	5	B. grade
3...	16	
4...	25	

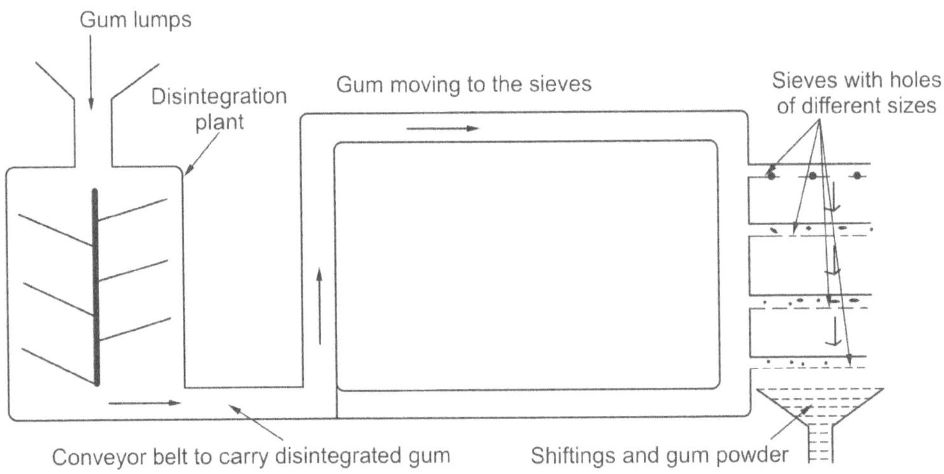

Fig. 8.5 Diagram of mechanical disintegration plant and Grading.

The gum passing up to 4 mesh size sieve is A grade and (5-16) B grade. Inspection of gum grades particularly A and B meant for food and pharmaceuticals is done on shining aluminium plate to detect rodent excreta and hairs at the time of final grading. There are electrically operated disintegrating, plants with arrangements for grading gums into different sizes and colours. Such plants crush about 5 to 6 quintals of raw gum per hour. Another machine which has a decorticator with fans for wind blowing is used for separating bark adhering to gum. The bark being lighter is winnowed off. Fine particles of gum blown are however recovered by straining through sieve. In manual processing normally from 75 bags of raw gum (5625 kg.), 50 bags (3750 kg) of A of all the 3 grades, 22 bags (1650 kg) of B of all the grades, 1.5 bags (112.5 kg) of powder (course and fine bark and gum) are obtainable. Further driage and losses are 1.5 bags (112.5 kg).

Storage of gum: The raw gum is generally stored in old used gunny bags (capacity 70-75 kg). The processed gum however, is stored and marketed in first quality gunny bags (80 kg). Gum should be stored under cool and dry conditions. Gum in powdered form is to be stored in well closed containers to prevent absorption of moisture. Graded and ungraded gum must be stored separately in clean hygienic premises and prevent it from being mixed with rodent excreta, and bird droppings. Depots and godowns should be rat-proof and rain-proof.

As it is an edible product, sanitary conditions in storage must be ensured and for this purpose it may be necessary to prescribe inspection of export consignments. Gum bags as they come from centre of production should be checked as per samples of each grade maintained under proper supervision and each packet should be labelled with grade it contains. Gum should be preferably packed in water proof containers and the containers, sprayed with an insect and rodent repellent.

Constituents and properties of Karaya gum: Structurally gum Karaya is a complex partially acetylated polysacharide of extremely high molecular weight reported to be in the range of 95,00,000. Hydrolytic studies have yielded the following quantitative components: D.galacturonic acid (43 parts), D.galactose (14 parts) and L.rhamnose (15 parts). Work by Rao and Sharma (1957). Confirmed that karaya gum is composed of L.rhamnose, D.galactose and D.galacturonic acid in the apparent molecular ratio of 4:6:5. When hydrolysed with phosphoric or sulphuric acids, the volatile acidity number varying from 13.4 to 21.3 (expressed in term of acetic acid) is obtained. The volatile acidity number of gum tragacanth is 2 to 2.94. On long standing, the gum develops a distinctive odour of acetic acid: Dissolved in water it has an acidic reaction with its pH (1% solution) about 4.6.

It does not dissolve in water completely to give a clear true solution but absorbs water rapidly and swells to form viscous colloidal sols. The rate of this hydration is dependent upon the mesh size of the gum with finer material hydrating much more rapidly than course grained material. Trimethylamine has been identified in the hydrolysis products in minor quantities. Normally a 3-4% concentration is the maximum that can be obtained by common cold water hydration giving a heavy gel-like paste of uniform smoothness and texture. Heating, lowers the viscosity but increases the solubility of gum. Heating, particularly under pressure gives smooth homogenous translucent (syrup like solution) colloidal dispersions with concentrations as high as 20-25%.

From the industrial point of view, it is the viscosity that is most important in the case of Karaya gum. Karaya gum will also form viscous sols in alcoholic solutions of up to 60% alcohol.

Specific properties and uses of gum karaya: Gum karaya is an important raw material in food, pharmaceutical, cosmetic, paper, textile and other industries. With increasing experience and knowledge of uses, gum karaya has proved superior in certain specific applications to other gums and as such it has found a market of its own. Excellent water-absorbing and water holding properties make karaya gum very useful in the preparation of ice cream, sherbets and ice-pops (ice-lollies).

The gum acts as a stabilizer and prevents formation of large ice crystals and "bleeding" of water (syneresis). If the gum is not used, the flavour and sugar from ice-pops would be sucked out leaving just a block of ice. The gum is used in the proportion of 0.2 to 0.4%. As an emulsifying agent with water absorbing and retentive properties, it acts as a binder in the preparation of meat products e.g. Bologna.

The gum, because of its cohesive properties, gives the final products a good smooth appearance. It helps in the processing and storage of meat products. It is used in less than 1% strength. It is effective in retarding staling of bread and other baked goods. As such it is used in the manufacture of doughnut mixes for the institutional trade. The preferred additive of gum karaya is about 0.1 to 0.9%. Gum karaya makes Baker's citrous-oil emulsions fairly stable and non-moulding. But slight creaming is noticed after about three months of standing. Because of its water absorbing properties gum karaya works as a stabilizer and finds its use in mereingue powders for topping of baked goods and in whipped cream products. The gum is used up to 0.8% concentration in cheese spreads. Because of its property to increase viscosity of oil-water emulsion and thereby preventing or slowing separation, it is used in French dressing and salad-dressing. Owing to its property to increase viscosity it is used in sauces, condiment bases, ketchups, sweet pickles, laxatives, denture-adhesives. The gum is effective as an emulsifier in hair wave, lotions and hair-set concentrates because of its cold water swelling and emulsifying properties. Because of its efficient binding properties of cellulose fibres it is used in the manufacture of long fibre, light weight papers e.g., condenser tissues and fruit wrap-tissues and usual rates of addition of gum are 4 to 6 kg per tonne of pulp. The textile gum solution is used principally as a thickening agent for printing dye in direct colour printing on cotton fabrics as well as for textile sizing. Small quantities of karaya gum are also used in the manufacture of dry cell batteries as well as insect repellent emulsifiers. It has been reported to treble the effectiveness of nicotine sprays in cigars when gum up to 0.2% is added.

Requirements of a standard gum: The material shall be in the form of tears of variable size or in broken irregular pieces. It possesses a slightly acetous base. The gum swells in water and in 60% rectified spirit.

The material shall contain not more than 0.5, 1.5 and 3.0 per cent of bark and other foreign organic matter for Grade I, II and III respectively (Indian Standard 5025-1969).

The material, for food and pharmaceutical industries, shall be free from contamination of animal filth, particularly rodent excreta and rodent hair. This can be tested as per I.S. 5025-1969, macroscopically against a white shining background and also microscopically.

Following are some optional requirements subject to the agreement between the purchaser and the supplier. (For method of test, appendix A clauses 4 to 10 of I.S. 5025-1969 may be seen).

1. Moisture percent	By weight max-20
2. Test for starch	Dissolve 2-6 gm of iodine and 3 gm of potassium iodine in 100 ml obtain iodine solution. Then heat 1 gm of gum with 100 ml of water to boiling until the material is dispersed. Cool the solution and add a few drops of iodine solution. The material shall be taken to have passed the test if no blue colour is produced.
3. Maximum acid insoluble ash	1% by weight on dry basis
4. Chlorides	Appendix A column 7 and 8
5. Sulphates	
6. Swelling properties min	125 i.e. swelling of gum 62.5 times.
7. Water absorption (Min.)	50 i.e. for every gram of gum 25 ml of water

Natural regeneration: Natural reproduction of the species is adequate only in inaccessible areas which are protected from grazing and where the trees are not tapped intensively for gum. It is tolerant to adverse soil conditions. As such it is found in hill tops, exposed ridges, rocky crevices, eroded slopes and similar other places. The seeds get lodged in the above places and germinate. However complete protection from grazing is necessary for the seedlings to get established. It is believed that the trees are weakened by heavy tapping for gum and hence trees do not produce fertile seeds. It seems, therefore, essential that tapping intensity should be regulated so that the capacity of tapped trees for production of viable seed is not affected.

Toona ciliata Roem.

Artificial regeneration: Seeds may be collected from April to June from forests. Nothing definite is known regarding the viability of seeds it has been ascertained that untreated seeds give best results in germination. Germination has been found to range from 32 to 77% with 18 plant per cent. It takes about 11 to 53 days for completion of germination. Seeds may be sown in polythene bags early in May soon after their collection. If watered, germination takes place in 10 to 15 days. Seedlings reach 15-20 cm height by July when they are planted out. In areas where the period of rainfall is short it would be desirable to plant one year old seedlings in the field. Plants raised in nursery beds give rise to vigorous seedlings with carroty roots.

The final spacing of plants may be from 8 to 9 m. Weedings may be done as and when necessary. As plant is palatable, cattle and chital damage the seedlings. Protection from browsing is essential for its success.

37. *S.villosa* Roxb.; Sterculiaceae.

Na : Assam: Udal, Odol; Bihar: Ganjher; Garo: Ubak; Hindi: Udal, Udar; Kanerese: Savaya, Chauri, Bilidall; Mal.: Vakka; Marathi: Sardal, Sarda, Kuthada; Oriya: Kodalo: Tamil: Muratham, Arnivakenar.

Hab : Pl. see edible plants.

The tree is the source of an important gum. This gum like karaya belongs to the tragacanth group. Since its properties are similar to Karaya gum it has the same uses. Tapping method and processing details are also similar to gum karaya.

38. *Toona ciliata* Roem. (Syn. *Cedrela toona* Roxb. ex Rottle. & Willd.); Meliaceae.

Na : Assam: Poma; Bengali & Hindi: Tun, Mahanim; Eng.: Cedrela tree, Red cedar; Kanerese: Mandurike; Mal.: Malarveppu; Marati: Kuruk; Sanskrit: Tunna; Tamil: Santhanvembu, Tunumaram; Telugu: Nandichettu.

Hab : It is a tree found in the Punjab, Assam, Bihar, the Western Ghats, Maharashtra, Madhya Pradesh and the Nilgiris. It is the source of a gum which is used as an adhesive.

Woodfordia fruticosa (L.) Kurz

39. *Woodfordia fruticosa* (L.) Kurz (Syn. *W.floribunda* Salisb.: *Lythrum fruticosum* L.); Lythraceae.

Na : Bengal: Dawai, Dhai; Bihar: Icha, Dhenti, Phuldawai, Dhai phul; Gujrathi: Dhavdi; Hindi: Dhawi, Dhauta, Jaju; Jammu & Kashmir: Thawi, Thai; Kanerese: Tamrapushpi; Oriya: Jaliko, Harwari; Punjab: Thawi, Dhawi; Mar.: Phulsatti; Dhaiphal; Sanskrit: Dhalaki, Agnijwala; Telugu: Jargi, Serinji, Gaddaisinka.

Hab : It is a common ornamental shrub growing wild in the tropical as well as sub-tropical forests in the country

It is the source of gum which can be used as a substitute for gum tragacanth.

Section - 2

8.2 Resins and Oleoresins

Resins are insoluble in water but usually dissolve readily in alcohol, ether, carbon disulphide and certain other solvents. They are chemically related to the terpenes or the essential oils. With heat they soften and then melt to a clear sticky fluid and are resistant to decay.

8.2.1 Origin of Resins

All the natural resins are of vegetable origin with the exception of lac, and similar insect exudations. The resins may be found in any part of the plant or may occur only in the inner tissue.

Usually resins do not exude unless there has been some injury. For this reason tapping has to be resorted to in order to obtain a free production of resin from the plant. Resins are secreted in plant tissues in special cavities or resin ducts. These ducts are connected with one another and when one of these passages is severed the resin flows to the wound. The method of formation of resin is not fully known but it is believed by some that it forms through polymerization and reduction of carbohydrates particularly starches. As regards its function in the life of the plant it is generally held that it may not be of any use in the growth of the plant and may assist in preventing desiccation to the injured tissue. The three most important families for the bulk production of natural resins for the commercial purposes are the Pinaceae, Leguminosae and Dipterocarpaceae. The first one includes the pines yielding rosin and turpentine. The leguminosae include the copals. The Dipterocarpaceae include dammars. The family Burseraceae yields aromatic resins also known as oleoresins such as frankincense *Boswellia serrata* Roxb. (salai) and myrrh *Commiphora mukul* (Hook. ex Stocks) Engl. Another family well known for its scented and medicinal resins is the Umbelliferae with such products as asafoetida (Hing). Some resins like Kauri are a fossilized product of trees long dead. It is derived *from Agathis australis* in New Zealand.

8.2.2 Properties of Resins

The natural resins as a group are very heterogeneous. The fact, that all are insoluble and unaffected by water, serves to distinguish them from the gums. They are soluble in a large number of organic solvents. Usually the hardest resins give the hardest and most durable films in varnish marking. Some of the resins may be quite hard and brittle like copals while others may be quite soft or semi-fluid like pine resins. .The colour of resins may again vary from colourless through various shades of amber and brown to pink. The natural resins do not usually have a definite melting point. It may vary according to the species. Essential oil is sometimes an important constituent of a resin. The fossil resins contain practically no essential oil and the hard resins very little.

8.2.3 Main uses of Resins

The use of natural resins in medicines, in varnish or as protective coatings or sources of perfume go back to very early times. The present day uses are numerous and they are employed by a large number of different manufacturing industries.

The hard resins like copals and some of the dammars are used in the manufacture of varnishes, paints and lacquers. Resin is employed in the paper industry for sizing paper to impart lustre and weight and to hinder the absorption of ink or moisture. It is also used in soap making, boot polishes, also as a lubricating agent. Resins are important constituents in the manufacture of linoleum. Sealing wax is made from shellac.

Turpentine oil is extensively used in paint and varnish industry. It is used in polishes of various kinds and as a basis for the manufacture of synthetic camphor. Varnishes made from natural resins fall into two main categories – oil and spirit varnishes. Among the spirit varnish resins, shellac is the most important.

8.2.4 Kinds of Resins

Following are some of the important resins found in India.

1. ***Dammars :*** They are soluble in turpentine, coal-tar and hydro-carbons. Good dammars are hard and almost transparent. They are derived mainly from species of *Canarium, Hopea, Shorea, Vateria* and *Balanocarpus.*
2. ***Myrrh:*** Myrrh refers to an aromatic exudation of a number of species belonging to the genus *Commiphora.* In India it is derived from *Commiphora mukul* (Hook. ex Stocks) Engl.; Syn. *Balsamodendron mukul* Hook. *ex* stocks. It is called Indian myrrh, Indian bdellium or gum-guggul. It is used as incense and as a fixative in perfumery and in medicine.
3. ***Frankincense:*** It is gum oleoresin obtained from the species of *Boswellia* of which, B. *serrata* Roxb. ex Colebr. occurs in India. This aromatic gum oleoresin is known as salai gum or Indian olibannm. The gum oleoresin is obtained on tapping of the tree. To appreciate the method of tapping of *Boswellia serrata.* (salai) it is necessary to know the distribution of oleoresin ducts in the stem of the tree.

Wood structure in relation to gum-oleo-resin exudation: (Fig. 8.6) Resin canals in *Boswellia serrata* are formed very early in developing shoot of the trees, however, in the pith development of canals takes place a little later.

They are present in all the parts of the stem viz. bark, wood and pith and even in the leaf tissues being most abundant in bark portion. As the cambial growth proceeds new resin canals are continuously developed in the parenchymatous tissue cut off by the cambium. The occurrence distribution and arrangement of these canals are best understood from the study of the different tissues of the main bole.

The bark of a mature tree is fairly thick about 1 cm in thickness and greenish yellow or greyish green. The T.S. of shoot near the apex showing resin exfoliating in thin flakes. The perioderm, which forms the outer most layers, often consists of 10-12 rows of cork cells. The tissue coming next in transverse section is phelloderm which contains a number of vertical resin canals and these are very unevenly distributed ramifying in longitudinal direction.

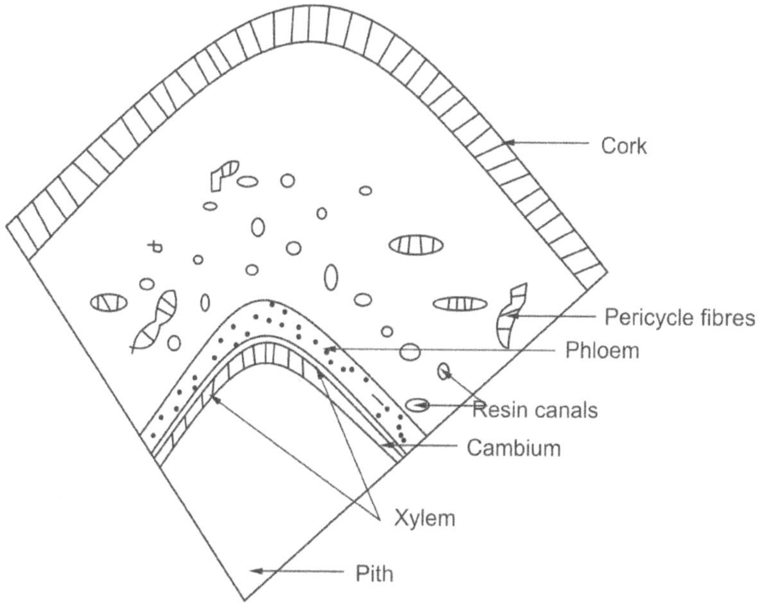

Fig. 8.6 T.S. of shoot near the apex showing resin canals.

These canals are also connected with the canals of the phloem towards the cambium. In this inner phloem, resin canals vary in shape and size in a transverse section and form a constant feature of the parenchymatous layer. In the tangential section these vertical resin canals show up as long channel which ramify and sometimes anastomose.

Boswellia serrata Roxb. ex Colebr. – A three dimensional view of the trunk showing the resiniferous system in the different parts of the tree. (Fig. 8.7).

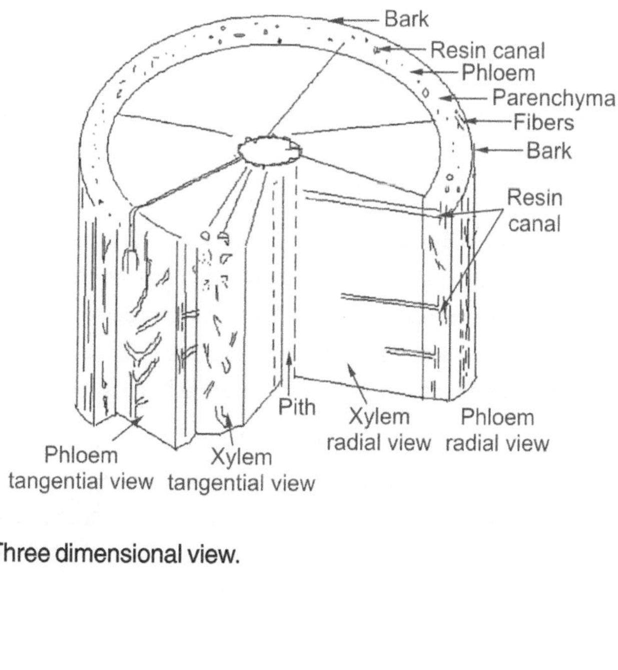

Fig. 8.7 Three dimensional view.

Although vertical resin canals are absent, horizontal ones are present in some of the wood rays. They run for long distances and do not branch like the vertical canals in the phloem. As they follow fairly straight course, they may occasionally be noticed also in the rays in transverse sections. These are connected with those of the phloem through the cambium but majority of the transverse canals do not run for long distances in the phloem and mostly end near the cambial region. It may be mentioned here that the number of horizontal canals increases from outer to inner phloem. Further, close to the cambium the number more or less corresponds to that found in the wood which is usually more than double than that of outer phloem. A few of these horizontal canals have been found to traverse the entire radius of the wood being connected with the vertical canals of the pith. These vertical canals of the pith which begin to form as a result of the onset of cambial growth have round shape and are smaller in size than those occurring in the phloem. They run only for a short distance vertically without branching and end abruptly.

From a detailed description of different types of resin canals met with, it is not difficult to construct the resiniferous system of this tree. It will be noticed that the distribution of resin canals in three different planes of the trunk vary considerably. In transverse section vertical canals are small, circular to oval in shape and are distributed in the parenchymatous layers of the bark.

The orifice of the canal is 70 to 80 μ. In the tangential plane, these canals are profuse and conspicuous, appearing as long tubular much branched anastomosing system. Besides these vertically aligned canals, there are also noticeable some horizontal resin canals as small circular structure in the phloem rays. In the radial section it will be noticed that the vertical canals of the bark have been cut in short lengths due to the ramifying nature of the canals. In some places, horizontal resin canals are also seen to be connected with the vertical ones of the bark. Further while most of the horizontal canals originate at different distances from the pith, a few are connected with the vertical canals of the pith.

Changes in wood after blazing: The blazing or exposing the resin canals of the inner bark results in the reduction in thickness of the bark from original width of 1 cm to 0.6 cm. However, the original thickness is regained after a rest of about 3 years. The number of vertical resin canals in the bark appears to increase as a result of injury.

Whether this is due to the development of new canals or more branching of the existing canals can not be said definitely. However, near the cambial region number of vertical resin canals were not observed to differ appreciably from the uninjured bark. On the other hand the horizontal canals both in the bark and in the wood have been noticed to increase in number appreciably as a result of wounding.

In bark the average number of horizontal canals increases from 4 per square centimetre in the untapped tree to as many as 10 per square centimetre in the tapped one. In the case of wood near the cambium the average number has been found to increase from 10 to 18 per square centimetre as a result of tapping.

Suggestions for making blazes: It has been stated earlier that the incidence of resin canals is maximum in the bark region. The tangential section of this portion shows that these canals are noticed to run vertically and are seen to branch and ramify profusely. As these are also connected with the horizontal resin canals coming from the wood, any blazing done to expose these canals (i.e. lying in the bark) will draw out all resinous contents present in the resin canals of bark, cambium and wood. Further, since the bark is very thin about 1 cm in thickness and the maximum yield of gum-oleo-resin is obtainable only from the bark portion, any deep blazing is not only unnecessary but may also adversely affect the growth of the tree. Therefore, the final depth of the blaze should in no case be more than 0.5 to 0.6 cm.

Tapping Method

Selection of trees: Trees of and above 90 cm at Breast Height may be tapped. Trees of lower girth classes are not to be tapped since any damage done to them may adversely affect their growth. Middle aged trees yield more gum. Selected trees are marked and numbered.

Season of tapping: Tapping season commences from October and continues until the end of May.

Method of tapping: So far no definite method of tapping of *Boswellia serrata* for gum-oleo-resin has been evolved. There have been prescriptions for blazing the trees all around or leaving a little space but this does not seem to be successful in the long run as the trees are likely to be girdled off resulting in their death. Hence following tentative method by blazing is suggested to obtain salai guggul from trees avoiding the risk of girdling the trees.

Blazes (Fig. 8.8)

Initial work: The initial blaze of 20 cm wide and 30 cm in length or height may be made in November on trees at 15 cm above ground level with sharp axe having 7.5 cm. wide blades or with saluli. Since the mature bark is 1 cm. thick and the pinkish layer of phloem is seen only by going up to 0.6 cm depth, where gum canals lie, it is necessary to make the blazes 0.6 cm deep in the bark. Deeper blazing is not likely to be of any use and rather it may damage the tree and delay the process of healing.

Number of blazes will depend on the girth of the trees and initially for a tree of 90 cm girth, one blaze may be made and for every increase of 50 cm. One more blaze may be planned for light & continuous tapping. Added blazes may be made horizontally leaving approximately equal space between the blazes. The blazes should not have any loose fibre.

The lower surface of the blaze should be slightly sloping outwards to avoid lodging of guggul in the blazed pocket in case initial blazing is done by axe. The surface should be smooth to get pure guggul.

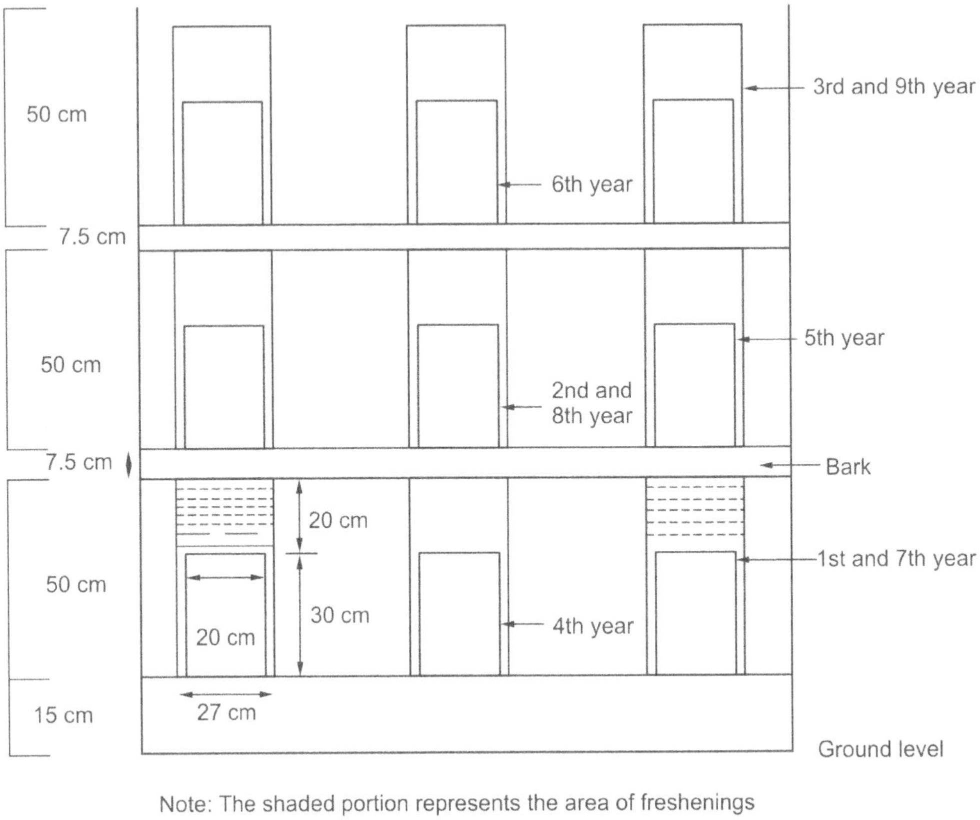

Fig. 8.8 Boswellia serrata Roxb. ex Colebr. – Plan diagram for light and continuous tapping.

Freshening of the blazes: *G*uggul starts oozing out soon after blazes are made which may be collected initially after a month when the blazes may also be freshened. Subsequent collections and freshenings may be done at fortnightly intervals till end of May. For freshening, a specially designed tool called *Saluli* maybe used. In each freshening the lower surface is not to be freshened. The edges may be slightly scraped with *Saluli* so that only 3.8 cm is increased on either side in width at the end of tapping season. This means about 0.3 cm may be scraped off on either side in width in each freshening. In the length or height about 1.6 cm may be freshened every time towards the upper edge. Thus at the end of one season the final size of the blaze in length or height will be (30 cm + 20) 50 cm. The surface of the blaze may be slightly scraped each time taking care that depth is not increased.

Following are the details of space to be covered per blaze

Initially (a) in width 20 cm

(b) in length or height 30 cm

In subsequent freshening.

Initially (a) in width 7.5 cm

(b) in length or height - 20 cm

Total size of each year's blaze.

Initially (a) in width or horizontally - 27.5 cm

(b) in length or height or vertically - 50 m

For the sake of convenience following schedule has been drawn for making suitable number of blazes in trees of different girth classes.

Blazing in subsequent years: The bole is divided into three zones, each one being tapped for one year for continuous tapping. Thus the three zones are covered in three years. For making another horizontal row of blazes in subsequent years 7.5 cm, space may be left above the blazed portions. The blazes of the subsequent year's horizontal row should be alternating or staggering with the previous year's blazes i.e. the previous and the fresh blazes will not be in the same longitudinal row. The size, shape etc., of the blaze before and after freshening may be the same as for the first year. Similarly in the third year the same method will be followed. In the fourth year blazing may be continued in the zone where it was done in the first year. But this time the blazes may be in between the previous ones. With this method the blazed region gets complete rest for two years and one year's partial rest.

Schedule of number of blazes as calculated on the basis of girth

Girth of the tree	No. of blazes	Total space to be covered horizontally by the number of blazes indicated in the previous column	Minimum space left unblazed on the tree horizontally	Minimum space available between blazes for tapping in the next cycle
90-140 cms	1	27.5 cms	62.5 cms	62.5 cms
140-190 cms	2	55.0 cms (27.5x2)	85.0 cms	42.5 cms
190-240 cms	3	82.5 cms (27.5x3)	107.5 cms	35.0 cms
240-290 cms	4	110.0 cms (27.5x4)	130.0 cms	32.5 cms
290-340 cms	5	137.0 cms (27.5x5)	152.5 cms	30.0 cms

Note : Space between the blazes has been calculated taking the minimum of the girth class.

This cycle could be repeated making use of all the trees available simultaneously. In the 7th year tapping can be reverted to the blazed area of initial exploitation i.e. blazes may be made overlapping the first year's blaze. Thus the blazed portions will get complete rest for six years during which period the wound would get healed up and will resume normal activity. After blazing if the space between the two adjacent blazes is not sufficient, for further blazing, it should not be taken up but stopped

Collection: After making initial blazes salai guggul may be collected after a month. Collection may be done by a scraper keeping a tray having a semi circular edge around the blazed surface. Subsequent freshenings and collections may be done fortnightly.

Yield: Average annual yield per tree is estimated to be about 1 kg.

Rosin or Colophony: It is one of the two important distillation products of the oleoresin of pine trees. The other product is turpentine oil. In India the species tapped is *Pinus roxburghii* Sar.

(a) *Tapping method:* French cup and lip system of continuous light tapping is employed. Areas under tapping are protected from fire. The trees which have been marked for felling are tapped heavily, in other words are tapped to death. Trees below 35 cm diameter are not tapped. Trees up to 60 cm diameter have only one and those above 60 cm two channels. Heavily tapped trees have more than 2 channels. At the time of commencement of tapping, the dimension of the channel is 10 cm long and 10 cm broad and 1 cm deep. Five freshenings are made in a month. At each freshening maximum extension in length is 0.7 cm to 1.0 cm and thus at the end of the year is about 60 cm. In subsequent years channels are extended by 40 cm every year. In making channel the outer bark is first removed and next with a mallet and a curved chisel a channel is made. A lip made of tin is nailed to the tree at a height of 16 cm (5 ½") from the ground. The lip is placed in a slanting position to allow the resin to flow down it just at about 3.5 cm (1 ½") above the cup. Some times a lip is inserted by making a cut in the tree. The cut made for the insertion of lip is not desirable as it interferes with the cambial activity and instead the lip is tucked to the tree with nails.

(b) *Freshening:* In order to open up the clogged ducts a thin shaving of wood from the channel is removed by an adze. Freshening should not be deeper than 2 cm. It is started from March and continued till November. Four to five freshenings are made every month. Increased freshenings for the same width and depth increases the yield of resin by 20 per cent. Depth of freshening has no appreciable effect on the yield. It is seen that application of 50 per cent solution of sulphuric acid mixture by weight to the exposed surface increases and prolongs the flow of resin. This may eliminate repeated freshening and reduce labour cost. Recent investigations have shown that chemicals 2,4-D 2,4,5-T (weedicides) if applied on the blazes in small proportions (about 2 per cent strength) increases the yield of resin by about 50 to 60 percent.

(c) *Collection and processing:* Resin is collected in buckets and brought to the central places and from where it is taken to the factory. Resin is distilled to get rosin and turpentine oil.

Competition from synthetic resins: In recent years there has been some decline in the use of natural resins in varnish industry on account of the competition from synthetic resins. To what extent these natural resins will be replaced by the synthetics it is difficult to foretell.

However production of natural resins at lesser cost accompanied by more careful grading and sorting may enable them to retain their position.

8.2.5 Resin and Oleoresin Yielding Plants

Pine tapping

1. *Ailanthus triphysa* (Dennst.) Alston (Syn. *A.malabarica* DC.); Simbarubaceae;

Ailamhus is a small genus of large trees distributed in South-East, China and Australia. Four species are indigenous to India

A.altissima (Mill).Swingle Syn. *A.glandulosa* Desf. of Japan and China is sometimes planted in Northern India as an ornamental tree. The wood of this species is characterised by ringporous structure.

Na : Hindi: Guggula dhup; Kanerese: maharukh, Dhup, Malmaddi; Mal.: Mattipal, Pongilyam; Marathi: Guggula-dhup; Tamil: Mattipal, Peru, Perumaram; Telugu: Maddipalu, Peddamanu.

The species is found in the Western Ghats from Kanara southwards up to 1,500 m elevation, and also in Ceylon and lower Burma. It is a large tree with tall cylindrical trunk. Sapwood and heart wood are not clearly demarcated. Bark is grey, rough and often shows bright reddish grains of resin. The tree yields an aromatic resin which is used for burning as an incense and also for making joy-stick (Agarbatti). The resin is also used in dysentery in local medicine and the bark in indigenous medicine as tonic and febrifuge. Collection of this resin is usually done from natural exudations. To induce the flow of resin, the bark is injured.

Ailanthus triphysa **(Dennst.) Alston**

Properties of gum resin: The pure gum-oleo-resin is very soft having the consistency of thick treacle (like mollasses). Alcohol readily dissolves the resin and on evaporation leaves it as a very viscous, transparent, light-brown, semi-liquid, which does not solidify even on exposure for many days.

Analysis indicates that gum-oleo-resin contains ash 2.15%, volatile oil 2%, loss at 100 °C 14.4%, matter soluble in 95% alcohol (resin) 71%, matter insoluble in 95% alcohol (non. resin) 14.5%, acid value 31.53.

Alcohol soluble resin is light brown, highly viscous and sticky. Its acid value is 29.2, sapondification value 64.2, easter value 35.0, and ash content 0.2%.

The resin falls in the Elemi class and could be used as a satisfactory substitute for gum elemi in varnishes and lacquers.

2. *Altingia excelsa* Noron.; Hamamelidaceae.

Na : Assam: Jululi; Hindi: Silaras.

Hab : It is a deciduous tree of Assam and Burma. The stem is the source of a resin styrax.

Altingia excelsa **Noron.**

3. *Anacardium occidentale* :L. Anacardiaceae.

Na : Eng.: Cashew; Hindi: Kaju.

Hab : Pl. see edible plants

Anacardium occidentale L.

The shell oil is the source of resinous matter used in the manufacture of varnishes, typewriter rolls, industrial flooring tiles, inks, oil cloth, paints, waterproofing paper, acid and alkali resistant cements, laminating resins and rubber-compounding resins. It is also used in the treatment of water required for boilers, air conditioning units, ice plants, textile industries.

4. *Anisoptera scaphula* (Roxb. Pierre) (Syn. *Hopea scaphula* Roxb.); Dipterocarpaceae.

Na : Hindi: Bailam.

Hab : It is a tree found in Assam. The stem yields resin on wounding.

5. *Boswellia serrata* Roxb. ex Colebr.; Burseraceae.

Na : Eng.: Indian frankincense, Indian olibanum; Hindi: Salai.

Hab : Pl. see medicinal plants.

On tapping the tree exudes gum-oleoresin from the bark which is known as salai guggal. Method of tapping has been given in the preceding pages of this chapter.

Nature and properties of gum oleo-resin: Gum-oleo-resin hardens slowly after exudation, retaining its golden colour and transparency. It varies in colour from transparent golden brown to dark-brown or dark greenish brown. The odour is like olibanum but fainter. It burns readily and diffuses an agreeable odour. Salai gum has the following average composition – moisture 10-11, volatile essential oil 8-9, resin 55-57, gum 20-23, insoluble matter 4-5%. These constituents can be separated. Its resin is very brittle, with a vitreous fracture and resembles colophony in odour and is tasteless. It is soluble in most organic solvents and does not dissolve completely in sodium carbonate solution even on prolonged boiling. Its softening point varies from 65-72 °C and the melting point from 73-78 °C.

Hydrolysis of the pure gum of salai yields mainly pentoses (65% as arabinose) with a high proportion of arabinose. Galactose and xylose are present only in small quantities. The gum also contains oxidizing and diastatic enzymes and 3.03% of total nitrogen. Commercial samples of gum contain varying proportions of resin and about 50% of water soluble material.

It is used as an incense and in the preparation of ointments, paints, varnishes. In large cities it is used as a fire-lighter for domestic angeties.

Good quality light coloured gum-oleoresin can be used in place of conventional canada balsam for mounting microscopic slides. For this purposes it has to he mixed in 50 per cent xylene.

Boswellia serrata Roxb – Branch

6. *Canarium bengalense* Roxb.; Burseraceae.

Na : Assam: Nerebi, Dhuna; Eng.: East Indiancopal; Hindi: Narebi, Dhuna; Lepcha: Narockpa.

Hab : It is a tall tree found in the evergreen forests in Assam and Bengal.

The resin which exudes in large quantities from wounds and fissures in the bark is clear and amber coloured and it rapidly hardens to a brittle mass and is soluble in turpentine. It is aromatic and is burnt as incense.

***C. mukul* (Hook. Ex Stocks) Engl.**

7. *C.sikkimense* King; Burseraceae.

Na : Bengal & Nepal: Goguldhup; Bhutan: Pah; Hindi: Gokul dhup; Lepcha: Narokpa.

Hab : It is a tall tree found in Sikkim and West Bengal.

The stem yields an aromatic amber-coloured resin which is used as an incense. On exposure the resin becomes hard and brittle.

8. *C.strictum* Roxb.; Burseraceae.

Na : Eng.: Black dammar; Hindi: Kala dammer.

Hab : Pl. see medicinal plants.

Stem on wounding or injury yields a resin called black dammar in trade. A common method of obtaining the dammar is to burn brush wood round the base of the tree so as to scorch the bark and outer layers of wood. The resin begins to ooze and continues to flow. Sometimes vertical incisions are made in the bark and then fire is ignited against them in order to induce the flow of resin. On exudation the resin hardens to a bright shining black coloured mass which breaks, with a glassy fracture. It is used in medicine as a substitute for burgundy pitch in making plasters.

9. *Commiphora caudata* (Wight & Arn.) Engl.; (Syn. *Balsamodendron caudatum* Wight & Arn.); Burseraceae.

Na : Eng.: Hill mango.

Hab : Pl. see medicinal plants.

Bark on wounding yields gum oleo-resin which is used as an incense, in medicine and for embalming.

10. *C.mukul* (Hook. Ex Stocks) Engl. (Syn. *Balsamodendron* (Hook. ex Stocks); Burseraceae.

***C.mukul* (Hook. Ex Stocks) Engl. Branch**

Na : Bengali, Hindi & Gujrathi: Guggul; Eng.: Indian bdellium tree; Mar. & Kan.: Guggule; Sanskrit: Guggulu, koushikaha, Devadhupa; Tamil: Maisakshi guggul; Telugu: Guggul.

Hab : It is a small tree or shrub with spinescent branches found in arid tracts in Rajasthan, Maharashtra and Karnataka. It is also planted as a hedge.

The stem yields a fragrant gum oleoresin, called Indian myrrh, Indian bdellium or gum-guggal in trade. Guggal is collected in cold season. It is obtained by making oblique cuts on the stem with a sharp knife or an axe. The gum-resin is scraped off the wound with the knife and collected in a tin. It is used as an incense, as a fixative in perfumery and in indigenous medicine.

Dipterocarpus alatus **Roxb. - Tree**

11. *C.roxburghii* (Arn.) Engl.; (Syn. *Balsamodendron roxburghii* Arn.); Burseraceae.

Na : Hindi: Gugal.

Hab : It is a dwarf tree found in Rajasthan, Gujarat, Madhya Pradesh and also eastwards. It yields an aromatic gum-oleoresin which is used as an incense.

12. *Croton aromaticum* L. (Syn.*C. lacciferus* L.); Euphorbiaceae.

Hab : It is an aromatic shrub found in Andhra Pradesh. It produces a gum-resin which could be used in varnishes.

Dipterocarpus alatus **Roxb. Branch**

13. *Daemonorops kurzianus* Hook.f. (Syn. *D.grandis* Kurz. *Calamus grandis* Kurz.) Palmaceae.

Na : Eng.: East Indian Dragon's blood; Hindi: Hiradukhi, Aprang; Malayalam: Rotanjarima; Mar. & Guj.: Hiradakhan; Tamil: Kondamurae, Rattam.

Hab : It is a slender lofty climbing cane palm of the Andaman Islands.

The resinous exudate obtained from the fruits is used for colouring lacquers and varnishes.

D.tuberculatus **Roxb. - Tree**

14. *Dipterocarpus alatus* Roxb.; (Syn. *D.costatus* Gaertn.); Dipterocarpaceae.

Na : Eng.: Gurjun; Hindi: Gurjun.

Hab : It is a tall tree of the Andaman Islands.

Stem yields a resin which is used in varnish and paint industry.

15. *D.indicus* Bedd.; Dipterocarpaceae.

Na : Eng.: Gurjun; Hindi: Gurjun.

Hab : It is found in the Western Ghats. Resin obtained from the stem is used in the manufacture of varnishes and lithographic inks. It is also used for smearing on wood work, in caulking and varnishing boats.

D.tuberculatus **Roxb. Branch & fruits**

16. *D.tuberculatus* Roxb.; Dipterocarpaceae.

Na : Eng.: Gurjun; Hindi: Gurjun.

It is the source of resin used in varnish industry.

17. *D.turbinatus* Gaertn. f.; Dipterocarpaceae.

D.turbinatus **Gaertn. f.**

Na : Eng.: Gurjun; Hindi: gurjun.

Hab : It is found in the Andaman Islands.

It is the source of an oleoresin known as Gurjan oil. For extracting the resin, oblique pyramidal shallow cavities are made in the trunk and fire is ignited near them to accelerate the flow of oleoresin.

Tapping is done during the hot and dry months. Gurjun oil has many local uses. It is said to have medicinal properties. It is used for caulking and varnishing boats and for making torches. It is also used for smearing on woodwork.

18. *Garcinia morella* Desr.; Guttiferae;

Na : Assam: Khujithekera; Bengali & Marathi: Tarnal; Eng.: Indian gamboge; Hindi: Gota ghanba, Tarnal; Kanerese; Harctala, Devanabali, Jarise; Malayalam: Chigiri, Daramba, Karukkampuli, Pinnarpuli; Tamil: Makki, Solaippuli; Telugu: Pasupuvarne, Revalchini.

Hab : It is an evergreen small tree found in the Khasia hills and the Western Ghats.

It is the indigenous source of gamboge. Gamboge is a resinous substance which is obtained from the bark on wounding. The exudate is collected in bamboo cups. Gamboge is reddish yellow or brownish orange in colour. It is odourless and tasteless or slightly acidic. It forms a yellow emulsion with water. It can be used in the preparation of water colours and gold coloured spirit varnishes for metals.

19. *Gardenia gummifera* L. f.; Rubiaceae.

Na : Eng.: Gummy cape Jasmine; Hindi: Dikamali.

Hab : Pl. see medicinal plants.

It is the source of an oleoresin which is known as Dikamali gum in trade. It is secreted on the buds and young shoots. It has the flavour of asafoetida and is used as a flavouring agent in food and is also used for medicinal purposes.

20. *G resinifera* Roth; (Syn. *G.lucida* Roxb.) Rubiaceae.

Gardenia gummifera **L. f.**

Na : Eng.: Brilliant Gardenia; Gujarathi, Hindi & Marathi: Dikamali.

Hab : It is a handsome shrub or small tree found in the forest in central and South India. Sometimes it is cultivated in gardens.

Hopea odorata **Roxb.**

It is the source of an oleoresin known as Dikamali gum in trade. Gum-oleo-resin is secreted on the buds and young shoots. It has medicinal value.

21. *Hopea odorata* Roxb.; Dipterocarpaceae.

Na : Andamans: Thingan, Rimda; Kanerese: Bilitirupu, Kallurala; Malayalam: Urappimpasa; Tamil: Urappuppicin; Trade: Thingan, White thingan.

Hab : It is a tall tree growing to 45 m in height and is found in Bengal, Assam and Andaman Islands.

When the stem is injured a fragrant oleoresin exudes, which on exposure to air solidifies in the form of nodules. They are brittle. The trees are generally not tapped but natural exudation only is collected.

The resin is hard and it is known as rock dammar. It is used in paints and varnishes. In indigenous medicine it is used in the preparation of ointments for sores and wounds.

22. *Jurinea macrocephala* Benth.; Asteraceae (Compositae.)

Na : Hindi & Common name: Dhoop.

Hab : It is an herb without aerial stem and is found in the western Himalayas from Kashmir to Kumaon in the sub alpine to alpine zone.

The roots are highly aromatic and yield oleoresin which is used as incense. The roots as such are also used as incense.

23. *Kingiodendron pinnatum* (Roxb.) Harms; (Syn. *Hardwickia pinnata* Roxb.); Caesalpiniaceae.

Na : Coorg: Choupaini, Kolavu; Kanerese: Yennemara; Mal.: Shurali, Kiyavu, Kodapalla; Tamil: Madayan, Sambrani, Kodavu, Kodapalai.

Hab : It is a large tree fairly common in parts of Maharashtra, Karnataka, Kerala and Tamilnadu. .

The tree on tapping yields dark red oleoresin which is used for preserving wood work. Trees of 1.5 metres girth are usually tapped for the oleoresin. Stem is bored right up to the pith. The incision is kept sloping downwards from the pith to the bark. A lip made of tin is fixed below the bore and the liquid is collected in Kerosene oil tin. If the liquid is thin it flows out entirely in one day. When the flow stops the bore is plugged with a piece of wood. Trees are tapped all the year round except during monsoon.

Liquidambar orientalis **Mill.**

24. *Liquidambar orientalis* Mill.; Hamamelidaceae.

Na : Eng.: Styrax, Oriental sweet gum; Hindi: Silarus

Hab : The tree is native of Asia Minor and has spread to other places.

The tree produces an oleoresin known as styrax which is used in fumigating powders, pastilles, soaps and in various perfumes.

***Pinus gerardiana* Wall. - Tree**

25. *Melanorrhoea usitata* Wall.; Anacardiaceae.

Na : Eng.: Bunnese lacquer tree; Hindi: Kheu, Suthara; Manipur: Kheu; Trade; Thitsi.

Hab : It is a large deciduous tree found in Manipur, Bengal and Assam. On tapping, the stem yields oleoresin or natural varnish known as Bunnese lacquer or thitsi. A "V" shaped incision is made in the bark and a receiver is kept at the pointing end of "V". After ten days when exudation stops fresh incision is made & freshenings done periodically. Oleoresin is used as a water proofing paint for boats, household vessel and paper and cloth sizing.

26. *Pinus gerardiana* Wall.; Pinaceae.

Na : Eng.: Neosia pine, Edible pine; Hindi: Chilgozah.

Hab : Pl. see edible plants.

Resin obtained from this species is a rich source of pinene which could be used for synthesis of camphor. Since the species is found scattered in nature and also the yield of resin being low it is not tapped for resin. It has greater value for its seeds.

***Pinus gerardiana* Wall. Seeds**

27. *P.insularis* Endl. (Syn. *P.khasia* Royle ex Parle.). Pinaceae.

Na : Bengal: Saral; Eng.: Khasia pine, Benguetpine; Hindi: Digsa, Taru; Khasi: Ding-se, Dieng-kysi; Lushai: Far.

Hab : It is a medium sized pine tree found in the Eastern temperate Himalayas especially in the Khasia and Jaintia hills, Manipur and Naga hills.

The resin obtained from the stem is used for the manufacture of turpentine and resin. The turpentine oil is very rich in á pinene and as such it can be a good source of synthetic camphor.

28. *P.roxburghii* Sar.; (Syn. *P.longifolia* Roxb.); Pinaceae.

Na : Eng.: Long leaved pine, Three leaved pine; Hindi: Chir, Salla.

Hab : Pl. see essential oils.

It is found in the lower Himalayas and Madhya Pradesh. Resin obtained from the stem on distillation yields turpentine oil and (natural forest) resin.

***P.roxburghii* Sar.**

(Method of tapping of this pine for resin is given in the preceding pages of this chapter.) Turpentine oil is used for the manufacture of camphor and also in varnish industry. Rosin or colophony is used in paint, sizing paper and cloth, boot polishes and in the manufacture of linoleum, sealing wax, oil cloth, lubricating compounds and several kinds of inks and also for soldering.

P.wallichiana **A.B.Jackson**

29. *P.wallichiana* A.B.Jackson; (Syn. *P.excelsa* Wall.; *P.griffithii* M Clell); Pinaceae.

Na : Bhutan: Tongschi, Lamshing; Eng.: Indian blue pine, Five leaved pine; Himachal Pradesh: Lim; Hindi: Kail; Kashmir: Yiro, Kaiar, Kail; Kumaon: Raisalla, Lamshing, Byans, Dolchilla; Lepcha: Neet-kung.

Hab : It is found in the temperate to sub-alpine regions generally in the Western Himalayas.

The tree on tapping yields resin which could be used for distillation of turpentine oil and rosin. The pinene content in turpentine oil is very high and as such it can be a good source of synthetic camphor. But owing to inaccessibility, short tapping season and low yield of resin, the species is not fully exploited.

Podophyllum hexandrum **Royle**

30. *Podophyllum hexandrum* Royle (Syn. *P.emodi* wall. ex Hook.f.x. Thoms.); Berberidaceae.

Na : Eng.: Indian Podophyllum; Hindi: Papra, Vakasa.

Hab : Pl. see medicinal plants.

The resin obtained from rhizomes and roots is used as a cholagogue purgative. Recently it has been found to affect the growth of malignant tumours.

Semecarpus anacardium **L.f**

31. *Semecarpus anacardium* L.f. (Syn. *Anacardium orientale* L.); Anacardiaceae.

Na : Eng.: Marking nut; Hindi: Bhilawa,

Hab : Pl. see edible plants.

A black resin obtained from the pericarp of the fruit is used as a varnish for baskets and wickerwork.

Semecarpus anacardium **L.f**
Branch with fruit

32. *Shorea robusta* Gaertn.f.; Dipterocarpaceae.

Na : Eng.: Sal; Hindi: Sakhya, Sal, Karshya.

Hab : Pl. see edible plants.

It is the source of an oleoresin known as Ral or Laldhuna or sal dammar. When incisions are made in the trunk of the tree the oleoresin oozes from the wounds. On exposure to air it hardens. Therefore in order to get a regular supply, the blaze has to be frequently freshened by scraping off the hardened resin. A good mature tree yields about 10 pounds of the oleoresin. Natural exudation is frequently found on the tree when they are injured by insects or owing to other damages. Sal dammar is nearly opaque and brittle. It is used as an incense and fumigant. It is also used in paints and varnishes and for caulking boats. It is said to have medicinal properties. The sal dammar on distillation gives "chua" oil which can be used in perfumery.

33. S.tumbuggaia Roxb.; Dipterocarpaceae.

Na : Hindi: Kala damar; Malayalam: Tampakam, Vanbogar; Tamil: Tambagom, Karuppudamar, Cangu; Telugu: Thamba, Nalladamara.

Hab : It is found in South India.

The stem is the source of a resin which is used in marine yards as a substitute for pitch and also as incense.

34. *Vateria indica* L. (Syn. *V.malabarica* Blume); Dipterocarpaceae.

Shorea robusta **Gaertn.f**

Na : Bengal: Chundrus, Safedamar; Bombay: Ral; Coorg: Biledupa, Velthapaini; Eng.: White damar, Indian copal tree, Piney resin, Malabar tallow tree; Hindi: Bada sal, Safed damar, Ral, Dhup; Kanerese: Dhupadamara, Mundadhupa, Saldhupa, Bileguggila, Biledammara; Mal.: Vellakunturukkum, Peinimaram, Perumpiney, Payani, Payin; Sanskrit: Ajakarna; Tamil: Vellaidamar, Vellaikuntricum, Pineimaram, Dhupmaram; Telugu: Telladamaru, Dhupadamaru; Trade: Yellaipiney.

Hab : The tree grows mostly in moist evergreen forests but frequently extends into the semi-evergreen and occasionally into deciduous forests adjoining rivers. It is found in the Western Ghats and foothills from North Kanara (Karnataka) to Kerala through the Ghat forests of South Kanara, Coorg, Malabar and Coimbatore. Apart from Ghats it is also found on the Deccan Plateau and along the Arabian Sea Coast, the altitudinal range being 60 to 1220 meters. It is a shade and moisture loving tree. It is usually of sporadic occurrence but in favourable localities, especially near streams in evergreen forests where there is plenty of moisture with good drainage it tends to form nearly pure patch. The resin obtained from the stem is used as incense and in paint and varnish industry. When the bark is wounded in the beginning of the dry season the resin begins to ooze within 3 or 4 days and continues flowing for two or three months. Sometimes fire is ignited round the base of the tree so as to scorch the bark which then splits and the resin exudes. The latter method is injurious to the plant.

35. *Vatica lanceaefolia* Blume; Dipterocarpaceae.

Na : Assam: Morhal, Morakur, Monal; Eng.: Vatica; Hindi: Morhal; Khasi; Dieng-so-karina.

Hab : It is a small evergreen tree found in Assam ascending to an altitude of 900 m.

The bark on wounding yields a white aromatic oleoresin which is used as incense and for flavouring tobacco.

9

Fatty Oils

9.1 Nature of Oils

Fatty oils, also known as fixed oils, are compounds of glycerol (glycerin) and fatty acids (glycerides). Unlike the essential oils, they do not evaporate or become volatile. They are insoluble in water but are soluble in organic solvents. When they are liquid at normal temperature they are known as fatty oils whereas those, which become solid, are called fats. Various uses of the different fatty oils depend on their chemical and physical composition and behaviour. Some oils on exposure to air absorb oxygen and form a firm tenacious film or skin. They are known as drying oils. This property of the drying oil is therefore of great importance in paint, varnish and linoleum industries. They are unsuitable for lubricating purposes. The semidrying oils absorb oxygen slowly and only in limited amounts. They form film only after long exposure. They are used in the manufacture of candles, soaps and also as illuminants. The non-drying oils and solid fats are obviously suited for use as food and lubricants. When these compounds break down they yield fatty acids and glycerine and usually develop a rancid odour and taste.

The waxes are found on the epidermal layers of leaf and fruits and because of their impervious nature they check water loss by transpiration in plants. They are harder and have a higher melting point than the fats. They do not become rancid and are less easily hydrolyzed. In chemical composition they are similar to fats and are esters of monohydric alcohols rather than glycerides.

9.2 Occurrence of Oils

Occurrence of oil in plants is practically limited to the fruit and, in the majority of cases, the seeds or seed kernels are the sources of oil though some fruits are provided with an oily pericarp or outer pulp surrounding the seed. Generally the albuminous seeds are rich in oil content. Drying and semi drying oils are of more frequent occurrence in plants of temperate regions while non-drying oils and fats predominate in tropical species.

9.3 Collection and Processing of Seed

There is a wide gap between the consumption and production of fatty oils production being far less than the requirement. Consequently there has been a consistent rise in the price of oils. To augment the supplies, efforts are being made to exploit forest oil seeds. The non-edible oil seeds of forest origin like mahua *Madhuca longifolia* (Koenig) Mac. Var. *longifolia; M.longifolia* (Koenig) Macb.var.*latifolia* Roxb.; *neem Azadirachta indica* Linn.; Kusum *Schlichera oleosa* (Lour.) Merr.; Kokam *Garcenia indica* A.juss.; *Shorea robusta* Gaertn.; Karanj; *Pongamia pinnata*; Kakan; *Salvodora oleoides* Dcne; Nahor, *Mesua ferrea* Linn. and Undi *Calophyllum inophyllum*, Linn. Most of these spp. excepting sal are scattered. Collection of fruits is done by climbing up the tree and cutting of the branches bearing fruits. Some times they are collected when they are fallen on the ground. Thereafter, the outer skin or fruit pulp is removed. The seed is spread out on the floor for drying. Moist seeds develop heat during the process and require agitation from time to time. The seeds are conveniently stored in gunny bags after they are completely dry. Long storage particularly in moist conditions results in the deterioration in quality. The kernel is very much susceptible to fungal damage though the shell remains apparently unaffected.

9.4 Extraction and Processing of Oil

On a commercial scale oil is obtained from the seed by expression or by extraction by solvents, the former being a common practice in the villages. There are three traditional methods of expressing oil from seed (i) Boiling method; (ii) Chappa method; (iii) Bullock Ghani method. In the boiling method the de-pulped seed is boiled in water for some time and then dried. Thereafter it is roasted on a hot pan until the shell becomes crisp and friable. The whole seed is then crushed into powder in a Dhenki (husking appliance). The powder is boiled with water for some time after which the oil floating at the top of water is separated by decantation. The husk and shell are devoid of oil. They reduce the value of oil cake for feeding purposes and in most cases the quality of oil is also affected. Therefore it is desirable to remove them before extraction of oil. However in the other two methods the outer shell of the seed is removed by rolling a stone or a wooden plank over a thin layer of the seed kept on a hard surface and winnowing off the husk. The separated kernel is powdered in a Dhenki. In the Chappa method the powdered kernel is steamed over a vessel of boiling water and pressed between the two limbs of Chappa. In the Bullock-Ghani method the powdered kernel is sprinkled with a little hot water and pressed in a Bullock-Ghani. Very often the crude oil contains impurities such as fine meal in suspension, water, mucilage etc. On storage they settle down and the oil is separated by decantation. Lately the fatty oils are extracted by solvents in solvent extraction plants. The details of this process are given under Shorea robusta Gaertn. f. in this chapter. The residues obtained on extraction of oil from seeds or kernels are rich in proteins and are valuable as cattle feed and fertilizers.

9.5 Uses

Fats and fatty oils are put to a variety of uses. Since many of the oils of forest origin have an unpleasant odour, colour and taste, they are not generally used for edible purposes. They find uses in the manufacture of soaps, candles, paints, varnishes, linoleum, for preparation of grease and as illuminants. However with modern technological developments such oils are being rectified by bleaching, deodorisation and hydrogenation and converted into edible oils. Some oils have been credited with medicinal properties.

9.6 Fatty Oil Yielding Plants

1. *Acacia nilotica* (L) Del. (Syn. *Acacia arabica* Willd.) Fam. Mimosaceae.

Na : Hindi: Babul;

Hab : Pl. see gums and resins.

Babul seeds are good source of proteins. 19.6-26.4%. The seeds also contain oil 3.3-6.2% which can be recovered by solvent extraction. De-oiled babul seed meal has a ready market as animal feed. Seed oil is used in soap industry. However the problem in babul seed oil production is that the seed coat is hard and difficult to break.

2. *Actinodaphne hookeri* Meissn.; Polypodiaceae.

Na : Eng., Hindi & Marathi: Pisa; Mal.: Malavirinya; Tamil: Thali; Uriya: Jharchampa.

Hab : It is a moderate sized evergreen tree found in Western Ghats and also in the forests of Orissa.

The seeds yield about 48 per cent fatty oil. It is also the source of lauric acid. Oil is used for soap making and production of lauric alcohol. The plant can be propagated by seed.

Acacia nilotica (L.) Del.-Pod

3. *Aleurites fordii* Hamsl ; Euphorbiaceae and

4. *A.montana* E. H. Wils.; Euphorbiaceae. *A. fordii* Hams.

(China wood oil, Tung oil tree) grows abundantly in the cooler parts of central and Western China and ascends up to 1500 m. It is also found in the Southern Shan States of Burma. In India a few plantations were started in the Tea estates of Assam, Bengal, Bihar, Coorg and other areas of Karnataka. In its natural habit, it is essentially a hillside plant and is said to flourish in rocky areas and on poor soils. It requires a cool climate, but is very frost tender. In India, the species has been grown in moderately fertile well-drained and slightly acidic soil. It cannot stand water logging even for a short duration and will not grow in alkaline soil with excess of lime and phosphate. An elevation of 750-1050 m and a rainfall of about 1250 mm are best suited to its growth.

A.montana E.H. Wils. (Tung wood oil tree) occurs in the subtropical parts of Southern China and in some of the Southern Shan States of Burma. Like *A.fordii* this is also a hillside species, but it can thrive in warmer climate and can stand heavier rainfall, provided the area is well drained. Its growth is good even in slightly acidic soil. Experimental cultivation in different parts of India has shown that this species is better suited to the Indian climatic and soil conditions than *A.fordii, Aleurites montana* has been found to thrive better in Assam, Bengal and in the area around Ranchi in Bihar. However, there are yet no large-scale plantations in India. *A.montana* has done very well in the alluvial soil of Assam. A rainfall varying from 1750 to 2750 mm and a maximum temperature of 36 °C and a minimum of 6 °C have not proved unfavourable to its growth. It has done equally well in Karnataka at an elevation of 750-900 m and an annual rainfall of 1500 mm.

Aleurites fordii **Hamsl - Branch with fruits**

A.montana **E.H.Wils. Tree**

A.montana **Branch with fruits**

Annona Sqamosa L..

Argemone mexicana **L.**

It is frost tender. This species also gives a much better yield of fruits than *A.fordii.*

Industrial Uses: *A.montana* seed oil is not at present a commercial product though it has properties similar to those of the oil of *A.fordii*. Tung oil which is usually a mixture of the oil of *A.fordii* and *A.montana* (about 10%) trade does not make any distinction between these due to their similar properties. Tung oil, which is a quick drying oil, has been used for caulking boats and junks, painting of woodwork and in admixture with natural resins in preparation of varnishes. It is also used for water proofing wood, paper and fabrics. It is at present extensively used, after suitable treatment, in the manufacture of paints, enamels, lacquers, etc. It has become an important raw material in the linoleum and water proofing industry and in the manufacture of electrical insulating varnishes and in the treatment of aeroplane fabrics and in the automobile industry (Brake linings). The oil when burnt in an insufficient supply of air yields carbon, an ingredient of Indian or Chinese ink.

5. *Annona Sqamosa* L..; Fam. Annonaceae.

Na : Eng.: Custard apple; Hindi: Sitaphal.

Hab : Pl. See edible wild plants.

It grows wild in many parts of A.P., Karnataka and Maharashtra. It is also grown in farmlands for its high priced fruits. The fruit contains large number of seeds. Kernel is 68.5% of seed and the oil yield is 39.5% of kernel. De-oiled meal can be used as cattle feed only after removal of toxic principles.

6. *Aphanamyxis polystachya* (Wall.) Parker; (Syn. *Amoora rohitika* Wight and Arn.); Miliaceae.

Na : Beng.: Tikataraj, Pittaraj; Eng.: Sophaga; Hindi: Harin hara; Kanerese: Mullumunthala; Mal.: Chemmarom; Mar.: Rohada; Sanskrit: Rohitika; Tamil: Malampuluvan; Telugu: Chawa-manu; Trade: Amoora.

Hab : It is an evergreen tree found in eastern parts of Uttar Pradesh, Bengal, Assam and the Western Ghats.

The seeds contain 78% of kernel, which yields 47% of semidrying oil, which is used for illumination and soap industry.

7. *Argemone mexicana* L. Papaveraceae.

Na : Bengal: Sialkanta; Eng.: Mexican prickly poppy; Hindi: Shial Kanta; Kanerese: Datturi; Mal.: Ponnummatam; Sanskrit & Telugu: Brahmadandi; Tamil: Kudiyoeti.

Hab : It is a common prickly herb of the tropical regions.

Its seeds can yield 22-36% bitter non-edible semi-drying oil, which can be used as an illuminator and to a lesser extent as a lubricant.

8. *Azadirachta indica* A. Juss.; *(Syn. Melia azadirachta* L*)*; Meliaceae.

Azadirachta indica – A Juss. tree

Na : Eng.: Margosa tree, Neem; Hindi: Neem.

Hab : It is common all over the tropical regions in the country. Pl. see edible plants.

It is multipurpose tree. Neem oil and leaf extract have insecticidal and medicinal properties. Azadiractin extracted from seed kernel (yield 1.5 to 4.8%) has assumed great commercial importance as insecticide. High yielding clones can be propagated through tissue culture. It grows very well in different agro-climatic conditions exhibiting its wide genetic base. Being an oilseed, viability is very short. Within a week of shedding of fruits there is gregarious germination. Therefore soon after collection, the seeds are required to be sown in the nursery and one year old seedlings are transplanted at 2 × 2 metre spacing initially and later on it can be 5 × 5 metres. Yield of seed in 5th year is 15 kg. From 20th year yield may be 75 kg annually. Larger trees yield more.

Azadirachta indica – A Juss . fruits

The yield of oil from the kernel is 35 to 40 per cent. The estimated potential of neem seed oil is 83,000 tonnes. The oil is yellow in colour and bitter in taste and is used in skin diseases. It is semi-solid oil. It is extensively used in soap industry. It is also used by the poorer classes as an illuminator. Neem cake is used as manure and pesticide.

9. *Baliospermum montanum* Muell.Arg. *(Syn. B.axillare* Bl.*)*; Euphorbiaceae.

Na : Hindi & Sanskrit: Danti.

Hab : It is an erect under-shrub fairly common in Doon Valley in Uttar Pradesh and also in other parts of sub-Himalayan tract from Kashmir to Khasi Hills.

The fruit is a 3-lobed capsule. The seeds resemble those of the caster plant. The oil from the seed resembles croton oil and is a drastic purgative. It is said to have medicinal properties and is used locally in external application in rheumatism.

10. *Bombax ceiba* L. *(Syn. B.malabaricum* DC; *Salmalia malabarica* (DC.) Schott & Endl.*)*; Bombacaceae.

Na : Eng.: Red silk cotton; Hindi: Simul, Semur.

Hab : Pl. see medicinal plants.

The seeds yield pale yellow oil, which deposits "stearine" on standing. It could be used for edible purposes, for soap making and as a substitute for cotton seed oil. The yield of oil is about 20 per cent. The oil cake has a mild and not unpleasant taste and would make a good type of cattle feed. Further it contains very little or no Gossypol, the toxic principal of cotton seed.

Bombax ceiba L.

Butea manosperma (Lamk)Taubart.

11. *Butea manosperma* (Lamk.)Taubart. (Syn. *B.frondosa* Koenig ex Roxb.) Fam.Fabaceae, (Papilionaceae.)

Na : Eng: Flame of the forest; Hindi: Palas.

Hab : Pl. see gums.

Plant is abundantly found in Madhya Pradesh, Maharashtra, Andhra Pradesh, Bihar, Gujarat, Orissa, Chattisgarh and also in other tropical regions. Collection of seeds was started in 1987. Yield of seed near the villages is low since the branches are cut for collection of leaves for platter. The seeds are separated from pods manually. Seeds contain 17 to 19% oil. Palas oil is used for making soap. De-oiled cake is used as manure.

Caesalpinia bonducella Flem - branch

12. *Caesalpinia bonducella* Flem.; Caesalpiniaceae;

Na : Hindi: Katkaranj, Karanju.

Hab : Pl. see medicinal plants.

The seeds yield pale yellow tasteless oil which can be used for illuminating purpose. The yield of oil from seed is about 20 per cent.

13. *Calophyllum elatum* Bedd. (Syn. *C.tomentosum* T. Anders. in part, non wight); Guttiferae.

Na : Bombay: Pun; Eng.: Poonspartree, Sirpoon tree; Hindi: Bobbi, Sirpon; Kanerese: kuve, Bobbi; Malayalam: Kattupunna, Malampunna, Punnappai; Marathi: Nagani; Tamil: Kattupinnai.

Caesalpinia bonducella Flem -Fruit and seeds

Hab : It is a tall tree found in the evergreen forests in Western Ghats from North Kanara in Karnataka to Kerala.

The seed kernels yield 70% of reddish brown oil, which is used as an illuminator.

14. *C.inophyllum* L.; Guttiferae.

Na : Eng.: Indian laurel, Laurel wood; Hindi & Bengali: Surpan, Sultan champa; Kanerese; Vuma, Honne; Mal.: Punna; Marathi: Undi, Surangi; Sanskrit: Nagachampa, Punnaga; Tamil: Punnai, Pinnay; Telugu: Pouna; Trade; Poon.

Calophyllum inophyllum L.

Hab : It is a medium sized evergreen tree found in South India and Andaman Islands.

The seed kernel (43-52% of the fruit) yields 50 to 73 per cent of dark green viscous oil. It has unpleasant odour and taste. It has toxic constituents. Therefore it is not used for edible purposes. It is however used for soap making and as an illuminator as hair oil. Cake is used as manure. It is propagated by seed. It is light demander and sensitive to frost.

15. *Cannabis sativa* L; Cannabinaceae.

***Cannabis sativa* L.**

Na : Eng.: Hemp; Bengali, Gujrathi & Hindi: Ganja, Bhang, Charas, Siddhi, Jia; Kanerese: Bangi; Sanskrit: Bhanga Vijaya; Tamil: Ganja, Bhangi; Telugu: Ganjai, Kalpam-chettu.

Hab : It is an under shrub found both wild and cultivated. It is a weed in parts of Uttar Pradesh forests and also eastwards.

The seeds yield by expression 32-35% of a greenish yellow drying oil, which can find use in paint and varnish industry and in the manufacture of soft soap.

16. *Capparis grandis* Lf.; Capparidaceae

***Capparis grandis* Lf.**

Na : Guj.: Dhuti; Hindi: Puchaonda; Kanerese: Revapi; Mal.: Waghutty; Marathi: Kauntel; Tam.: Nakkulinjan; Telugu: Oridonda.

Hab : It is a small tree found in Rajasthan and Western Ghats.

Fruit yields oil, which is used as an illuminator.

17. *Catunaregam speciosa* (Thunb.)Triveng. Syn. *Randia dumetorum Lamk. Gardenia dumetorum Roxb*.; Rubiaceae.

***Catunaregam speciosa* (Thunb)Triveng.**

Na : Eng.: Emetic nut; Hindi: Mainphal.

Hab : Pl. see medicinal plants.

The seeds yield yellowish-green fat, which is of the consistency of butter.

18. *Celastrus paniculatus* Willd.; Celastraceae.

***Celastrus paniculatus* Willd.**

Na : Hindi: Malkangni.

Hab : Pl. see edible plants.

The crushed seed yields brownish yellow pungent oil. On standing it deposits "stearine". It is said to have medicinal uses in skin diseases.

19. *Daphne papyraceae.* Wall. ex Steud. (Syn. *D.cannabina* Wall.); Thymelaeceae.

***Daphne papyraceae*. Wall. ex Steud.**

Na : Bhutan: Dayshing; Eng.: Nepal paper plant; Hindi: Satpura, Mahadev-Ka-phul, Setburva, Setharosa; Punjab: Niggi, Jeku.

Hab : It is an evergreen shrub found in the temperate regions in the Himalayas. It is abundant in Garhwal forests in Uttaranchal as an under shrub in *Quercus* forests.

The seeds yield about 36 per cent oil, which is greenish yellow in colour.

20. *Diospyros peregrina* (Gaertn.) Gurke; (Syn. *D.embryopteris Pers*;); Ebenaceae.

Na : Eng.: Indian persimmon; Hindi: Gab.

Hab : Pl. see edible plants.

The seeds yield a brownish yellow oil of peanut-oil odour. It is locally used in dysentery and diarrhoea.

21. *Diploknema butyraceae* (Roxb.) H.J. Lam. (Syn. *Bassia butyraceae* Roxb. *Madhuca butyraceae* (Roxb.) Macb. Family; Sapotaceae.

Na : Bengal: Gophal, Hindi Chiura, Phulwara, Phulel; Trade: Indian butter tree, Phulwara.

Elaeis guineensis Jacq.

Hab : The tree is commonly met with in the sub-Himalayan tract and in the outer Himalayas from Kumaon eastwards to Sikkim and Bhutan up to an altitude of 1500 m. It is also been recorded from Andaman Islands where it is frequently found in semi-deciduous and evergreen forests. It is a fast growing tree found chiefly along the sides of ravines in hills and in shady valleys. The fruit is a berry, blackish in colour, with a thick, soft, sugary pericarp and a characteristic sweet odour. The pericarp forms nearly 70% of the weight of the berry. The kernel forms about 70% of the weight of the seed (weight of 100 seeds is 78 g). The fat known in the trade phulwara butter is extracted by crushing either the seeds as a whole or the separated seed kernels into a cream-like paste and squeezing out the oil through a cloth bag. The yield of oil is 42-47% of the weight of the seeds or 60-67% of the weight of the kernels. It has the consistency of ghee and when properly prepared has a white colour not unpleasant taste and odour, and good keeping qualities.

Phulwara butter is used mostly as a substitute or as an adulterant of ghee. It is used also as a substitute for cocoa butter in chocolate manufacture. It may be employed in the manufacture of soap and candles, and for illuminating purpose. It burns without emitting smoke or unpleasant odour. An ointment prepared from it is said to be useful for rheumatic pains. *D. butyracea* oil-cake, like other mowra cakes, contains saponins and finds use as manure. Percentage of saponin probably falls within the range 29-31 reported for mowra cakes in general. The cake is used as a substitute for soap and as fish poison. It is also used as worm killer.

22. *Elaeis guineensis* Jacq.; Palmae

Na : Trade: Oil palm, African oil palm.

Hab : Oil palm is a native of West Africa and is now extensively grown in Malaysia, Nigeria, Indonesia, Republic of Zairi and Ivory Coast and to a limited extent in India, Ghana and south and Central America. It is considered as the highest oil-yielding crop per unit area capable of producing up to 8 tonnes of oil per hectare. In India it is mainly grown in Andhra Pradesh, Karnataka, Tamilnadu, Kerala, Goa, Gujarat, Orissa and Andaman as they receive higher rainfall.

***Climate and soil requirements*:** Oil palm is a tropical species and grows well in localities where the annual temperature varies between 20 °C and 35 °C with well distributed rainfall of 2500 to 4000 mm per annum. Since drought affects the yield, supplementary irrigation in summer may be necessary to realize economic yield. It does not tolerate shade.

The palm grows well in sandy loam to alluvial soils. Considering perennial nature of palm and shortage of edible oil production in the country, its high yielding nature, it is worthwhile trying as a plantation crop in forest areas. Since suitable climatic and edaphic conditions are available in Assam, Kerala, Andaman and Western Ghats.

***Brief Description*:** It is a moderate sized palm with compact crown. In young palms the crown will have as many as 50 leaves. The leaf bases adhere to the stem for about 12 years or longer. Thousands of adventitious roots spread from a massive basal bole. Oil palm bears male and female flowers in separate bunches in the same palm (monoecious). In the axil of leaves there is bud, which may develop into a male, female or occasionally bisexual inflorescence. After 2 to 3 weeks the outer spathe splits and the inflorescence pushes its way through. Occasionally a bisexual inflorescence is produced between the two male and female inflorescences. Flowers open from the tip of the inflorescence and completes opening in three days,. The stigma has three hairy lobes, which is moist and is ivory or orange coloured during receptive stage. Male flowers begin to open from bottom to the top of the inflorescence. Most of the pollen is shed during 2 or 3 days of starting opening and is completed within 5 days. The fruit is ovoid or elongated bulging at the top. It may be 2-5 cm, long weighing 3-30 grams. The fruit consists of outer thin skin known as exocarp, the middle oil bearing mesocarp and the inner part the endocarp. The endocarp consists of shell together with the kernel inside forming the seed. Although kernel contains oil but the yield is less and hence, only mesocarp the middle portion is used for extraction of oil. Oil palm bunch may weigh from a few kg to a quintal the average being 30 kg. Well-set bunches bear 1000 to 3000 fruits. Ripening is usually from tip downwards. A bunch takes about 5 to 6 months for ripening. Since oil palm is a cross-pollinated crop, assisted pollination is found necessary. For collection of pollen the male inflorescence is bagged before the flowers open. When the flowers open the inflorescence along with the bag is separated from the palm and dried for 24 hours at 30 to 40 °C. The pollen, which is shed in the bag, is sieved and stored in moisture free containers. It remains viable for 30 days or longer, if stored in a refrigerator. Pollination is affected by applying the pollen into the receptive stigma with a puffer. The fruits ripen in 5 to 6 months from the date of pollination. Several species of weevils visit male and female flowers both and affect pollination.

***Varieties*:** There are 3 varieties of oil palm. The first being dura which has a thick shell with low to medium mesocarp content (35-55%). It is not useful for commercial purpose. Second variety is tenera, which is obtained by crossing dura (female) and pisifera (male). It has a thin shell usually measured 0.5 mm to 4 mm and a medium to high mesocarp. This is widely cultivated all over the world due to higher mesocarp content and resultant oil out-turns. The third variety is pisifera. The fruit does not contain shell with pea like kernel inside. Some times kernel is also absent. Therefore, propagation of the species is difficult.

***Seed germination*:** Fruits are separated from the bunch and are extracted by scraping exocarp and mesocarp. Alternatively the fruits are retted in water for ten days. Thereafter the fruits are gently pounded to remove the adhering mesocarp. After removing adhering fibre seeds are then dried by spreading them on clean and dry floor for two days. Such seeds can be stored for 3 to 4 months at 27 °C. Under normal conditions the seed takes two years to germinate. For quicker germinations seeds are soaked in water for 5 days changing the water daily. Thereafter the seeds are spread out for drying for 24 hours. The dried seeds are put in polythene bags and closed tightly with a rubber band and placed in a germinator maintaining temperature of 40 °C. After 80 days the seeds are taken out from the germinator and polythene bags and soaked in water for five days changing the water daily and dried in the shade for two hours. After this the seeds are filled into bags and kept in a cool place to maintain the moisture content. Germination commences in about 10 to 12 days and germination is 90-95 per cent.

***Raising nursery*:** The poly bags preferably black of 400-500 gauge and 40 × 35 cm are used. After filling the bags with good fertile soil one or two seeds are dibbled into it. Watering the bags alternate day is necessary. During rainy season no watering is necessary. A fertile mixture containing 15 g N, 15 g P_2O_5 and 6 gms K_2O at 8 gms of these in 5 lit of water for 100 seedlings may be applied when the seedlings are more than two months old.

***Planting*:** Oil palm is planted in the field in triangular system at spacing of 9 meters accommodating 140 palms per ha. Planting is done with the onset of regular shower of monsoon. Polythene bag is torn open and with ball of earth it is planted in the pit and levelled. The pit can be 50 cubic cm. Young plants are to be protected from rodents.

Weeding: The plants should be kept free from weeds.

***Manuring*:** Manures may be applied only when soils are poor. Supply of sufficient quantities of green leaf or compost is advantageous especially where the soil is poor in organic matter.

***Irrigation*:** Irrigation at the rate of 50 lt of water per day per plant is required to be applied during the dry season.

***Leaf Pruning*:** Dead and diseased leaves and inflorescence up to three years are required to be removed. Thereafter when the plants are yielding, about 40 leaves per plant can be retained.

Harvesting: First harvest can be had after three and a half or four years after planting. When a few ripe fruits start falling off, bunches can be removed. When the plants are still small with the help of chisel the stalk of the bunch is cut and pushed down. When the palm becomes taller the stalk of bunches are cut with the help of knife or sickle tied to the pole and bringing them down. If they are still taller, it may be necessary to climb them for harvesting.

***Yield*:** Under average management an yield of 12 tonnes of fresh fruit bunches per ha per year can be expected. Under good management with irrigation 4 to 6 tonnes of oil per ha per year can be expected. The average oil yield is 2.5 tonnes per ha per annum. The tenera fruit has on an average 35% palm oil, 35% residue, 7% kernel and 25% shell.

***Diseases*:** In nursery stage, the plants are affected by leaf fungi Anthracnose and with the application of fungicide, Captan the disease can be controlled. In spear-rot disease the bases start yellowing and drying. Such affected parts are cut and removed and destroyed by burning. Some times bunch failures are noticed. There is no set of fruit followed by drying or rotting of bunches. The disease may be because of excess pruning, natural shading or lack of proper pollination.

***Pests*:** Rhinoceros beetle and red palm weevil cause damage to the plants. Many birds feed on the mesocarp of oil palm fruits. Rats, porcupines, squirrels, monkeys, wild boar and dogs are the common mammalian pests.

***Processing*:** For plantation not exceeding 40 ha, a hand operated hydraulic press will be sufficient. It can process about 3.5 tonnes of bunches per day. The fruits should be processed within few hours after harvest to obtain good quality oil. Oil mills capable of processing 3 tonnes of FFB per hour are available. A minimum of 200 ha of plantation is required to make such a unit viable.

***Uses of Palm Oil*:** Palm oil is used as frying oil. It can be blended with any other edible oil.

23. *Garcinia echinocarpa* Thw.; Guttiferae.

Na : Mal.: Para; Tamil: Madul.

Hab : It is a, handsome tree found in Kerala and Tamilnadu.

The seeds yield 49.6% of oil, which solidifies slowly at about 26°C to a soft brown fat. It is used as an illuminator and for soap making.

24. *Gindica* (Thou.) Choisy; Guttiferae.

Na : Eng.: Kokam butter tree, Mango steen oil tree; Hindi: Vishambil, Kokam.

Hab : Pl. see edible plants.

The seeds yield (23 to 26 per cent on the weight of seed and 44% on the weight of kernels) hard fat, which is used for edible purpose having low

***Garcinia indica* (Thou) Choisy fruits**

digestibility, which can be improved by blending it with hydrogenated oil. It is highly priced for use as cocoa butter substitute in chocolate and confectionery. It is also used for candle and in soap making.

Propagation is by seed. The flowers are unisexual and plants can be monoecious or dioecious. Hence, good clones should be propagated vegetatively. Tree starts yielding fruits at 10 or 12 years. Japan is the main buyer of this oil.

25. *G.morella* Desr.; Guttiferae.

Na : Eng.: Ceylon gambogi; Hindi: Gotaghanba.

Hab : Pl. see edible plants.

The seeds yield 30% of fat similar to kokam butter from *G.indica* (Thou.) choisy, which is used in cooking and confectionery as a substitute for ghee.

26. *Givotia rottleriformis* Griff.; Euphorbiaceae.

Na : Eng.: White catamaran tree; Kanerese: Bilitale, Punki, Pulkeer, Bhutale; Mar.: Polki; Tamil: Vendalai, Vandarlei, Pundarlai, Vendarirbudali; Telugu: Tellapoliki, Poniki.

***Givotia rottleriformis*, Griff**

Hab : It is a moderate sized tree found in deciduous forests in South India. The under surface of leaf is woolly. Seeds are the source of lubricating oil. Wood is white and light in weight. Hence, it is used for toy making. Natural regeneration is scarce as the seeds are eaten away by animals and birds. The seeds are soaked in water for 3 to 4 days to remove epicarp. Mesocarp is hard and it has to be cracked by secature or soaking in water by alternate wetting and drying for 15 days. They are then treated with GA 1000 ppm for 72 hours and sown. Germination is 40%. Germinated seeds are sown in the field. Semi-hard wood cuttings of 1-2 cm diameter and 15 to 20 cm length with one or two leaves are treated with fungicides, Bavistin 1% and dipped in root initiating hormone IBA 3000 ppm and planted. They sprout within 3-4 days.

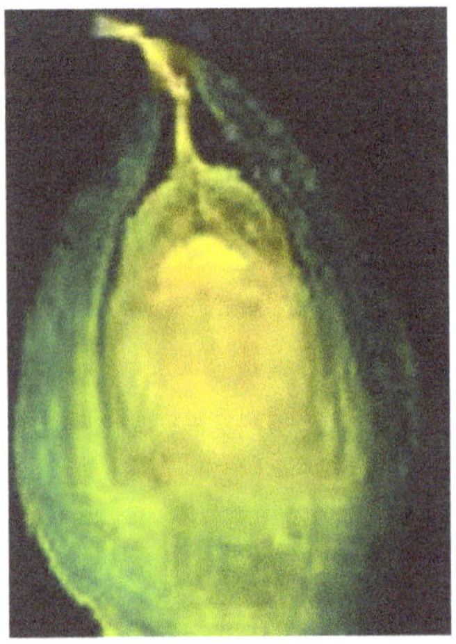

***Gnetum ula* Brogn. - Fruit**

27. *Gnetum ula* Brogn.; Gnetaceae.

(Na. & *Hab :* Pl see edible plants).

The seed kernel (30% of the wt. of the seed) yields 14.2% of fixed oil. The oil is used as an illuminator and to a small extent for edible purpose.

28. *Gynocardia odorata* R. Br.; Flacourtiaceae.

Na : Hindi: Chaulmogra.

Hab : Pl. see medicinal plants.

The seed kernels (67% of the wt. of seed) yield 27% of a pale yellow drying oil.

29. *Hevea brasiliensis* (G.B.K.) Muell. Arg; Fam. Euphorbiaceae.

Hevea brasiliensis **(G.B.K.) Muell. Arg**

Na : Eng.: Caoutchouc tree, Hevea rubber, Para rubber.

Pl. see rubber plants.

The rubber tree is cultivated mainly in Kerala, Karnataka and Tamilnadu producing para rubber. The dry seeds contain about 19.3-24 per cent oil. The oil content of dry kernel is 38-50 per cent. Rubber seed oil is soft, semi-drying. It is used mainly for soap making in Kerala and Tamilnadu. Some quantities of the oil are also consumed by paint and varnish industry. The cake is used as cattle feed and also as manure.

30. *Heynea trijuga* Roxb.; Meliaceae.

Na : Assam: Theng-are-arong, Rolang-phang; Beng.: Kapiakushi; Bihar: Chinnanji; Bombay: Limbara; Kanerese: Kora, Limbara; Kumaon: Banritha; Lepcha: Migtiok kung; Mal.: Korakkadi, Kurukkati; Marathi: Gundira; Tamil: Karaivilangu, Karai.

Hab : It is a small handsome tree found in Assam, West Bengal, Kumaon and the Western Ghats.

Seeds are the source of oil, which is used for illumination purposes.

Hippophae rhamnoides **L - Branch with fruits**

31. *Hippophae rhamnoides* L.; Elaegnaceae.

Na : Eng.: Sea-buck thorn; Hindi: Dhurchuk, Chuma, Tarwa; Ladakh and Lahoul: Sinna, Tasru; Punjab: Neichak, Kalabisa, Sirma.

Hab : It is a shrub or small tree found in the North Western Himalayas.

The seeds yield about 12 per cent oil, which is yellow in colour.

32. *Holarrhena antidysenterica* Wall.; Apocynaceae.

Holarrhena antidysenterica **Wall.**

Eng.: Easter tree, Ivory tree; Hindi: Kurchi, Kadva indar jau, Kutaja.

Hab : Pl. see medicinal plants.

The seeds yield 19 to 30 per cent oil which in greenish yellow in colour. It has penetrating odour mild taste and free from alkaloids. It is used locally as an anthelmintic.

33. *Holoptelea integrifolia* (Roxb.) Planch.; Ulmaceae.

Holoptelea integrifolia **(Roxb.) Planch. - Branch with fruits**

Na : Guj.: Kanjho, Waola; Hindi: Papri, Kanju, Banchilla, Chilbil, Dhamna, Begana; Kanerese: Thavasai, Rasbija, Kaladri; Nilavahi; Kumaon: Paper, Kanju; Madhya Pradesh: Karanji, Chirhol, Karanjalam; Mal.: Aval; Marathi: Vavli, Papara; Oriya: Dauranja, Turuda; Punjab: Rajain, Khulen, Arjan; Sanskrit: Chirabilva; Tamil: Aya, Ayil, Kanchi, Vellaya; Telugu: Nemilinara chettu; Trade: Kanju, Indian Elm.

Hab : It is a large tree commonly found in the tropical and sub-tropical forests in the country.

The seeds (husk 27.5%, Kernel 72.5%) yield 37.4% (53.27% in dry kernel) of yellow oil

34. *Hydnocarpus kurzii* (King Wartt) Syn.*H.heterophyllus*; *Taraktogenos kurzii* King;.; Fam.: Flacourtiaceae.

Na : Eng.: Chaulmogra; Hindi: Chaulmogra.

Hab : Pl. see medicinal plants.

The seeds yield fatty oil known as Chaulmogra oil. This oil is known to have medicinal properties and is used in skin diseases and leprosy. It is also used in the manufacture of soap. The oil cake is used as manure and is not used as cattle feed because of toxic principles found in it.

35. *Hydnocarpus pentadra;*Syn. (Buch.-Ham.) *Oken H. laurifolia* (Dennst.) Sleumer;. *H.wightiana* Blume; Flacourtiaceae.

Na : Hindi: Kava.

Hab : Pl. see medicinal plants.

The kernels (60-70% of the weight of seed) yield 63.25% oil which is used in indigenous medicine in the treatment of leprosy, skin diseases, rheumatism, sprains, bruises, sciatica and also as an illuminator. Seed cake is used as manure and as insect repellent.

36. *Jatropha curcas* L; Fam.: Euphorbiaceae.

Na : Eng.: Physic nut; Hindi: Safed arund, Bagbherenda, Jamalgutta; Telugu: Nepalam, Dola.

Hab : Pl. see medicinal plants.

Jatropha is a small sized, soft-wooded tree of arid to semi-arid areas in Central America, Mexico, South America. It is found in Brazil and other Latin American countries. It has spread to other parts of the world. In India, it has been grown in Rajasthan, Gujarat, Maharashtra, Tamilnadu, and Andhra Pradesh and in many other places. It is not festidious about soil or climate. It can come up in areas of low rainfall as low as 200 mm and can tolerate higher rainfall of 1200 mm. It is not browsed by cattle and can tolerate desiccative conditions of summer. As such, it is ideally suited for afforesting waste lands, barren and denuded areas. The seeds are rich in oil content about 35-40% (in seed) and 50-60% (in kernel). The seed oil is being used in soap industry and as an illuminator, but of late it has assumed great importance as a potential source of bio- fuel because of shortage and soaring prices of mineral fuel. Its calorific value is almost the same as that of diesel. The oil on refining can be used as a bio-fuel in the form of extender to the HSD to the tune of 20%. It may cut the drain on the exchequer and reduce import of mineral fuel. It is a non-conventional renewable energy source. It is eco-friendly, since it burns without emitting much smoke. The smoke emitted is odourless and non pungent and less harmful. The oil cake can be used as a bio-fertilizer. As such the plant deserves extensive cultivation in the country.

***Propagation*:** The plant can be raised by seed as well as cuttings. The plant starts flowering in April/May. It bears unisexual flowers. When fruits turn yellow, they are separated and dried. Since the seeds are oil-bearing their viability is not long. As such, they cannot be stored for longer durations. They can be sown in poly bags of size 18 × 13 cm filled with good mixture of soil. Before sowing, the seeds may be soaked in water to get speedy germination. About 3 kg seed is required for raising a hectare of plantation. They germinate, in about a week's time. In about two to three months, the plants are ready for transplanting. The proper time for planting would be the commencement of monsoon. In the nursery stage, the weeds are required to be removed. The poly bags are torn at the bottom and planted in the field in 30 cm cube pits at an espacement of 3 × 3 m (about 1100 plants per hectare) The object of having wider espacement, i.e. 3 × 3 m (or) 1100 plants per hectare is to allow good root formation with abundant lateral branching. Close espacement of 2 × 2 m will help in vertical growth of plant. In the year of planting, two or three weeding-cum-soil working may have to be carried out. In the second year weeding and shrub cutting may have to be done. Casualty replacement may also be carried out, if found necessary. Soil and moisture conservation measure will boost up the growth. If it is a sloping ground contour trenching will also help. At a later stage, if the plants are found to be congested, light thinning by removing the weaklings may be necessary.

Jatropha Curcas **L.with fruits**

Direct sowing in the filled up-pits can also be done with the commencement of monsoon. Two or three seeds may be dibbled in each pit, ultimately retaining one seedling in each pit. Semi- hard wood cutting of more than thumb thickness and 20-30 cm in length may be planted in the nursery at an espacement of 15 cm. Regular watering is done till the cuttings strike roots and get established. It may take about 25-30 days for the cuttings to strike roots and get established. The time of planting, of cuttings, care should be taken to see they are not planted upside down. The rooted cuttings are planted at the same spacing of 3 × 3 m with the commencement of monsoon. Weeding and other operations may be done as in the case of seedling plantations. Manuring and periodic irrigation during desiccating summer months will increase the yield of crop.

***Damage to the crop*:** Since the branches and stem are soft wooded and brittle, there may be some damage from cattle which may be prevented.

The plants may start yielding fruits in the second year in the case of cutting planting and third year in the case of seedling planting when about 1 kg seed per plant can be collected. At the age of 5 or 6 years, the seeds up to 4 kg per plant can be obtained. Normal life of the plant may be about 30 to 40 years depending on maintenance.

37. *Kingiodendron pinnatum* (Roxb.) Harms (Syn. *Hardwickia pinnata* Roxb.); Caesalpiniaceae.

Hab : Pl. see resins & oleoresins.

Oil obtained from the seed is useful for soap and varnish manufacture.

Litsea glutinosa (Lour.) C.B. Robinson - Branch with fruits

38. *Kokoona zeylanica* Thw.; Celastraceae.

Na : Eng.: Kokum.

Hab : It is a tall tree found in the Western Peninsula and South India.

Seeds are the source of oil, which is used for illumination purposes.

39. *Litsea glutinosa* (Lour.) C.B. Robinson (Syn. *L.chinensis* Lamk.; *L.sebifera* Pers.) fam.: Lauraceae.

Na : Assam: Heluka, Bagnara; Bengali: Kukurchita, Garur, Ratun; Guj.: Maeda lakari; Hindi: Garbjaur, Medh, Maida lakari; Madhya Pradesh: Menda; Mar.: Maidalakdi; Lepcha: Suppatnyok; Oriya: Jaisanda; Punjabi. Medasak, Chandra, Medalakri; Tamil: Mushaippeyetti, Elumpurukki, Uralli; Telugu: Naraalagi, Naramamidi.

Hab : It is an aromatic evergreen shrub or small tree found in the Punjab, Himachal Pradesh, Assam, Bengal and S. India at lower altitudes. The fruit is globose in shape and black or purple in colour.

The seeds yield about 35 to 51 per cent of fat, which is useful for making candles. The fat is of disagreeable taste. It is stated that one tree is sufficient to provide material for the production of 500 candles a year. The fat contains laurin. Local people use this oil for skin diseases. The bark is used as an adhesive in preparation of joysticks.

40. *L.monopetala* (Roxb.) Pers. (Syn. *L.polyantha* Juss.) Fam.: *Lauraceae.*

Na : Assam: Muga, Hoanlu; Bengal: Barakukurchita; Hindi: Meda, Katmarra, Patoia, Kakuri; Lepcha: Sunyok-kung, Sapotkung; Mar.: Ranamba, Rapamba; Oriya: Baghoari, Kulya, Bastura; Punjab: Rian, Gwa, Harein, Medalakdi, Karkawa; Tamil: Maidalagadil, Muchaippeyetti, Picinbattaw; Telugu: Naramamidi.

Hab : It is found in tropical forests in the East, North arid Central India.

The seeds yield 21 to 33 per cent fat, which can be used for candle making.

41. *L.zeylanica Nees.* Syn. *L.umbrosa;* fam.: Lauraceae.

Hab : It is found in the temperate regions of Central Himalayas.

The yield of oil from seed is 61 per cent. The oil is useful for candle making.

L monopetala (Roxb.) Pers.

42. *Madhuca indica* J.F. Gmel. (Syn. *M.latifolia* (Roxb.) Macb.; *Bassia latifolia* Roxb.); Fam.: Sapotaceae.

Na : Eng.: Mahua tree; Illipe butter; Hindi: Mahua, Moha.

Hab : Pl. see edible plants.

It is one of the important fat yielding species. The fat content in the kernel of the seed is up to 50 per cent. It has many industrial uses. Fat is used for soap making. Locally it is used as an illuminator and a hair fixer. It is also used by the local people for edible purpose when it is fresh. It is used as an adulterant for Ghee. For this purpose it is clarified with buttermilk to mask off the odour and give a pleasant taste.

Madhuca indica **J.F. Gmel – Kernel**

43. *M.longifolia* (L.) Macb. (Syn. *Bassia longifolia* L.); Fam.: Sapotaceae

Na : Eng.: Mowra fat; Hindi: Mohua. Names in Indian languages are the same as those of M.indica J.F. Gmel.

Hab : It is common in monsoon forests of western Ghats usually along the banks of rivers and streams.

The seed kernel yields about 51 % of fatty oil. The fresh oil obtained is yellow in colour with not an unpleasant taste. But commercial oils are generally greenish yellow in colour with an offensive odour and disagreeable taste. Fatty oil is used for soap making as an illuminator and, even for edible purposes, when fresh by the tribals and other villagers. It is sometimes used as an adulterant for ghee. This fat also is clarified with buttermilk to mask off the odour. Mahua oil is used in manufacture of Vanaspathi. Mahua cake is used as manure and as vermicide. After treatment meal can be used as cattle and poultry feed.

Madhuca longifolia **(L.) Macb.**

44. *M.malabarica* Bedd. Parker (Syn. *Bassia malabarica Bedd.*); Sapotaceae.

Hab : It is a small to medium sized tree found in the West coast and Western Ghats usually along banks of streams.

The fatty oil obtained from the seed is used for soap making and for edible purposes. It is also used as a hair fixer.

45. *Mallotus philippinensis* Muell. Arg.; Euphorbiaceae.

Na : Assam: Jorat, Losan; Bengali: Kamala; Gujrathi: Kapilo; Hindi: Kamala, Rohini, Sindur; Kanerese: Kunkumadamara; Lepcha: Puroakung; Mal.: Kuramadakku, Manjana; Mar.: Shendri; Oriya: Kapilogundi, Kunkumo, Sinduri; Tamil: Kapli, Kungumam, Kurangumanjanatti; Telugu: chendiramu, kunkuma, Sinduri.

It is a small tree found commonly along the foothills of Himalayas from Bengal Westwards and in Orissa, Madhya Pradesh and Maharashtra.

Fruit is a capsule covered with reddish brown glandular pubescence. The pale white seed kernels (60 per cent of weight of seeds) yield on solvent extraction about 35 per cent of dark brown to pale yellow oil or semi-solid fat. By expression method yield of oil is low. It is a drying oil as such it forms a good substitute for tung oil in the formulation of rapid drying paints and varnishes. It can be employed in coating compositions giving either glossy or wrinkled finishes.

Mallotus philippinensis **Muell. Arg.**

Mangifera indica **L - tree**

Mangifera indica **L Fruit**

Melia azedarach **L**

Mesua ferrea **L**

Michelia champaca **L**

Mimusops elengi **L**

46. *Mangifera indica* L; Anacardiaceae.

Na : Eng.: Mango; Hindi: Am, Amra.

Hab : Pl. see edible plants.

The kernel of the seed (stone) yields about 15 per cent of a white solid fat which is odourless and has a mild taste comparable with that of cocoa fat. It has good export market. Pant University has developed hand and power operated decorticators for separation of kernel from the stone. Oil is used for edible purpose and soap making. De-oiled kernel meal is used as cattle feed.

47. *Melia azedarach* L; Meliaceae.

Na : Eng.: Pride of India, China tree, Persian lilac; Hindi: Bachain, Drek.

Hab : Pl. see medicinal plants.

Seed yields about 40 per cent of drying oil with agreeable odour. It is suitable for the manufacture of soap and hair oil.

48. *Mesua ferrea* L; Guttiferae

Na-Eng.: Ironwood; Hindi: Nagakesar, Nagchampa, Nageshwar champa.

Hab : Pl. see essential oils.

The kernel of the seed yields about 60 to 77 per cent of oil. The oil is semi-hard to soft. Oil is dark in colour. It has unpleasant odour. After treatment it can be used for soap making, lubricating and as an illuminant. Cake contains toxic constituents. Therefore, it can be used only as manure. Tree regenerates naturally from seed. It requires shade when young. It cannot stand drought conditions in early stage.

49. *Michelia champaca* L; Magnoliaceae.

Na : Assam: Titasopa, Bolnabat, Shap; Bengali & Hindi: Champa, Champaca; Guj.: Champa, Raechampa, Pitochampa; Kan.: Sampige, Kala-sampige; Mal.: Champakam; Mar.: Pivalachampa, Kudchampa, Sonachampa; Tamil: Shembuga, Chambugam; Telugu: Champakamu; Trade: Champak.

Hab : It is a large tree found in the evergreen forests in Assam, Bengal and South India. It is also grown in gardens for its aromatic flowers. The seeds yield about 32 per cent butter-like fat, which is said to have medicinal properties.

50. *Mimusops elengi* L; Sapotaceae.

Na : Eng.: Spanish cherry; Hindi: Maulsari.

Hab : It is a tree found in Central and South India. It is grown also for its fruits. Pl. see edible plants.

The seed kernel yields 16 to 25 per cent oil, which is edible and is useful for illuminating purpose.

51. *Moringa oleifera* Lamk. (Syn. *M.pterygosperma* Gaertn.; *Guilandina moringa* L); Moringaceae.

***Moringa oleifera* Lamk.**

Na : Eng.: Drumstick tree, Horseradish tree; Hindi: Sahinjna, Soanjna.

Hab : Pl. see medicinal and edible plants.

The seeds yield 21 to 24 per cent oil, which is clear, rather thick at ordinary temperature and almost colourless. It is commonly known as Ben oil and is used locally for edible purposes, illumination and cosmetics.

52. *Myrica nagi* Thunb. (Syn. *M.rubra Sieb.* & Zuce.); Myricaceae.

***Myrica nagi* Thunb.**

Na : Eng.: Bayberry, Box myrtle; Hindi: Kaiphal.

Hab : Pl. see edible plants.

The myrica fruit is of the size of a pea and covered with a snow white crust of fat, which is permeated with brown or black spots. The commercial wax has a green colour, the uppermost layers of which get bleached on exposure to sun and air to a whitish mass. The wax is useful for manufacture of candles and soaps.

53. *Myristica canarica* Bedd.; Myristicaceae.

Na : Hindi: Pindi.

Hab : Pl. see essential oils.

Seeds yield fatty oil, which is useful for making candles.

54. *M.malabarica Lamk. Myristicaceae.*

***M.malabarica* Lamk. - Fruit**

Na : Eng.: Malabar nutmeg, Bombay mace; Hindi: Rampatri, Pathiri; Kanerese: Kanagi; Mal.: Kattujattika, Ponnampanu; Telugu: Adavijajikaya,

Hab : It is a moderate sized tree found in the evergreen forests of Western Ghats in Karnataka & Kerala.

Seed kernels yield 15-16 per cent fat, which is used as an illuminator and locally for medicinal purposes.

55. *Olea cuspidata* Wall. (Syn. *O.ferruginea* Royle); Oleaceae.

***Olea cuspidata* Wall.**

Na : Garhwal: Bairbanj; Hindi: Kahu, Kan, Kao.

Hab : It is a medium sized tree fairly common in subtropical to temperate Himalayas in Uttara Khand up to an altitude of 2400 m.

Both pulp and seeds yield oil (pulp 31-32% and seeds 15%). The yield of oil from kernel is 25 to 28 per cent. The entire fruit is used for extraction of oil. The oil resembles olive oil. It is sweetish without possessing the characteristic taste of olive oil.

Perilla frutescens **(L.) Britt.**

56. *Palaquium ellipticum* (Dalz.) Engl.; (Syn. *Bassia elliptica* Dalz.; *Dichopsis eiliptica* Benth.); Sapotaceae.

Na : Eng.: Indian-gutta-percha tree; Hindi: Halganne; Kan.: Panchonta, Hadasale; Mal.: Pella, Pali, Choppala, Panchendi; Mar.: Panchoti; Tamil: Katillupei, Palvadinjan; Trade: Pali.

Hab : It is a large tree found in the Western Ghats, North Kanara and Tamilnadu up to an altitude of 1200 m. Fruit is a one-seeded berry. The seeds yield oil useful for lighting and soap making.

57. *Perilla frutescens* (L.) Britt. (Syn. *P.ocimoides* L); Labiatae.

Na : Eng.: Perilla; Hindi: Bhanjira, Bhasinda.

Hab : Pl. see spices.

The seeds yield 35 to 51 per cent fatty oil. It is a valuable drying oil known as Perilla oil. It is used for varnishes, paints, printers ink, waterproof clothes and also locally for edible purposes.

Pongamia pinnata **(L) Pierre - Tree**

58. *Pongamia pinnata* (L) Pierre (Syn. *P.glabra Vent. Derris indica);* Fabaceae (Papilionaceae.)

Na : Eng.: Pongam; Hindi: Karanj, Karanja.

Hab : Pl. see medicinal plants.

Pongamia pinnata (L) Pierre is a middle sized tree popular for its oil seed. The plant also yields small timber and leaf for mulching and green manure. The tree grows naturally along the stream banks and in varied soils. It is largely grown as avenue and shade giving plant in social forestry programme. The demand for pongam (karanj) oil has shot up in bio-diesel industry and as such its cultivation is required over large areas to cater to the needs of industries and rural people.

Nursery Technique**:** The seedlings that are intended to be utilized for planting are raised initially in 12 × 18 cm poly bags and later shifted to 25 × 30 cm polybags and maintained for one year in trenches to grow to 1m in height. Adequate watering is required for quick growth. The seedlings are taken out from trenches before dispatching to planting sites and allowed to recover from shifting shock.

Pongamia pinnata **(L) Pierre - Fruit**

Planting Technique**:** Pongamia seedlings raised from one year in polythene bags will be planted at an espacement of 4 × 4 m in 45 cm cube pits. Weeding and soil working may be taken up to prevent congestion and root competition. Lindane Dust and SSP shall be applied at the time of plantings as basal dressing. Application of fertilizer, watering during dry season in the early stages of growth may be necessary. Soil and moisture conservation measure like digging trenches in sloping areas will greatly help the growth.

Maintenance**:** In the first year maintenance casualty replacement along with one (or) two weedings cum soil working are taken up. In the subsequent years, only tending, maintenance of inspection paths, watch and ward and other contingent works are required. At a later stage, when there is congestion in plants thinnings may be carried to help formation of good crown.

Pterospermum acerifolium **Willd.**

***Yield*:** Though the plant starts flowering from third year onwards, it is expected to yield economic returns from sixth year. The yield of fruit may increase with age and development of the crown. A well grown tree may yield about 0.2 mt of fruit. The seeds yield 30-35% yellowish, brown oil. The life of the plantation may be fifty years if the plants are not pollarded or cut back for leaf or other purposes.

***Economics*:** Seed oil of Pongamia at present is used in soap industry, as an illuminator. It is said to have medicinal properties and is also used in treating skin diseases. The oil cake is used as bio-fertilizer, fungicide and insecticide. Filtered and semi-processed oil, up to 20% can be used along with diesel in diesel engine without impairing the efficiency of engines. With proper processing pongamia oil has great potential to substitute diesel.

59. *Prinsepia utilis* Royle; Rosaceae.

Na : Garhwal: Bhekor; Hindi: Bhekal, Bhekoi, Bekkra, Cherara, Dhatila, Jhatela, Karanga, Krungora, Mhat; H.P.: Arund, Bekkli, Garandu, Gurinda, Kharungura, Phulwara, Talua; Jaunsar: Bhekoi, Bhek; Khasi: Sohmanrit, Dieng sia-soh-khar.

Hab : It is a fairly common thorny shrub thriving in open situations in the temperate regions of central and western Himalayas.

The seed kernels yield 37 per cent of semi-drying, pale yellow fatty oil. It resembles the oil derived from cottonseed. It is said to have medicinal properties and is used locally in skin diseases and also for illuminating purposes.

60. *Pterospermum acerifolium* Willd.; (Syn. *P.aceroides* Wall); Sterculiaceae.

Na : Assam: Hatipeala, Morra, Moragos; Beng.: Kanakchampa, Muskanda; Hindi: Kanak champa, Kaniar, Kathachampa, Muchkund; Jaunsar: Mayeng; Khasi: Dieng-khongswet, Diengtharomasi; Lepcha: Nambong; Oriya: Kanakochampa; Telugu: Matsakanda; Trade: Hathipaila. .

Hab : It is an evergreen tree commonly found in Maharashtra, Karnataka, Manipur, Assam, North Bengal, Bihar and Uttar Pradesh.

The seeds yield about 22 per cent of pale yellow oil.

61. *Quercus incana* Roxb.; Fagaceae.

Na : Eng.: Whiteoak; Garhwal: Phanat; Hindi: Banj; Jaunsar: Inai, Bani; Kashmir: Silasupari, lri, Shiddar; Kumaon: Ban, Bang.

Hab : It is a moderate sized tree found in the temperate regions of central and western Himalayas.

Fruit is an acorn. The kernel on extraction with petroleum ether gives 16 per cent of yellow fatty oil.

Quercus incana **Roxb.**

Rhus succedanea **L**

62. *Rhus succedanea* L; Anacardiaceae.

Na : Eng.: Japanese wax tree; Hindi: Kakra singi.

Hab : Pl. see medicinal plants.

The seeds yield 20 to 30 per cent fat, which can be used in the preparation of polishes and waxing floors. It is also used in the manufacture of candles.

63. *R..vernicifera* DC.; (Syn. *R..verniciflua* Stokes); Anacardiaceae.

Na : Eng.: Japanese varnish tree, Lacquer tree.

Hab : It is native of China but is found in Kumaon and Garhwal regions at altitudes between 900-1800 m.

Seeds are the source of fat used in polish and varnish industry and also in the manufacture of candles.

Ribes rubrum **L**

64. *Ribes rubrum* L; (Syn. *R.sylvestre* (Syme.); Saxifragaceae.

Na : Eng.: Red currant, Hindi: Dak.

Hab : Pl. see edible plants.

Seeds yield 16 to 25 per cent yellowish-brown oil, which has good drying properties. A film exposed to air-dries up quickly to a colourless skin. It is quite suitable in paint and varnish industry.

65. *Salvadora oleoides* Delle.; Salvadoraceae.

Na : Hindi: Badapilu, Khakan.

Hab : Pl. see edible plants.

The seeds yield greenish oil, which is useful for soap making.

It can be propagated by seed and root suckers. Collection of seed is made in Rajasthan and Gujarat. Since the trees are lopped heavily for fodder, collection of seed is hampered. Kernel contains 42-48.8% oil. Oil is used for soap making. Cake is not palatable but mixed with other cakes it is used as cattle feed.

Sapium sebiferum **(L.) Roxb. - Branch with fruits**

66. *Sapium sebiferum* (L.) Roxb.; Euphorbiaceae.

Na : Bengali: Momchina; Eng.: Chinese vegetable tallow; Hindi: Vilaiti or pahari Shisham, Pippalyang; Kanerese: Meenadabattimara; Oriya: Rollojita.

Hab : It is an ornamental tree grown chiefly in North Western India.

The seeds yield Stillingia oil; whereas the fruit flesh gives the Chinese tallow. It is useful in the manufacture of soaps, candles, paints and varnishes.

Schleichera oleosa **(Lour.)**

67. *Schleichera oleosa* (Lour.) Oken (Syn. *S.trijuga* Willd.) *Pistacia oleosa* Lour.; Sapindaceae.

Na : Eng. Lac, tree, Honey tree; Hindi: Kusum.

Hab : Pl. see medicinal plants.

The seeds yield 30-38% valuable oil, which at ordinary temperature is yellowish, white and buttery with, fairly bitter almond smell. Yield of oil from kernel is 51-62%. Kusum oil also called maccasar oil is unique among other vegetable oils as it contains 3.6 – 3.9% of glycerin while normal other oils contain 9-10% glycerin. It is employed as an illuminator by the villagers and is also suitable in soap manufacture. It is said to be very efficient and stimulating agent for the scalp both in cleansing and promoting the growth of hair. It is said to have other medicinal uses. It is used for skin diseases. Seed cake is used as manure and cattle and poultry feed.

Schleichera oleosa (Lour.) Seeds

It is propagated by seed. Direct sowing gives good results. It produces suckers profusely. They can also be dug out and planted.

68. *Semecarpus anacardium* L. (Syn. *Anacardium orientale* L); Anacardiaceae.

Na : Eng.: Markingnut tree; Hindi: Bhilawa.

Hab : Pl. see edible plants.

Semecarpus anacardium L.

The kernels yield about 20 per cent oil of very pleasant taste. Kernels are extensively used in confectionery. The oil is however not extracted for edible purposes.

69. *Shorea robusta* Gaertn. f.; Dipterocarpaceae.

Na : Eng.: Sal; Hindi: Sakhya, Sal

Hab : Pl. see edible plants.

Sal fat was used by tribals for edible and lighting purposes, in parts of Madhya Pradesh, Bihar and Orissa since long. The De-oiled residual meal after drying and pounding into flour was eaten with Mahua flowers. Though the possibilities of utilising sal fat in soap making and in confectionery as a substitute for Borneo tallow obtained from *Shorea stenoptera* Burck was recognized as early as in 1939 and 1942 respectively, commercial exploitation of sal seed for production of fat was considered uneconomical due to low fat content of about 12.5 to 18.5 per cent in sal kernel. Owing to enormous increase in the prices of oil, commercial collections of sal seeds for production of fat was started in 1968 and since then there has been steady increase in the utilisation of sal seed and presently sal fat is being produced on a commercial scale. The area of sal forests in India is estimated to be 1,14,379 sq. kilometres occurring in states of Punjab, Himachal Pradesh, Uttar Pradesh, Bihar, Madhya Pradesh, Orissa, West Bengal, Assam, Tripura and Srikakulam districts in Andhra Pradesh.

Shorea robusta Gaertn.f.- Tree

Flowering and fruiting in sal: The time of flowering and fruiting in sal varies from locality to locality depending on climatic conditions. High temperature results in early flowering and fruiting and in dry and warm localities the fruits fall early. The low temperatures at the time of beginning of flowering generally result in a delay in flowering and fruiting. The flowers open full from the end of February to well on in April and fruits ripen in June/July and fall as soon as they ripen.

Shorea robusta –Gaertn.f.- Flowers

***Collection of fruits, their decortication and storage*:** Collection of sal fruits extends from the time of ripening and falling of fruits to the commencement of monsoon. After the onset of rains fruit collection is given up as it becomes difficult to dry them and also the fallen seeds either deteriorate or start germinating. Thus the collection period extends from 4 to 6 weeks only. The seeds collected after some showers of monsoon get infested with fungus and insect pests on storage. Sal fruits are collected by hand picking and by sweeping from the forest floor after cleaning the ground. The collected fruits are generally brought to the drying centres where they are spread in sun for few days to dry up. Sufficient drying is necessary for efficient decortication. Dry fruits are decorticated either manually or mechanically. In the traditional method of decortication by hand the dried fruits are spread on ground to a thickness of about 10 cm and beaten with sticks to dewing them. The de-winged seeds are then spread again on a hard surface and a small wooden plank or a flat stone piece is run over to and fro pressing over the seeds. The dried brittle shell cracks open and separates from hard kernels. The husk being lighter is then separated from the heavier, kernels by winnowing. This method of decortication gives about 17.7% wings, 30% husk consisting of shells and adhering fragments of calyx; 50% kernels and 2.3% losses by weight. Mechanical decorticators both hand and power operated are also existing which are found to give 97 to 98% decortication with 10% and 5% broken kernels and fines respectively. The kernels after decortication are stored in gunny bags. One gunny bag holds about 75 kg of sal kernels. By decortication there is volume reduction by 50 to 60% and weight by about 83% per unit volume of fruits. As the collection season is short and also as rains commence soon after the fall of fruit, it is necessary to have good storage facility. The seed is to be stored in dry and rodent, proof go-downs. During storage the kernels are susceptible to attack by insects. A treatment with aluminium phosphide is reported to protect the kernels from insect damage. If the seeds are dried and stored properly there is no adverse effect on their quality even when stored for a period of about 6 months. There is variation in the yield of fruits depending on the climatic and edaphic factors and size and spread of the crown and the intensity of fruiting. The intensity of fruiting is largely dependent on whether it is a good, moderate or poor seed year. Based on the above, fruit (dry) yield per tree can be taken as 13.7 kg in good seed year, 9.7 kg in moderate seed year and 3.1 kg in poor seed year. Taking an average of all the age crops and site qualities the number of fruiting trees is estimated at 181 per hectare. The stand tables however have been worked out in sal forests with 100% stocking. This density is seldom achieved in natural forests. For practical purposes a density of 0.4 is considered optimum in sal forests, which gives 72 fruiting trees per hectare. Considering 7,200 fruiting trees per sq. kilometre with fruit yield of 13.7, 9.7 and 3.1 kg per tree for good moderate and poor fruiting years, kernel yields (50% of the weight of fruits) 14% fat (on the weight of kernels) and taking 50% of the total sal forests (half of 114379 sq. km. say 57190 sq. km.) as available for exploitation, the estimated potential yield of fat in good fruiting year is 3,95,000 tonnes, in a moderate fruiting year 2,80,000 tonnes and in a poor fruiting year 89,000 tonnes.

Extraction of fat, its refining and bleaching: The traditional method used by the tribals to extract fat was to de-husk the seeds, pound the kernels and boil them in water and skim off the fat that floats on the surface of water. On cooling, the fat sets hard like 'ghee'. But this however gives a low fat yield. Expeller method is also not practicable in this case. The only method suitable for the purpose is extraction with organic solvents. In the solvent extraction method kernels are reduced to a coarse powder preconditioned by cooking with moisture followed by flaking and finally extracting with food grade hexane solvent. Flaked kernels with normal hexane in cold yield 15.6% to 18.5% fat. There are two types of solvent extraction plants, namely, batch-type and continuous type. The crude fat is greenish yellow due to the presence of pigments in it. Removal of free fatty acids and pigments is essential. Bleachability of sal fat is not quite normal. It bleaches white on exposure to sunlight but the process takes longer time than that required for a commercial turnover. However, on a commercial-scale bleaching is done with activated earth and activated carbon. For removal of free fatty acids caustic soda solution is used.

Utilisation of sal fat and de-oiled meal: Sal fat is used in soap industry. Owing to its relatively high melting point i.e., 34.5 to 36.5 °C and large content of low chain saturated fatty acids it forms good substitute for coconut oil, which is one of the important vegetable oils. Sal fat is reported to form a good substitute for imported borneo tallow and cocoa-butter. After further processing, it is used for edible purposes. The de-oiled sal meal is rich in protein, starch, sugars etc. Presence of tannin (6-13%) in de-oiled sal seed meal is posing a major problem in its utilization at increased level in animal feed. The de-oiled sal meal is used in mixture up to 20% level in cattle feed without any adverse effect. However, with higher dosages of over 30% it may have some adverse effect owing to some tannin content in it. Similarly, in poultry feeds at 5% level it has no adverse effect on birds. However, sal seed tannins which could be extracted from de-oiled meal by hot water extraction could be used in leather tanning. The de-oiled meal starch 30-50% can also be used as a substitute for maize starch as binder ingredient and also as starch based food products and can also be used as manure in agricultural or forest crops.

70. *Simondsia chinensis* (Link.) C.K.Sehneid; Fam.: Buxaceae.

Hab : Jojoba is a small to moderate sized ever green tree growing in arid regions of South Western America - Arijona and Mexico (Baja California).

It comes up in scanty rainfall areas. It can grow to a height of 3 to 3.5 meters with crown spread of about 1.8 meters. It can withstand prolonged drought conditions. Seed of the plant contains 50% valuable fatty oil, actually a liquid wax highly priced as a quality lubricant in place of the now scarce sperm whale oil. It can also be used in cosmetics, lotions, shampoos and conditioners. As such the species requires large-scale planting in arid and dry tropical areas in India as a measure of afforestation, and also to get commercial benefits. It can be propagated by seeds, cuttings and also clones. The plant flowers in April. The seeds are sown in poly bags. After regular

***Simondsia chinensis* C.K.S Jojoba male flowers**

shower of monsoon well grown seedlings are planted in 30 cubic centimetre pits at a spacement of 2 m × 2 m. Water and soil conservation methods may be adopted. Although plants survive without additional water even in dry areas but by planting in moderately good rainfall areas having well drained soils, yield of fruits can be increased. Manure also helps in increasing the production. Jojoba is a unisexual and dioecious plant. Occasionally both male and female flowers are borne on the same plant. It may be necessary to maintain male and female plant ratio in the plantation. The nut is edible, tasting somewhat like hazel nut but it is bitter until it is processed.

71. *Simorouba glauca* Ed.FGilman and Deniis G.Watson.

Na : Eng.: Aceituno tree, Paradise tree.

Hab : It is an evergreen tree of semi-arid tropics of Central America particularly El-Salvadar. The tree was first introduced at NBPGR station at Amaravati (Maharashtra). It has also been planted in Karnataka and Tamilnadu and introduced in A.P., Orissa, etc. It has remarkable natural re-generation capacity and ability to grow even as arid, eroded and degraded soils. As such plant is required to be cultivated in a big way in degraded areas in afforestation programmes. The flowers are unisexual and the plants can be monoecious are dioecious. Flowering is annual, beginning in December and continuing up to February. Fruit is a drupe. It turns black in pink variety and greenish yellow in green variety. They are harvested in April/May. Seeds contain 60% shell and 40% kernel. Oil content of kernels is 60.3%. Kernel oil is solid hard fat. Refined oil is edible and is used for cooking and making biscuits and soap industry. It can be used as bio-diesel after processing. The kernel cake is used as manure. To be used as animal feed toxic principles are required to be removed.

***Simorouba glauca* – Ed FGilman and Dennis G Watson male tree**

It grows in all types of neutral and well-drained soils. It comes up in areas where rainfall is as low as 500 mm and as high as 4000 mm. Jackals, porcupines and squirrels devour the fruits and help in dispersal of seed. Fleshy pulp adhering to the seed is removed by rubbing and washing. The seeds are dried. They have viability of 2-3 months. If they are kept in airtight container, they remain viable up to 50 per cent even after one year. Germination period is 20-30 days as it has hard seed coat. Alternate wetting and drying for 3-4 days with dipping in hot water for half an hour may hasten up germination. There is almost 100% germination. Bigger sized poly pots are required. One or two seeds are sown in poly pots soon after collection and dried. Regular care of the seedlings by watering, weeding and prophylatic spraying may be undertaken. With the first regular shower of monsoon June/July seeds are sown so that the seedlings get established by end of monsoon. They may be planted in pits of 45 cubic cm after tearing of the bag and at espacement of 5 m × 5 m. Before planting the pits may be filled with topsoil and 2 kg compost 20 gms. Phosphorus and 20 gms Nitrogen. Water conservation measures may be adopted. Timely weeding will help the plant for effective pollination and good bearing. Planting of high yielding monoecious plants (pollen donors) at distance of 60 meters from each

***Simorouba glauca* - Ed FGilman and Dennis G Watson male Fruiting tree**

other among female plants is advised or some selected female plants may be grafted with a few branches of high yielding monoecious scions. Wind and honeybees are pollinating agents. The plants are not browsed by cattle and also have the capacity to recover from injuries. In commercial plantations where the plants are growing vigorously final spacing may be about 7-7.5 meters. In the initial stages inter-cropping can be done. Application of fertilizers and manure will boost up the yield of fruits. There are no serious diseases of simorouba plants.

Tamarindus Indica L - branch

Harvesting: When fruits are ripe, branches are shaken or beaten mildly and fallen fruits are collected. Immediately they should be de-pulped and seed dried. After decortication the oil is extracted from kernel. The plant comes to fruiting in 6-8 years and stabilizes at 10th year and remains productive for 50-60 years. The yield of oil from a hectare of plantation at 10th year may be 1-2 tonnes besides 1-2 tonnes of oil cake. A good plantation may yield even up to 3 tonnes of oil.

Tamarindus Indica L. Fruit

72. *Tamarindus indica* L.; Fam.: *Caesalpiniaceae Common* name: tamarind.

Hab : Pl. see edible plants.

Tamarind seed kernel (70% of seed) contains 6-8% of soft oil, which can be extracted. De-oiled kernel cake is used as cattle feed.

73. *Terminalia bellerica* (Gaertn.) *Roxb. Myrobalanus bellerica* Gaertn.; Combretaceae.

Terminalia bellerica (Gaertn.) Roxb. – fruit

Na : Bengali: Bhairah; Eng.: Bahera, Belleric myrobalan, Bastard myrobalan; Hindi: Bahera, Vibhitaka; Malayalam: Tani; Marathi: Bahed; Oriya: Bhara; Tamil: Tani; Telugu: Tani.

Hab : It is a large tree growing wild practically throughout the tropical forests in the country. Stem is buttressed when old.

The kernel of the seed yields yellow oil, which is agreeable in taste and is used by the poor people as a substitute for Ghee in some parts of Madhya Pradesh. It is also applied to the hair. The oil content in the seed is about 25 per cent.

T.chebula Retz. - branch

74. *T.chebula* Retz.; Fam.: Combretaceae.

Na : Assam: Silikha; Bengal: Haritaki; Eng.: Yellow myrobalan, Chebulic myrobalan; Gujrathi: Hardo; Hindi: Harir, Harda, Haritaki; Marathi: Hirda; Oriya: Haridra; Punjab; Harar; Tamil: Kadakkai; Telugu: Karakkaya.

Hab : It is found practically throughout the dry tropical forests in the country.

On removing the astringent pulp of the fruit a hard stony seed remains, which weighs 37 per cent of the fruit. The seed yields about 36 per cent oil, which is yellowish in colour. It has pleasant odour and agreeable taste.

T.chebula Retz. Fruit

Theobroma cacao **L**

75. *Theobroma cacao* L; Fam.: Sterculiaceae.

Na : Eng.: Cacao, Cocoa or chocolate tree.

Hab : It is a small tree native of tropical America but now cultivated along the Malabar Coast in Kerala and in the Nilgiris and Pulney hills in Tamilnadu, Orissa and certain parts of Eastern India.

The cocoa butter obtained from the seeds is used in cosmetics and in the preparation of chocolates. Cocoa plants come up very well under shade and in open. In the 3rd year 85% plants fruit.

76. *Thespesia populnea* (L) Soland. ex corr.; Fam.: Malvaceae.

Na : Bengal: Dumbla; Eng.: Portia tree; Gujrathi: Paraspeepla; Hindi: Paras pipal, gajadanda, porush; Kanerese: Hoovarase, kandarola; Jogiyarale; Malayalam: .Poovarasu; Marathi: Parsachajhada; Oriya: Gunjaosto, Porosopippali; Sanskrit: Gardha bhanda; Tamil: Poovarasam kallal, cheelanthi; Telugu: Gangaravi, Munigangaravi.

Hab : It is a small tree found along the coastal regions of Konkan, Bengal and the Andaman Islands. The seed oil is used as an illuminator by local people.

77. *Triumfetta bartramia* L (Syn. *T.rhomboidea* Jacq.); Fam: Teliaceae.

Na : Assam: Agra; Bengali: Banokara; Gujrathi: Jhipato; Hindi: Chikti; Kanerese: Kadubende; Marathi: Jhinjira; Oriya: Bojoramuli, Jotojuti; Sanskrit: Jhinjharita; Tamil: Ottupullu, Puramutti; Telugu: Chirusitorika; Uttar Pradesh: Bhora.

Hab : It is a common small shrub found throughout tropical and sub-tropical India up to 1200 m in the Himalayas.

The seeds yield greenish-yellow and odourless oil

78. *Vateria indica* L (Syn. *V. malabarica* Blume); Fam.: Dipterocarpaceae.

Na : Eng.: White damar, Indian copal tree, Malabar tallow tree; Hindi: Bada sal, Safed damar, Ral

Hab : Pl. see oleoresins and resins.

The seeds yield oil, which is used for illumination and in the preparation of soap and candles. Locally it is used for edible purposes.

The dried kernels, on solvent extraction, yield 20 to 22 per cent of pale-yellow fat having a tallow-like consistency, which turns white on standing. The fat is known by a number of names, such as Malabar tallow, Piney tallow; Dhupa tallow. It can be extracted by boiling the kernels in water for

Thespesia populnea
(L) Soland. ex corr - branch

2 to 4 hours. The softened kernels are then ground to a paste, which is again boiled in water for 4 to 6 hours and allowed to cool overnight when the fat floats on the surface and is skimmed off. It can also be extracted by solvent extraction method.

The chief characteristics of the fat are given below:

Specific gravity (20 °C)	0.9120
Refractive index (25 °C)	1.4556
Saponification value	190.4
Iodine value (Hanus)	40.0
Acid value	1.4
Unsaponifiable matter	0.8%

Collection, drying and storing of these seeds is rather a difficult problem because the fruits ripen during the rainy season and deteriorate soon after falling to the ground. Kernels must be separated from the husk and extracted for the fat as early as possible. It is planted as avenue tree. There is good natural regeneration. Tree loves shade and moisture in early stage and is sensitive to drought and fire.

Wrightia tomentosa **Roem. & Schult.**

79. *Vernonia anthelmintica* Willd.; Fam.: Asteraceae (Compositae.)

Na : Hindi: Kaliziri.

Hab : It is a common herb in the Dun Valley.

The seeds yield about 18 per cent fatty oil. With it are mixed essential oil, resin and tannin. The oil is considered to have powerful anthelmintic properties.

80. *Vibernum coriaceum* Bl.; Fam.: Caprifoliaceae.

Na : Khasi: Sohjia hynlam, Dieng soh law; Kumaon: Kalatitmalia, Titmulia, Tita, Karwa.

Hab : It is an evergreen shrub found in the temperate regions in Central and Western Himalayas.

Seeds yield yellow coloured fatty oil, which is used as an illuminator and also for edible purpose by the Nepalese.

81. *Wrightia tomentosa* Roem. & Schult. (Syn. *W.mollissima* Wall.; *Nerium tomentosum* Roxb.); Fam.: Apocynaceae.

Na : Hindi: Dharauti, Duchi.

Hab : Pl. see edible plants.

It is a small deciduous tree found throughout warmer parts of India. The seeds yield thick red coloured oil, which is said to have medicinal properties.

Xylocarpus moluccensis **(Lamk.) Roem**

82. *Xylocarpus moluccensis* (Lamk.) Roem; (Syn. *Carapa moluccensis* Lamk); Fam.: Meliaceae.

Na : Eng.: Carapa; Hindi: Pussur.

Hab : It occurs in the coastal regions of Andaman Islands.

The Carapa oil obtained from the seeds is used for manufacture of soap and as an illuminator.

10

Tanning Materials

10.1 Nature and Kinds of Tannins

Tannins are complex organic structures often with very large and high molecular weights built up from the elements carbon, hydrogen and oxygen. Some tannins contain smaller amounts of other elements such as nitrogen and phosphorus. Tannins are found in abundance in plants. In fact, any vegetable produce can be said to contain some tannin. They are mostly uncrystallizable colloidal substances with astringent properties and with the ability to precipitate gelatin from solution and to form insoluble compounds with gelatin yielding tissues. It is this property which enables them to convert hides and skins into leather. The tannins are generally classified into two groups. The hydrolyzable tannins can be hydrolyzed by acids and enzymes and comprise the gallotannins (from galls) and ellagitannins which produce "bloom" on leather from myrobalan, divi-divi and oak bark. These correspond to pyrogallol tannins. The condensed tannins are not hydrolyzable and yield "phlobaphenes" or reds; they comprise the cutch (catechu tannin) mangrove tannin etc; these correspond to the catechol tannins. In the cycle of organic degradation the tannins get converted into sugars and gums.

10.2 Occurrence of Tannins in Plants

The tannins are more prevalent among the Angiosperms or higher plants especially in certain Dicotyledonous families than they are among the lower plants. Among the Dicotyledons there are many families in which tannin occurs very freely, the important ones being Leguminosae (wattle) Anacardiaceae (*Rhus* spp.), Combretaceae (myrobalan), Rhizophoraceae (mangroves). Myrtaceae (eucalyptus). Plants rich in tannin occur both in temperature and in tropical or sub-tropical climates but the frequency of such species is more in the tropics. Tannin may occur in almost any part of the plant –root, stem or trunk, bark, leaves, fruit and even plant hair like growths. It may occur either in isolated individual cells, in groups or chains of cells or in special cavities or sacs. Tannin is found in solution in the cell sap and not in the cell membranes. As the cell ages and loses its protoplasmic contents, the tannin commonly becomes absorbed in the cell wall. In dead tissue, tannin often accumulates in considerable quantity. Tannins may occur in all parts of the plant but the content may vary. Tannin content in any species may depend on the climate, soil, age, position of the bark on the tree and even on the season of collection.

10.3 The Function of Tannins in the Life of the Plant

The function of tannins in plants is not well understood. Some contend that it is a by-product of metabolic process while others think that it has a definite function in the plant. Some argue that the presence of tannin in leaves and green tissue would, owing to its astringency, prevent damage from browsing animals or creatures. Some consider that one of the functions of tannin in the plant may be to assist in preventing infection by fungi. In more recent years the belief has grown that tannins may play some part in the formation of cork or corky tissue. Tannins and dyes are very closely found in the body of the plant and it is also considered that the tannins get transformed to produce coloured pigments.

10.4 Collection of Tanning Materials

Collection is a seasonal business especially in the case of fruits and leaves. The proper season for collection of fruits varies; in some cases the ripe fruit is best for collection and in others the full-sized unripe fruit. The season for collection of leaves is even shorter than fruits. The only leaf which has been collected to some extent is *Anogeissus latifolia* Wall (Dhawa). In this case young leaves have higher tannin content (25 % to 46 %) than the large leaves (22 % to 32 %). In the case of trees, the bark is often not removed until they are felled and have been logged for timber or firewood. It is better to collect the bark immediately after the trees are felled. For the purpose of supplies of tanning materials the forest areas are leased out to the contractors. Lease period is usually one year. Method of collection depends on the nature of the tanning material. The fruits like chebulic myrobalan are collected by shaking the tree and picking the fruits from the floor. They are then thoroughly dried in the sun. In the case of bark the trees or the branches, as the case may be, are cut. Long incisions are made on the logs or twigs with an axe. With the blunt side of the axe or the hammer, pounding is done on the bark. The bark gets loosened from the stem. It is thus separated and dried. Some times the standing trees are stripped of the bark.

10.5 Extraction of Tannin

Manufacture of industrial tannin extract based on forest produce, is of fairly recent origin. Generally, those substances which combine with the active elements in the hide are termed as tannins, whereas the other substances, which are present in the solution, but in no way undergo any change, are termed as Non-tans. The non-tans form an inevitable part of the extract and consist mainly of sugars and gums. Although a wide variety of indigenous tannin containing plants are found in our country all of them are not exploited due either to the unsatisfactory colour of the extract, to their relatively slow growth and or to their comparatively low tannin content. Wattle bark has been found suitable in these respects. Wattles are native to Tasmania and South Australia but have been introduced in other countries particularly in Natal, South Africa, Kenya, Uganda, Sri Lanka and South India in the Nilgris. In tannin industry, by 'wattle' it is normally meant Acacia mearnsii dewilld (Syn. *A. mollisima* Willd. A. *decurrens* var. *mollis* Lindl.). In the past, 3 other species of Acacia were also associated with the wattle industry, namely, *Acacia decurrens* Willd. syn. (*A.decurrens* var. *normalis* Benth.) or Green wattle; *Acacia dealbata* Link. syn. *A.decurrens* Willd. var. *dealbata* (or Silver wattle) and *Acacia Pycnantha* Benth (or Golden wattle). Mature wattle bark contains an average of 35% tannin. It has been established that the tannin content even within a tree, varies, being maximum in the thick basal bark, gradually decreasing with increasing height. The tannin content usually increases with age.

10.6 Short Description of the Process

Matured wattle trees are stripped or peeled of bark. Normally, the trees strip with ease and almost continuously in areas of high rainfall. Stripping is done manually in the absence of any mechanical means. Bark from the bottom portions is peeled of from the standing trees and after felling, bark of the upper portions is stripped off. The material i.e., bark used for extraction of tannin has to conform to some basic specifications. It should, for instance, have a certain optimum growth, should have a definite dimension both in respect of length and width and should have definite moisture content. The stripped bark is sent directly to the processing factory. The colour of the dried bark is of paramount importance. Open-air drying is satisfactory under fine weather conditions. Continued exposure to moist conditions causes discolouration of the bark and may result in tan losses. Normally shade-drying is preferred. The black wattle extractives are readily soluble in hot water. During the commercial manufacture, strict analytical control is exercised both in respect of tannin content and the colour of the product. The solid extract sold in the market, should contain a minimum of 60% tannin, irrespective of the moisture content.

The bark, either green or dried, is fed to the chipping machines where it is cut into the required sized chips. These chips then are fed to the battery of autoclaves containing extracting medium i.e., water. The autoclaves have device to regulate temperature and pressure. Under proper temperature and pressure in the autoclaves the solubles in the bark are leached out and a weak overflow of the tan liquor is obtained. The liquor is then further processed, mainly to evaporate the water and to obtain a solid extract normally containing 18% to 20% of moisture. This evaporation is achieved in two different sets of evaporators, the first set achieving an evaporation leading up to the intermediate stage (50% solids) and this thick tan liquor is then further processed to the commercial solid extract in the finishers. Sometimes, where it is desired to manufacture the spray-dried powder, the intermediate stage liquor, instead of being fed to the finishers, is sent to spray drying unit to yield the powder. The solid extract is then traditionally bagged in double gunny bags, whereas the spray-dried powder is invariably packed in polythene lined gunnies.

The spent or the residual matter left in the autoclaves represents the waste material and is discarded. However, it can be used as fuel.

10.7 Tannin Content

From the tanner's point of view, high tannin content is a very desirable factor and certain minimum tannin content is essential. Too dark a colour precludes the use of materials with high tannin content; some factors like ellagic acid content make tanning material of low tannin content acceptable.

10.8 Adulteration

Worthless bark or twigs are sometimes mixed and in some cases it is even done to the extent of including stones to make the weight greater. Sometimes dust, leaves, etc., get into material when it is swept up after being spread out to dry or after being pounded for packing. Therefore every precaution should be taken to supply raw material of good quality.

Acacia decurrens **Willd. var. dealbata F. Muell.**

10.9 Uses of Tannis

Leather industry has always remained by far the largest user of tannins but certain other industries use them to some extent. Tannin is used in deep oil-well drilling in combination with caustic soda to reduce the viscosity of the drill mud. Tannin has long been used in the preservative treatment of fishing nets (mainly cutch) and in ink manufacture. Of late its use in the manufacture of inks has considerably reduced because synthetic materials have been partly substituted for it. Tannin is also used in boiler-water treatment, for the prevention and removal of boiler-scale.

10.10 Supply and Demand

In India tanning materials are obtained from trees or shrubs which are practically all wild, either growing in forests or waste lands or as relics of original forest in cultivated lands or planted sporadically as shade and avenue trees, etc. Except Wattle *Acacia mearnsii* de Willd (Syn. A.*mollisima* Willd.) which has been planted up in the Nilgiri hills no other species has been raised artificially on a large scale. Myrobalans are found extensively in the tropical forests in Central and South India and there is a considerable demand for the myrobalan extract in foreign trade. Leather industry occupies an important place in the country and large quantities of skins and hides of both processed and finished products are exported. There are over 725 tanneries requiring large quantities of tanning materials. In addition, the local village cobblers are also using the tanning materials. In India the demand for wattle extract alone is in the region of 20000 tonnes per annum and the present production is of the order of 4000 tonnes a year which highlights the wide gap between supply and demand and to reduce or close this gap, extensive planting programme of wattle in suitable localities is necessary, to meet internal and external demand.

10.11 Tanning Yielding Plants

1. *Acacia decurrens* Willd. var. *dealbata* F. Muell. (Syn.*A. dealbata* Link); Mimosaceae.

Na : Eng.: Silver wattle.

Hab : It is an evergreen tree attaining a height of 12 m and girth of 1.2 m. Its bark is silvery grey. It was introduced from Australia and is now being grown in the Nilgiri and Pulney hills at an elevation of about 1500 m. It is also introduced in the Himalayas at altitudes of 1600 to 2400 m but it suffers from snow break. It has lower tannin content than *A. mearnsii*. de mernsic Willd.

2. *A.mearnsii de* Willd. (Syn. *A.mollissima* Willd. *Syn.A.decurrens* willd. var. *Mollis* Lindl.); Mimosaceae.

Na : Eng.: Black wattle.

Hab : Black wattle an evergreen tree growing to a height of 20 to 21 m is native of Tasmania. It is the principal species among Wattles grown in South Africa and India. The bark of Black Wattle contains a high percentage of tannin. It is very much in demand as it imparts a high grade finish and colour to the treated leather. Consequently, extensive plantations have been raised in South India at elevations of 1500 to 2100 m during the recent past.

Acacia mearnsii **de Willd.**

Cultivation of Wattle on a large scale is confined to Tamilnadu, though sporadic plantings and plantations are also met with in other States. It is an average light demander. It can tolerate certain amount of lateral shade and is susceptible to wind damage.

***Climate*:** Black Wattle flourishes well between elevations of 1850 to 2500 m. It comes up in places where the mean daily shade temperature for the year is 14 °C and the average maximum temperature is 24 °C. It cannot tolerate severe summer. At lower elevations due to high temperature the growth is poor. The average minimum temperature should not be less than 1°C. The requirement of annual precipitation is 1250 to 2000 mm with about 100 rainy days. Heavy rainfall of over 2000 mm may deteriorate the quality of bark due to lichen covering. Winter frosts and cold winds during early part of the rainy season (June-July) adversely affect the growth rate. Localities experiencing severe frost are similarly not suitable. Generally speaking, wattles prefer moderate climate, exhibiting greater intolerance to the extremes – both of heat and cold. A well distributed rainfall of about 1500 mm is most suitable, but even higher rainfall is not disadvantageous provided it does not lead to waterlogged conditions.

***Edaphic requirement*:** Although Wattle can grow on a wide variety of soils it does best on a loose, deep, well-drained soil. Dark sandy loam or red loam on gentle slopes appears to provide optimum soil conditions. Soils of the Nilgiris are lateritic as such they are poor in plant nutrients. In such soils fertilizer application may be necessary.

***Method of propagation*: *Presowing treatment of seeds*:** The number of Black Wattle seeds per gram may vary from 55 to 80. Pre-treatment of seed is necessary due to its hard testa and for early and satisfactory germination. Water (about 5 to 7 times the volume of seed) is heated and when it starts boiling the fire from underneath is removed and the seed is put in the water and stirred. On no account boiling should continue after the seed is added. Water with seed is allowed to cool. Seeds are washed 5 to 6 times in cold water to remove the gummy mucilage and then dried in shade and stored in lead tinned boxes even for a year without affecting its germination capacity. Treated seeds give 53 to 60% germination as against 5% from untreated seed.

***Sowing*:** Plantation can be raised either by direct sowing in the area or by planting nursery raised seedlings. In the case of direct sowing the area is generally ploughed and prepared and seed is sown at about 2 to 3 kg per hectare with the first shower of monsoon in lines 3 to 3.5 m apart. It would be better to sow the seed in advance so that with the first shower of monsoon it starts germinating. Germination starts in about 8 days and is complete within a month. Direct sowing is resorted to in places where there is labour scarcity and planting target is high but the requirement of seed would be more. In case there is any gap in the monsoon after germination, there may be heavy mortality of seedlings. Similarly, by ensuing winter the seedlings will be too tender and are liable to be affected by frost, if any, if no cover is given. Hence transplanting of well grown seedlings appears to be advantageous.

***Nursery*:** Seeds are generally sown in nurseries 9 to 12 months prior to planting. In South India planting season is June-July and sowing is best done not later than September of the previous year. The seed rate is 0.5 kg per bed of 12 m × 1.2 m. Each bed will give sufficient seedlings for planting 2 ha of plantation at an espacement of 2.5 m × 2.5 m. Seeds in the nursery beds should be covered with hay and watered with a rose-can. Germination takes place in about a week and is complete in a month's time. To avoid the damage by rabbits, a rabbit-proof fencing is desirable till at least the seedlings become about 10 cm. In frosty localities frost-covers may have to be provided over the nursery beds. The best time of planting is June to mid-July depending on the outbreak of monsoon. The seedlings when they are 7 to 10 cm in height are lifted from the nursery bed 3 to 4 months prior to planting in the field. The roots are wrapped up in moss with little earth and tied with fibre. Alternatively they are pricked out to polythene bags. They are kept in shade till they are planted out. By planting time they are about 25 to 30 cm in height. Mossed seedlings are disturbed once a fortnight to prevent striking of roots in the ground.

***Planting*:** 0.3 m cube pits are dug at an espacement of 2.5 × 2.5 m in advance of planting. Planting of polythene bag raised seedlings or mossed seedlings are done only with the advent of monsoon particularly on a rainy day. The soils, where Wattles are planted in South India are normally lateritic and are deficient in plant nutrients. Hence Wattle fertilizer mixture N-6 to 7%, P_2O_3-10%, K20-10%, is mixed with the soil at 75 grams per pit at the time of planting. In some places along with wattle, taungya crop of potatoes is raised in the interspaces.

Casualties are beaten up periodically in the 1st year during the monsoon. In the 2nd year if the stocking is less than 75% casualties are replaced. Normally 3 weedings are done in the 1st year and 2 in the 2nd year, the last operation being combined with soil working. Where the incidence of grass and weed growth is more, more weeding operations may be necessary.

***Protection from injuries*:** The plantation is completely protected from fires. Young wattle up to a height of even 1m is susceptible to frost damage. When the seedlings are tender they are given frost cover during November to March where the winters are severe. In South India no serious incidence of insect pests is noticed. However, some damage occurs in nursery by cut-worms and cockchafer larvae, but this can be controlled by dusting with gamaxene. A considerable damage is caused due to browsing by wild animals to young plantations. Therefore fencing may be necessary. Sometimes because of fungal infection there can be water blisters.

***Growth*:** The growth varies depending upon the site quality. In 1st quality areas the plants may attain a height of 21 m with 23 cm in diameter while in quality III areas the height may be 15 m and diameter of 15 cm.

***Silvicultural system*:** The silvicultural system adopted is clearfelling followed by natural regeneration from 2nd rotation onwards.

***Rotation*:** The best rotation depends upon the site quality. In South Africa a rotation of 12 years is adopted for quality I and II sites and 8 to 10 years for poorer sites. In South India a rotation of 10 years is generally adopted.

***Second rotation crop*:** Black wattle produces prolific seeds annually and has the capacity to regenerate itself naturally. Advantage is taken of this habit. The area is clear felled and after extraction of utilisable material a light fire is run to eliminate weeds and stimulate germination of the hard-coated seed. This method restocks the area naturally. Since there is profuse natural regeneration it is necessary to space out the stems. A thinning is carried out in the 3rd year spacing the plants to 2 to 2.5 m apart. Along with this, unwanted growth is also removed.

***Stripping, drying, processing and packing of bark*:** Before felling the trees, a horizontal cut is made around the mature tree at about 1.2 m to 1.5 m above the ground and the sheet of bark is pulled off downwards. With this method bark found even on large roots is saved. Bark close to the ground is thicker and richer in tannin content. Therefore as much as possible of the bark close to the ground is pulled off before the tree is felled. The tree is then felled and the bark is stripped upwards upto immature bark – generally up to an inch in diameter of the stem or branch. After stripping, bark is chopped into lengths of about 1.2 m and bundled. In South India an average worker can strip about 150 to 200 kg of bark per day. The green bark should not be heaped up or kept bundled because it is infected by fungus. For drying, bark is spread on bushes keeping the outer surface exposed to the sun. The inner surface is not exposed to the sun as otherwise there is a possibility of decolourisation. On drying, bark becomes brittle and light-brown in colour. Bark can also be dried artificially in driers. The dried bark is chipped into lengths of 15 to 22 cm or chips of 5 × 5 cm and pressed into bales or packed in bags for disposal.

***Yield of bark and tannin content*:** Many factors such as locality, method of cultivation and climatic conditions have immense effect on the production of bark. As such the yields have been varying. Depending on the site quality a yield of 15 to 20 tonnes of bark besides 40 to 70 tonnes of fuel per hectare can be expected in about 9 to 10 years. The tannin content in the bark is normally about 35% with about 12.5% moisture content and 10.5% to 11.5% soluble non-tan. There is some percentage of colouring matter.

***Wattle bark as a tanning material*:** Wattle bark is one of the richest tanning materials in the world. Wattle tannin rapidly penetrates inside the hides and provides an excellent weight in finished leather and a good colour as well. Wattle tanned leather is very tensile and elastic. The requirement of tanning material in the country is far higher than actual production. As such large quantities of raw material as well as tannin extracts are being imported into the country. During 2002-03, 12384.314 tonnes of Wattle extract valued at Rs. 43,08,54,204 were imported.

***Uses of spent-wattle bark and wood*:** Spent Wattle bark can be used for manufacture of pulp for wrapping paper and straw board and also as fuel. Wattle wood besides being used as fuel and charcoal is also used as mine-props, tent poles and in building construction.

On destructive distillation of wood tar, high yield of acetic acid and methyl alcohol are obtained. The wood can also be used for manufacture of pulp.

A. leucophloea **Roxb. Willd.**

Acacia nilotica **(L.) Del.- Pods**

Acacia Pycnantha **Benth.**

Aegiceros corniculatum **(L.) Blanco**

3. *A. decurrens* Willd. var. *normalis* Benth. (Syn. *A.decurrens* Willd.); Mimosaceae.

Na : Eng.: Green Wattle

Hab : It is an ever green tree growing to a height of 12 m with leaves 7.5 to 10 cm long and olive green to dark grey bark. In India it is grown in the Nilgiris near Ootacamund & Kodaikanal.

The bark yields tannin which is used in tanning industry. It has undue redness in extract.

4. *A. leucophloea* Roxb. Willd. (Syn. *Mimosa leucophloea* Roxb.); Mimosaceae.

Na : Bengal: Safed babul; Guj: Haribaval; Hindi: Safed babul, Safed Kikar; Kan.: Bile jali; Mal.: Pattacharayamaram; Mar.: Hewar; Tam.: Velvayalam; Tel.: Tellatumma.

Hab : It is a small sized tree found in the Punjab, Rajasthan, Madhya Pradesh, Maharashtra and Andhra Pradesh.

The bark yields about 15 % to 16 % tannin on zero moisture basis which is similar to the Arabica tan obtained from Acacia nilotica (L.) Del. (Syn. A.arabica Willd.)

5. *A. nilotica* (L.) Del.(Syn. *A. arabica* Willd.);Mimosaceae.

Na : Eng.: Babul acacia; Hindi: Babul, Kikar.

Hab : It is a tree found in North, Western, Central and Southern India. (Pl. see gums)

The bark and pods yield tannin. The tannin content in the bark is about 17 % on zero moisture basis and that of soluble non-tans is about 11.1 %. Bark taken from a fully grown tree is found to be more astringent and better adapted for tanning purpose than the bark taken from a young and immature tree. Old bark yields darker colour. Bark from lower parts yield greater tannin content. Deep colour and high non-tannin content of the bark are serious disadvantages. Babul is good for heavier leathers. Pods contain about 22 % tannin. Soluble non-tans in them are about 29.5 %.

6. *A. Pycnantha* Benth.; Mimosaceae.

Na : Eng.: The golden wattle.

Hab : It is grown in the Nilgiri Hills.

The species is grown in Australia but not in S. Africa. In India it is not a favoured species as it contains less tannin than A. mearnsii de Willd.

7. *Aegiceros corniculatum* (L.) Blanco (Syn. *Rhizophora corniculata* L.) Myrsinaceae.

Na : Beng. Khalshi; Eng.: Goats horn mangrove; Hindi: Halsi; Mar.: Kanjala; Tamil: Narikandam.

Hab : It is an evergreen shrub or small tree found in the Sundarbans and the Andaman Islands.

The bark is the source of tannin.

Alnus nitida **Endl.**

8. *Alnus nitida* Endl.; Betulaceae.

Na : Eng.; Himalayan black cedar; Hindi: Kunis Saroti, Utis; Kashmir: Chamb; Punjab: Sharol.

Hab : It is a large tree of temperate regions in north western Himalayas.

The bark yields a tan which is used locally for tanning purpose

Anogeissus latifolia **Wall.**

9. *Anoegeissus latifolia* Wall.; Combretaceae.

Na : Eng.: Gum ghatti; Hindi: Bakli, Dhaura.

Hab : It is a deciduous tree occurring wild in the drier parts in the country. (Pl. see gums).

The leaves are used for tanning purpose. The tannin content is about 19 %, whereas the content of soluble non-tans is about 1.5 %. Small young leaves have high tannin content and less colour than older green leaves. The bark also contains about 13 % tannin.

Avicennia officinalis **L.**

10. *Avicennia officinalis* L. (Syn. *A.tomentosa* Jacq.; *A. resinifera* Forst.); Verbenaceae.

Na : Eng.: White mangrove; Beng. & Hindi: Bina; Guj.: Tavariyan; Mar.: Tiwar; Telugu: Mada.

Hab : It is s tree of the Sundarbans, the Coromandel coast and the Andaman Islands.

The bark is locally used for tanning purpose.

Bauhinia malabarica **Roxb.**

11. *Bauhinia malabarica* Roxb.; Caesalpiniaceae.

Na : Eng.: Malabar mountain ebony; Hindi: Amli.

Hab : Pl. see edible plants.

Bark contains about 12 % tannin. It imparts a red colour to leather and is not of much value.

Bauhinia purpurea **L.**

12. *B. purpurea* L.; Caesalpiniaceae.

Na : Eng.: Camel's foot tree, Pink Bauhinia; Hindi: Kaliar, Lalkachnar.

Hab : Pl. see edible plants.

The bark yields a tan.

Bauhinia vahlii Wight & Arn.

Bauhinia variegata L.

Bischofia javanica Blume

Boswellia serrata Roxb. ex Colebr.. branch

Bridelia retusa Spreng.

13. *B. vahlii* Wight & Arn.; Caesalpiniaceae.

Na : Bengal: Sihar; Eng.: Camel's foot climber; Hindi: Maljhan, Jallur; Mar.: Chambul; Telugu: Adda.

Hab : It is found practically throughout the tropical and sub-tropical forests in the country particularly along the valleys and wetter parts.

The bark yields 17 % tannin and 19 % non-tans

14. *B. variegata* L.; (Syn. *B. candida* Roxb.); Caesalpiniaceae.

Na : Eng.: Mountain ebony, Variegated Bauhinia; Hindi: Kachnar.

Hab : pl. edible plants.

The bark is locally used for tanning

15. *Bischofia javanica* Blume; Euphorbiaceae.

Na : Bengal: Kainjal; Eng.: Bishop Wood, Java cedar; Hindi: Bhillar, Paniala; Kan.: Gobranerale; Mal.: Nira, Thirippu; Mar.: Boke; Tamil: Thondi; Telugu: Nalupumushti.

Hab : It is a large tree found in shady ravines and river banks of the Sub-Himalayan forests in the eastern region, Orissa and Western Ghats. The twig bark contains 16 % tannin and 13 % non-tan which is used by the local tanners.

16. *Boswellia serrata* Roxb. ex Colebr.; Burseraceae.

Na : Eng.: Indian frankincense; Hindi: Salai.

Hab : Pl. see medicinal plants and oleoresins.

The bark is the source of tan.

17. *Bridelia retusa* Spreng.; Euphorbiaceae.

Na : Hindi: Ekdania, Khaja.

Hab : Pl. see edible plants.

The bark is used as a tan.

18. *Bruguiera conjugata* (L.) Merr. (Syn. *B. gymnorhiza* Lamk.; *Rhizophora gymnorhiza* L.); Rhizophoraceae.

Na : Beng. & Hindi: Kankra; Tamil: Sigapukokandam; Telugu: Thuddaponna.

Hab : It is a tree of the tidal marshes of the Western peninsula, Bengal and Andaman Islands.

Tannin content in the stem bark varies from 31 to 42 %. Content increases with the age of the bark. Soluble non-tans in the bark are about 13 to 16 %.

Because of tannin content it is a valuable tanniferous mangrove. Leaves also contain about 13.5% tannin but the soluble non-tans are 21.5%. As such leaves are not useful for tanning purpose.

B.cylindrica (L.) Blume - Branch

19. *B. cylindrica* (L.) Blume; Rhizophoraceae.

Na : Hindi: Kakandan, Madarna; Tamil: Kakandan.

Hab : It is a small tree found in the tidal forests in Kerala and Tamilnadu.

The bark contains 15 to 19% of tannin and 18-20% of non-tans. It is used locally for tanning purpose. It gives a supple light brownish red leather.

***B.parviflora* Wight & Arn.**

20. *B.parviflora* Wight & Arn.; Rhizophoraceae.

Na : Telugu: Vurada.

Hab : It is a small tree found in the Andaman Islands.

The leaves contain 12% of tannin and are suitable for light leathers.

21. *Buchanania lanzan* Spreng.; (Syn. *B. latifolia* Roxb.); Anacardiaceae.

Na : Eng.: Cuddapah almond; Hindi: Chironji.

Hab : Pl. see edible plants.

The bark contains 13.4 % of tannins and 9.4 % of non-tannins. It yields dark reddish brown leathers of somewhat stiff and harsh texture.

***Buchanania lanzan* Spreng. - branch**

22. *Caesalpinia coriaria* (Jacq.) Willd.; Caesalpiniaceae

Na : Eng.: Dividivi, American sumach; Hindi & Telugu: Libidibi; Tamil: Tividivi,Inkimaram.

Hab : It is a woody climbing shrub commonly found in South India. It is grown in north western India and Bengal.

The dried pods (fruits) are used for tanning. The tannin content in pods is about 38 % on zero moisture basis, whereas the content of soluble non-tan is about 26 %. The tan is usually blended with other tanning materials. It yields heavy and firm leather. The yield of pods is 2.5 tonnes per hectare of plantation per annum

***Caesalpinia coriaria* (Jacq.) Willd.**

23. *C.digyna* Rottler; Caesalpiniaceae.

Na : Bengali: Umulkuchi; Hindi: Vakerimul; Mar.: Vakerimula; Oriya: Odlo; Telugu: Nunegacha.

Hab : It is a climbing shrub found in Bengal, Assam and the Andaman Islands. It grows in waste lands and prefers a well-drained sandy soil.

C.digyna Rottler

Cassia auriculata L.

C. fistula L. - Tree

C. fistula L. Branch with flowers

Pods known as Teripods yield about 28 % tannin on zero moisture basis and 27.6 % soluble non-tans. Pod cases yield about 54 % tannins. They impart light colour. Collection of pods from wild source is difficult because of its scattered nature.

24. *Careya arborea* Roxb.; Barringtoniaceae.

Na : Eng.: Patana oak; Hindi: Kalikatbhi, Kumbhi.

Hab : Pl. see edible plants.

Bark yields tannins.

25. *Carissa opaca* Stapf; (Syn. *C. spinarum* Auct. *pl.* non L.); Apocynaceae.

Na : Hindi: Karaunda; Kashmir: Garaunda; Punj.: Gan; Sans.: Karamadika; Tamil: Chiru, Kila; Telugu: Kalivi; Uriya: Anka, kot.

Hab : It is an ever green shrub commonly found in the drier regions.

Leaves contain about 9.6 % tannin and 23.7% soluble non-tans. Tannin content increases with the maturity of leaves. It is not a good tanning material to be of much importance.

26. *Cassia auriculata* L.; Caesalpiniaceae.

Na : Eng.: Avaram; Hindi: Avaram, Tadwad; Tamil: Avaram; Telugu: Tangedu.

Hab : It is a very common shrub in drier places particularly in waste lands. The plant is conspicuous by its yellow flowers. (Pl. see edible plants.)

Avaram bark is extensively used in tanning industry in South India. It yields about 20 % tannin on zero moisture basis and the soluble non-tans found in it is about 11.5 %. Tannin increases with the age of the plant. Maximum tannin content is found in 3 to 5-year-old plants. The tan penetrates quickly and produces a light coloured leather with elastic grain and good tensile strength.

27. *C. fistula* L.; Caesalpiniaceae.

Na : Eng.: Indian labernum, Golden shower; Hindi: Amaltas.

Hab : Pl. see medicinal plants.

Bark of C. *fistula* known as Suman is the source of tannin which is extensively used in South India. It is used in admixture with avaram bark. It yields about 12 % tan on zero moisture basis and 13.3% soluble non-tans. Its tan when used singly does not penetrate as quickly as that of C. *auriculata* L. and the result is that the leather is compact and dense.

28. *Casuarina equisetifolia* L. (Syn. C. *muricata* Roxb.); Casuarinaceae.

Casuarina equisetifolia L.

Na : Bengali: Jau; Guj.: Vilayati saru; Eng.: Casuarina, She-oak; Hindi: Janglisaru; Kanerese: Saromara, Chabaku; Mal.: Chavakku; Mar.: Saru; Tamil: Savukku; Telugu: Sarugudu, Chavuku.

Hab : It is a tall tree extensively grown in the coastal regions.

Bark is locally used for tanning purpose.

29. *C. suberosa* Ott. & Diet; Casuarinaceae;

Na : Eng.: Swamp oak.

Hab : It is grown in the Nilgiris.

The bark contains up to 18 % tannin of catechol type. It penetrates the hide quickly and produces soft leather of pale reddish brown colour.

30. *Ceriops decandra* (Griff) Ling How.;Syn.*C.roxburghiana Arn..;* Rhizophoraceae.

Na : Beng. & Hindi: Badagoran.

Hab : It is a shrub or small tree found in the tidal forests of Sundarbans, east coast of India and Andaman Islands.

Bark and leaves are used locally in tanning industry. Bark yields about 29 % tannin on zero moisture basis and about 9 % soluble non-tan. The leaves contain about 9-15 % tannin. Bark tannin imparts a red colour to leather which can be avoided if it is blended with myrobalan and babul bark tannins in suitable proportions.

31. *C. tagal* (Perr.) C.B. Robins. (Syn. C. *candolleana* Arn.);Rhizophoraceae.

Na : Beng. & Hindi: Goran; Sindhi: Chauri, Kirari; Tamil: Pandikutti; Telugu: Gedera.

Hab : It is a small tree found in the tidal forests of Sundarbans and the Andaman Islands.

The bark and leaves are used for tanning purpose. The leaves yield 15.5 % tannin and about 20% non-tans. Twig bark contains about 26 % tannin, whereas the bole bark contains 41 %. Tannin content varies with the age of bark. Leaves give strong, supple pinkish leather. They can be used in combination with the bole bark. The bole bark is highly valued in the tanning industry.

C. tagal (Perr.) C.B. Robins

Chukrassia tabularis A. Juss.

Chukrassia tabularis A. Juss.-Tree

Cotinus coggygria Scop.

Dichrostachys cinerea L. Wight & Arn.

32. *Chukrasia tabularis* A. Juss.; Meliaceae.

Na : Eng.: Bastard cedar; Hindi: Chikrasi.

Hab : Pl. see gums and resins.

Bark is astringent. Young leaves and bark contain 22% and 15% tannin respectively.

33. *Cleistanthus collinus* (Roxb.) Benth. ex Hook. f.; Euphorbiaceae.

Na : Bengali: Karlajuri; Hindi: Garari; Kan.: Badedarige; Tamil: Nilaippalai, Oduvan; Telugu: Kodisha.

Hab : It is a small tree very common in the forests in South and Central India and Bihar.

Bark, leaves and green fruits yield tannin. Tannin content in fruit is 14 % and in twig bark it is 21 %. Leaves contain 19% and bole bark 33% tannin. The species is of only local importance.

34. *Coriaria nepalensis* Wall.; Coriariaceae.

Na : Hindi: Makola, Mesuri; Kashmir: Balel, Tadrelu.

Hab : It is a shrub found in the temperate Himalayas.

Leaves contain about 20 % tannin and are used locally for tanning purpose.

35. *Cotinus coggygria Scop*. (Syn. *Rhus cotinus* L); Anacardiaceae.

Na : Eng.: Smoke tree, Wig-tree, Venetian. Sumac. Indian Sumac; Hindi: Tungla; Kashmir: Darengri; Kumaon: Gadtung; Punjab: Tunga, Tung. Manu, Paan.

Hab : It is a shrub of north western Himalayas. The leaves are used locally for tanning purpose. They should be collected after the rains from the middle of September to the end of December when they contain 18-22 % tannin on air dry basis. Bark yields about 8-21% tannin and it is high during rainy period.

36. *Dichrostachys cinerea* L.Wight & Arn.; Mimosaceae.

Na : Hindi: Vurtuli, Kuntai; Kan.: Odatare; Mar.: Sagum-kati; Tamil: Vidattalai; Telugu; Velturu.

Hab : Pl. see medicinal plants. Bark is used in tanning.

37. *Diospyros melanoxylon* Roxb. (Syn. *D.tupru* Buch.-Ham.), Ebenaceae.

Na : Eng.: Coromandel ebony; Hindi: Tendu, Abnus;

Hab : Pl. see edible plants.

Tannin content in different partsof the plant – bark, fruit and half ripe fruit is 19%, 15% and 23% respectively.

38. *Engelhardtia spicata* Bl.; Juglandaceae.

Ficus glomerata **Roxb.**

Na : Assam: Wagkru, Rumguch, Lalamiri, Lewa; Beng.: Bolas; Hindi: Silapoma; Lepcha: Sugreot-kung.

Hab : It is a large deciduous tree found in the Eastern Himalayas and Assam:

Bark contains 16% tannin and 14% non-tan and is considered a good tanning material.

39. *Ficus glomerata* Roxb.; (Syn. *F. goolereea* Roxb.); Moraceae.

Na : Eng.: Cluster fig; Hindi: Gular, Udumbara, Domoor.

Hab : Pl. see edible plants.

Bark yields 14 % tannin.

40. *Garuga pinnata* Roxb.; Burseraceae.

Na : Assam: Thotmola, Gendeliporna, Rohimala, Dieng-Khiang; Bengal: Jum, Dabdabe, Turnkharpat. Nilbhadi; Bihar and Orissa: Kandwer, Karur, Armudaru, Kekar; Guj.: Khusimb; Hindi: Kharpat, Ghogar, Kaikar; Kan.: Halabalagi. Amelli, Godda; Mal.: Kosramba, Kattukalinjan; Mar.: Kakad, Kudak, Kuruk; Oriya: Mohi, Sompotri, Armu; Tamil: Karrevembu, Arunelli; Telugu: Garuga.

Garuga pinnata **Roxb.**

Hab : It is a common tree in tropical forests throughout the country.

Leaf galls are used for tanning since they are rich in tannin.

41. *Heritiera fomes* Buch.-Ham. (Syn. H. *minor* Roxb.); Sterculiaceae.

Na : Eng.: Sundri; Hindi: Sundri;

Hab : Pl. see edible plants.

Bark and leaves contain about 10 % tannin. The leather obtained by the use of Sundri bark is reddish buff in colour, tough hut supple and free from cracks

42. *Hopea parviflora* Bedd.; Dipterocarpaceae.

Na : Eng.: Hopea, Iron wood of Malabar, White Kongu; Kan.: Tirupu, Bovige, Kiral boghi, Bovumara; Mal.: Thambagam; Tamil: Kongu, Vellaikongu. Pongu, Agil, Irumbugam.

Hab : It is a handsome tree found in the moist tropical forests in South India.

Heritiera fomes **Buch.-Ham.**

The tannin content in the bark ranges from 14 to 28 %, whereas soluble non-tan varies from 5 to 10 %. It is a very astringent tanning material with a slow speed of diffusion. Therefore it is not suitable alone but with other tanning materials like myrobalan and wattle it gives good tanning properties.

Kandelia candel (L.) Druce

Lumnitzera racemosa Willd.

Myrica nagi Thunb.

Oroxylum indicum Vent.

43. *Hymenodictyon excelsum* Wall.; Rubiaceae..

Na : Assam: Kodom, Ding-dolong-sir, Phurkundi; Beng.: Latikaram; Bombay: Kolakadu; Guj.: Amarchhala, Dondro; Hindi: Bhaulan, Kukurkat, Bhurkur, Kan.: Bandarnyenni, Vilari; Mal.: Litthilei, Nichan, Vella katampu; Mar.: Bhoursal, Dancelo; Madhya Pradesh: Bohar, Potur; Oriya: Bodoka, Konoo; Punjab: Kukarhat, Barthoa; Sans.: Bhrnmarchhalika, Ugragandha; Tam.: Vellei kadambu, Sagapu, Peranjob; Telugu: Dadippa, Dadiyetta, Burja; Trade; Kuthan.

Hab : It is a large deciduous tree found in Bihar, Central, Western and Southern parts of the country.

Bark yields tannin which is locally used for tanning purpose.

44. *Kandelia candel* (L.) Druce; (Syn. K. *rheedii* Wight & Arn.): Rhizophoraceae.

Na : Beng. & Hindi: Goria; Kan.: Kandale; Mal.: Cerukandal; Oriya: Rasunia; Tam.: Thuvar, Kandan; Telugu: Kandigala.

Hab : It is a shrub or small tree found in Andaman Islands and the tidal swamps of the western and Eastern parts in the country. Bole bark contains about 17 % tannin and 13.5% non-tans. It is suitable for heavy leather tannage.

45. *Lumnitzera racemosa* Willd.; Combretaceae.

Na : Beng.: Kripa; Mal.: Kalakkantal; Oriya: Tunda; Tamil: Tipparathai; Telugu: Kadivi, Tandra.

Hab : It is a small tree found in the Andaman and Nicobar Islands and the Sundarbans. Bark contains 15-19% tannin, whereas leaves and wood contain smaller quantities.

46. *Myrica nagi* Thunb. (Syn. *M. rubra* Sieb. & Zucc.); Myricaceae.

Na : Eng.: Bay-berry, Box myrtles; Hindi: Kaiphal.

Hab : It is a tree native of China and Japan but now it is found in the sub-tropical Himalayas, the Khasia Hills and Assam. (Pl. see edible plants)

Bark is a rich source of tannin and belongs to pyrogallol group. Tannin content in the bark is 32% with 10.7% soluble non-tans.

47. *Oroxylum indicum* Vent.: (Syn. *Calosanthes indica* Blume); Bignoniaceae.

Na : Hindi: Sonapatha, Sora, Arlu.

Hab : Pl. see edible plants.

The bark and fruits are used locally for tanning.

48. *Phyllanthus emblica L.;Syn. Emblica officinalis Gaertn. Euphorbiaceae.*

***Phyllanthus emblica* L**

Na : Eng.: Emblic myrobalan; Hindi: Amla, Aonla.

Hab : Pl. see medicinal plants.

The stem bark yields about 12 % of tannin on zero moisture basis, whereas the tannin content in the twig bark is fairly high about 21 % and soluble non-tan being 13 %. The leaves yield about 22 % tan and 17 % soluble non-tans. Although the leaves and twigs contain high tannin content but collection of these for commercial exploitation is difficult. Fruit also contains fairly high tannin content about 25.5 % but the soluble non-tan being very high about 38.7 % is unsuitable in tanning industry.

49. *Pistacia khinjuk* Stocks; (Syn. P. *integerrima* Stew ex Brand.); Anacardiaceae.

***Pistacia khinjuk* Stocks**

Na : Eng.: East Indian mastiche; Hindi: Kakra, Gurgu.

Hab : The tree is found in the north-western Himalayas. (Pl. see edible plants).

Galls on the leaves contain 20-27 % tannins. Leaves and bark contain 16 % and 14 % tannins respectively. They are used for tanning purpose.

50. *Polygonum bistorta* L.; Polygonaceae.

***Polygonum bistorta* L.**

Na : Eng.: Snakeroot, Snake weed.

Hab : It is a perennial herb found in the temperate Himalayas from Kashmir to Assam.

Its roots contain tannins up to 36 %.

51. *Punica granatum* L.; Punicaceae.

***Punica granatum* L.**

Na : Eng.: Pomegranate; Hindi: Anar.

Hab : Pl. see medicinal and also edible plants.

Fruit rind yields tannins. The dried rind contains 18-22 % tannins and 16-23% soluble non-tans. Tannin is of myrobalan type. The leather tanned with this is soft with slightly greenish-brown appearance, whereas the blend with other tannins has produced light grey colour. Stem bark contains 10-25% tannin, whereas the root bark contains 28%. The fruit rind, stem and root bark are widely used in tanning industry.

Quercus incana **Roxb.**

52. *Quercus incana* Roxb.; Fagaceae.

Na : Eng.: White oak; Hindi: Banj.

Hab : Pl. see fatty oils.

Bark yields 16 % tannin and 8 % of non-tan. Hide tanned with its extract acquires a light biscuit colour. Bark is extensively used for tanning purpose.

53. *Rhizophora candelaria* DC.; (Syn. *R. conjugata* Kurz non DC.); Rhizophoraceae

Na : Malayalam: Naya Kandal.

Hab : It is a small evergreen mangrove tree found in the Sundarbans and Andaman Islands.

The bark which contains 25-36 % of tannin is used for tanning purpose.

***R. mucronata* Poir.**

54. *R. mucronata* Poir..; Rhizophoraceae.

Na : Andamans: Bairada, Jumuda, Pyu; Bengal: Kamo, Bhom; Hindi: Bhora; Kanerese: Kandaale; Malayalam: Paniccha kandaal, Pikandaal, Venkandal; Marathi: Kandal; Oriya: Rai, Rohi; Tamil: Kandaal, Peykkandal, Sorapinnai; Telugu: Uppuponna, Adaviponna.

Hab : It is an evergreen mangrove tree found in the Sundarbans, along the Coromandel coast and the Andaman Islands.

The bark yields about 24 % tannin and about 10 % of soluble non-tan. It is used in local tanning industry.

***Rhodomyrtus tomentosa* Wight**

55. *Rhodomyrtus tomentosa* Wight; Myrtaceae.

Na : Eng.: Hill guava, Hill gooseberry, Downy Rose myrtle; Kanerese: Tavuti, Guddade nelli; Tamil & Mal.: Thaontay, Koratta, Thavatlu.

Hab : Pl. see edible plants.

The bark which contains 19 % tannin is locally used for tanning.

56. *Rhus mysurensis* Heyne ex Wight & Arn.; Anacardiaceae.

Na : Hindi: Dansam, Dasni.

Hab : Pl. see edible plants.

Bark contains 15-22 % tannins and 11-15 % non-tannins and is used locally for tanning purpose.

***R. punjabensis* J.L. Stew. ex Brand.**

57. *R. punjabensis* J.L. Stew. ex Brand.; Anacardiaceae.

Na : Hindi: Titri.

Hab : Pl. see edible plants. .

Leaves which contain 13 % tannins are used locally for tanning purpose.

58. *R. succedanea* L.; Anacardiaceae.

Na : Eng.: Japanese wax-tree; Hindi: Kakrasingi.

Hab : Pl. see medicinal plants.

The galls formed on the leaves are used for tanning. Leaves also contain 20 % tannins.

***Rhus succedanea* L.**

59. *Semecarpus anacardium* L.f. (Syn. *Anacardium orientale* L.); Anacardiaceae.

Na : Eng.: Marking nut tree; Hindi: Bhilawa.

Hab : Pl. see edible plants.

The nuts are used locally as a tan.

***Semecarpus anacardium* L.f.**

60. *Shorea robusta* Gaertn. f.; Dipterocarpaceae.

Na : Eng.: Sal; Hindi Sakhya, Sat, Karshya.

Hab : Pl. see edible & fatty oil yielding plants.

The bark of the stem and branches yields about 9 % tannin on zero moisture basis and about 8 % soluble non-tan. The tanned leather is reddish in colour. Large quantities of the bark can be available for tanneries.

***Shorea robusta* Gaertn.**

61. *Soymida febrifuga* A. Juss.; Meliaceae.

Na : Eng.: Indian red wood, Bastard cidar; Hindi: Rohan. Somi.

Hab : Pl. see medicinal plants.

The bark contains about 17.5 % tannin of catechol group and is locally used for tanning purpose.

***Soymida febrifuga* A. Juss. - branch**

62. *Spondias* pinnata (Koen.) Kurz. (Syn. S. *mangifera* Willd.; *Mangifera* Pinnata L.f.); Anacardiaceae.

Na : Eng.: Hog plum; Hindi: Ambara, Amra.

Hab : Pl. see edible plants.

The bark is locally used as a tan.

***Spondias pinnata* (Koen.) Kurz.**

63. *Symingtonia populnea* (R. Br.) van Steenis (Syn. *Bucklandia* populnea. R. Br. ex Grifff.); Hamamelidaceae.

Na : Beng. & Hindi: Pipli; Khasi: Ding-dan; Trade: Piple.

Hab : It is found in the Eastern Himalayas, the Khasia Hills and Darjeeling.

The bark contains 11 % tannin and 10 % non-tans.

Tamarindus indica L.

Tamarix aphylla (L.) Karst.

Terminalia arjuna (Roxb. ex BC.) Wight & Arn.:

Terminalia bellirica (Gaertn.) Roxb.

64. *Tamarindus indica* L.; Caesalpiniaceae.

Na : Eng.: Tamarind, Tamarindo; Hindi: Imli, Tentul.

Hab : Pl. see edible plants.

Bark contains about 7 % tannin.

65. *Tamarix aphylla* (L.) Karst.; (Syn. T. *articulata* Vahl.); Tamaricaceae.

Na : Bengal: Raktajhav; Eng.: Athel tamarisk; Gujarathi: Laljhav nujhada; Hindi: Laljhau, Farash; Punjab: Farash, Kharlei, Narlei; Tamil: Shivappu-atru-shavukku; Telugu: Erraerasaru, Enrasirishar.

Hab : It is a moderate sized tree commonly found in the Punjab and Uttar Pradesh.

The galls formed on the branches are used for tanning.

66. *T. troupii* Hole (Syn. T. *gallica* auct. non L.); Tamaricaceae.

Na : Eng.: French tamarisk; Hindi: Jhau.

Hab : It is a bushy shrub found in the Punjab and UttarPradesh.

The galls formed on the branches yield tannin.

67. *Terminalia arjuna* (Roxb. ex BC.) Wight & Arn.: (Syn. *T. glabra* Wight & Arn.; *Pentaptera arjuna* Roxb.; P. *glabra* Roxb.); Combretaceae.

Na : Assam: Orjun; Bengal: Arjhan; Eng.: Arjun, Terminalia; Guj.: Sadado; Hindi: Arjun, Maruthu, Kahu; Kanerese: Maddi; Mar.: Sanmaddat, Sadaru, Vellamarda; Oriya: Arjuno, Sahajo; Punjab: Arjan; Tamil: Vellamatta; Telugu: Yermaddi.

Hab : It is a large tree growing in South, Central and Northern India and also in the Punjab mainly along the river banks and moister places.

The bark yields about 17 % tannin on zero moisture basis and 12 % soluble non-tan. It is suitable in tanning industry. It imparts light brown colour.

68. *T. bellirica* (Gaertn.) Roxb.; (Syn. *Myrobalanus bellerica* Gaertn.); Combretaceae.

Na : Eng.: Belleric myrobalan, Bastard myrobalan; Hindi: Bahera, Vibhitaka.

Hab : PI. see fatty oils.

The whole fruit yields about 12 % tannin and 24 % soluble non-tan on zero moisture basis. The tannin is found mainly in the pericarp or flesh. Its tannin content is low and is a cheap substitute for chebulic myrobalan.

69. *T.chebula* Retz.; Combretaceae

Terminalia chebula **Retz.**

Na : Eng.: Yellow myrobalan, Chebulic myrobalan; Hindi: Harlr, Harda, And Haritaki.

Hab : Pl. also see fatty oils.

Myrobalans are the fruits of T. chebula, T. bellerica and Emblica officinalis commonly known as "Harra", "Bahera" and "Amla" respectively. The myrobalan of commerce is usually the dried fruit of Terminalia chebula and is known commercially as "chebulic myrobalan". However, detailed investigations into what passes as chebulic myrobalan showed that species other than T. chebula e.g., T. pallida, T. travencorensis and also probably T. citrina are found mixed with it. True chebulic myrobalan is the one which is most important for commercial exploitation because of the higher percentage and better quality of tannin in it.

***Distribution*:** The species is found naturally in the forests of Madhya Pradesh, Maharashtra, Orissa, Bihar, Andhra Pradesh, Tamilnadu, Karnataka, West Bengal and Punjab and to a lesser extent in either States like Uttar Pradesh and Rajasthan. Madhya Pradesh and Orissa are the major producing states.

***Yield*:** No correct estimate of the total production of myrobalan in the country is available. However, it is estimated to be about 80000 to 100000 tonnes of dried fruit annually. From information collected from North Mandla forest Division, Madhya Pradesh alone, it appears that about 15000 tonnes of fruits are extracted annually from there. The potential annual yield of myrobalan fruit in the country is in the neighbourhood of 80000 tonnes. The estimated available supplies of myrobalans are 40000 tonnes.

***Processing*:** Myrobalans freshly collected and dried immediately have yellowish colour and fetch a better price. The fruits when allowed to lie on the ground have darker colour with sometimes mould attack. Tannin content in such decaying fruits is also very low. Mould attack also sometimes occurs on the trees and this is mentioned as the major cause of poor quality of myrobalans.

Varieties and Grades: The different grades of myrobalans are at present known by the names of the areas from where they are exported. The grades are based on shape, colour, compactness of the nuts and freedom from insect attack. Following four main grades are known: 1. *Jabalpore* coming from Madhya Pradesh and partly from Orissa, (ii) *Bimlipatnam* coming from Andhra Pradesh and partly from Tamilnadu, (iii) *Rajpores* or *Bombay variety* coming mostly from Kolhapur and other parts of Maharashtra, (iv) *Salem* or *Madras variety* coming mostly from Tamilnadu. These grades are again subdivided depending on the colour etc. In Madhya Pradesh two distinct varieties of fruits Elongated and Oval shaped are known. Among

Terminalia chebula **Retz. - fruits**

the oval shaped, a few trees have distinctly smaller fruits while others large. Tannin content in oval fruits is also higher about 28 to 36 % while the elongated fruits contain 21 to 23 %. The rind stone ratio in oval fruit is also higher 1.52/1 to 1.84/1 as against 1.33/1 to 1.36/1 in elongated fruits.

***Trade*:** India being the major producing country of myrobalans has practically no imports of myrobalan from any other country. Exports are made as a whole and crushed nuts as well as in the form of myrobalan extract. Extract containing about 50 to 60 % tannin is obtained from the fruit which is exported in addition to its large consumption in the country. Indian Standards Institution has formulated specification (IS: 2716-1964) for myrobalan. From the export figures of 2002-03, it appears that roughly 590.1 tonnes valued at Rs. 2,86,84,991 of myrobalan fruit extract were exported. Although India is a major producer of myrobalan but there has not been much increase in its export. There may be many reasons for this trend. Firstly, like most other forest produce the bulk of myrobalan fruit is great when compared to its value rendering transportation difficult. On account of its cheapness it used to compete with Quebracho etc., on favourable terms in the world market. The Indian Leather Industry has developed considerably in the recent years and as such Indian tanners are fast expanding the business both internally and for export. Quite a few tanning extracting factories at Calcutta, Kanpur, Kolhapur and Kharagpur have been set up in the country. These tanners and manufacturers of tannin extract are buying myrobalans in greater proportions for utilizing in their industry.

***Uses*:** Myrobalan is used in leather/tanning industry. It causes mellowness in leather on account of its astringent fermentative and certain acid-forming properties which causes satisfactory blooming of leather. It is also used to fix the colour of the skin. It is used both raw and in extract form, in the tanning industry. The fruit is attributed with many medicinal properties and is also used in the manufacture of ink. The tree provides valuable wood for the manufacture of carts, rice pounders, boats and building materials.

70. *T. citrina* Roxb. ex Flem.; Combretaceae.

Na : Assam: Monala; Bengali: Haritaki; Hindi: Harira.

Hab : It is a large deciduous often buttressed tree found in Assam and the Khasi hills.

The fruits are used for tanning.

71. *T. tomentosa* Wight & Arn.; Combretaceae.

Na : Beng.: Asan; Eng.: Indian laurel; Guj.: Sadar; Hindi: Asan; Sain, Saj; Kanerese: Sadada; Marathi: Ain; Oriya: Sahaju; Tamil: Karramarda; Telugu: Maddi.

Hab : It is a large tree found all over the tropical and subtropical regions in the country.

The bark yields about 15 % tannin and about 6.5 % soluble non-tan on zero moisture basis.

Toona ciliata **Roem.**

72. *Toona ciliata* Roem. (Syn. *Cedrela toona* Roxb. ex. Rattl. & Willd.); Meliaceae.

Na : Eng.: Cedrela tree, Red cedar Toon; Hindi: Toon.

Hab : Pl. see gums and resins.

The bark yields tan.

Toona ciliata **- branch with flowers**

73. *Vateria indica* L. (Syn. *V.malabarica* Blume); Dipterocarpaceae.

Na : Eng.: White damar, Indian copal tree, Malabar tallow tree; Hindi: Bada sal, Safed damar, Ral.

Hab : Pl. see resins.

The fruit shell is used for tanning.

74. *Woodfordia fruticosa* (L.) Kurz.; (Syn. W: *floribunda* Salisb.; *Lythrum fruticosum* L.); Lythraceae..

Na : Hindi: Dhawi, Dhauta.

Hab : Pl. see gums.

The bark and leaves yield tannin.

Woodfordia fruticosa **(L.) Kurz.**

75. *Xylocarpus moluccensis* (Lam.) M .Roem.; Meliaceae.

Na : Hindi: Pussur.

Hab : It is a mangrove tree found in the coastal regions.

The bark is quite useful for tanning purpose. It yields about 24 % tannic acid and about 8 % soluble non-tan on zero moisture basis.

Xylocarpus moluccensis **(Lam.) M .Roem.; - Tree**

76. *Ziziphus nummularia* (Burm.-f..) Wight & Arn.; (Syn. Z. *rotundifolia* Lamk.); Rhamnaceae.

Na : Hindi: Jhadber.

Hab : Pl. see edible plants.

The bark yields tan.

Z.oenoplia **Mill.**

77. *Z.oenoplia* Mill.; Rhamnaceae.

Na : Hindi: Makoil, Makob

Hab : Pl. see edible plants.

The bark is used as a tan.

78. *Z xylopyrus* Willd. Rhamnaceae.

Na : Bombay: Ghot, Suti; Hindi: Kat-ber, Goth-ber, Kakar, Ghont; Kanercse: Challe, Mullukare; Marathi: Goti; Oriya: Got. Gotoboro, Kantabohul; Tamil: Kottei; Telugu: Gotti.

Hab : It is a small tree found in the dry tropical forests in India. It is an important lac host in Madhya Pradesh.

The average tannin content of properly collected whole fruits is about 16 % but the flesh alone averages up to about 21 % tannin with 24 % soluble non-tan on zero moisture basis. The high proportion of soluble non-tans and the presence of mucilaginous matter have rendered difficulty in the fuller utilization of fruits.

11

Natural Organic Colouring Materials

The use of natural dyestuffs dates back to the remote past. It is to be presumed that, the first steps in the art of dyeing resulted from the accidental staining of the skin or fabric by a vegetable material and this was followed by the use of their extracts for the purpose of dyeing. Mordenting was known in India at a very early date but these mordents consisted of naturally occurring sulphates of aluminium and iron. A knowledge of the use of these salts gradually spread from India to other parts. Owing to the easy solubility of the natural colouring matters they are readily absorbed by the fabric during the dyeing process. Natural dyestuffs are less permanent than those obtainable by other methods. The colouring matter can be found in the leaf, flower, fruit, seed, stem, bark or root. There was considerable demand for many of the natural organic colouring materials like madder *Rubia tinctoria* Linn., Manjeet R. *cordifolia* Linn., Kamala, *Mallotus philippinensis* Muell. Arg. and Indigo *Indigofera tinctoria* L. in the dyeing industry. But of late they have, in most cases, become obsolete on account of the advent of synthetic dyes. Natural dyes are normally costly owing to higher cost of collection as compared to synthetic dyes. They were used in textile industry and also for colouring paints, varnishes, leather, inks, paper, wool, medicines and foods. Among the natural dyes annatto, *Bixa orellana* L. appears to be the only one, which has great economic and commercial importance. It was once used for colouring silk and cotton but it is not a fast dye. Owing to its non-toxic effects it is employed for colouring edible oils, fats, cheese, margarine, chocolate, butter, ghee and the like products. The colouring of 'Vanaspati' '(vegetable ghee) by annatto may serve to distinguish it from the natural ghee. Madder, *Rubia tinctoria* Linn., was used for calico printing. It has a characteristic property of yielding a variety of colours with the aluminium, tin and iron mordents viz., red, pink, orange, lilac, black and also brown or chocolate by employing a mixture of aluminium and iron mordents. All these colours are fast to soap and light. Manjeet, *R. cordifolia* Linn. was once being cultivated for its red dye. The dye was prepared by boiling the crushed powder with water but sometimes it was merely left to steep in cold water. The

Acacia catechu (L.f.) Willd.

A. concinna DC. - branch

A. concinna DC. Pods

A. decurrens Willd.
var. *dealbata* F. Muell.

A.leucophloea (Roxb.) Willd.

solution then obtained was used generally to dye coarse cotton fabrics. Alum was used as a mordant. The flowers of *Butea monospermo* (Lamk.) Taubert yield orange coloured dye, which is not fast and is easily washed out. For the purpose of colouring, the material was steeped in a hot or cold decoction of the flowers. A more permanent colour was produced either by first preparing the cloth with alum and wood ash or by adding these substances to the dye-bath. The dye indigo *Indigofera tinctoria* L was produced by steeping the plant in water and allowing it to ferment. This was followed by the oxidation of the solution with air in a separate vessel. Kamala *Mallotus philippinensis* Muell. Arg. yields orange colour. It was used for dyeing silk and wool. To prepare the annatto dye from *Bixa orellana* Linn., the fruits are collected when nearly ripe. The seeds and pulp are removed from the mature fruit and macerated with water. Thereafter they are either ground up into 'annatto paste' or dried and marketed as annatto seeds. Sometimes when the seeds and pulp are macerated with water, the product is stained through a sieve and the colouring matter, which settles out, is collected, partially evaporated by heat and finally dried in the sun.

11.1 Natural Organic Colour Yielding Plants

1. *Acacia catechu* (L.f.) Willd.; Mimosaceae.

Na : Eng.: Catechu; Hindi: Khair, Katira.

Hab : Pl. see Gums.

The cutch obtained from the heartwood yields a brown colour, which is used in dyeing canvas, fishing nets and ropes.

2. *A.concinna* DC.; Mimosaceae.

Na : Beng.: Banritha; Eng.: Soap pod plant; Guj.: Chikakai; Hindi; Shikekai, Kochi; Kanerese; Sige; Mal.: Chikaka; Mar.: Shikakai; Tamil: Shikai; Telugu: Shikaya.

Hab : It is a prickly scandent bush commonly found along the west coast and other parts of South India.

The bark yields brown dye.

3. *A. decurrens* Willd. var. *dealbata* F. Muell. (Syn. *A. dealbata* Link); Fam.: Mimosaceae.

Na : Eng.: Silver wattle.

Hab : Pl. see Tanning materials.

The bark yields brown dye. The bark is mostly used for tanning purpose. It imparts brown colour to the leather.

A. nilotica **(L) Del.- tree**

4. *A.leucophloea* (Roxb.) Willd. (Syn. *Mimosa leucophloea* Roxb.; Mimosaceae.

Na : Hindi: Safed kikar, Ronj, Arinj, Nimbar.

Hab : Pl. see Gums and resins.

Leaves and bark yield red dye.

A. nilotica **(L) Del.- pods**

5. *A. nilotica* (L) Del. (Syn. *A. arabica* willd.); Mimosaceae.

Na : Eng.: Babul Acacia; Hindi: Babul, Kikar.

Hab : Pl. see Gums and resins and tanning materials.

Bark and pods yield black dye.

Adenanthera pavonina **L. branch**

6. *Adenanthera pavonina* L; Caesalpiniaceae.

Na : Beng.: Raktakambal; Eng.: Coral wood, Bead tree; Hindi: Badi gumchi; Kan.: Manjetti; Mal.: 'Manjadi; Mar.: Thorligunj; Tamil: Anikundumani; Telugu: Pedda-guruvenda. .

Hab : It is a moderate sized deciduous tree found in South India, Andaman Islands and the Eastern sub-Himalayan tract. Propagation is done by seed sowing or by stem cutting. Nursery raised seedlings can be planted in the field at an espacement 4 × 4 m. Powdered wood yields a red dye: The dye is used by Hindus for making marks on the forehead.

Adenanthera pavonina **L. fruits & seed**

7. *Adhatoda vasica* Nees: (Syn.*Justicia adhatoda* L); Acanthaceae.

Na : Eng.: Malabar nut; Hindi: Basak, Arusa, Adulasa, Vasaka.

Hab : Pl. see Medicinal plants.

Leaves yield yellow dye.

Adhatoda vasica **Nees – Branch**

8. *Aegle marmelos* (L) Corr. (Syn. *Crataeva marmelos* L); Rutaceae.

Na : Eng.: Bael, Bengal quince; Hindi: Bel, Bilva.

Hab : Pl. see Edible plants.

Rind of the fruit yields yellow dye.

Aegle marmelos **(L.) Corr. branch & flowers**

Ailanthus triphysa (Dennst) Alston

9. *Ailanthus triphysa* (Dennst.) Alston; (Syn. *A.malabarica* DC.); Simarubaceae.

Na : Hindi: Guggula, Dhup.

Hab : Pl. see Resins.

The leaves are the source of a black dye.

Albizia odoratissima (L.f.)Benth

10. *Albizia odoratissima* (L.f.) Benth. (Syn. *Mimosa odoratissima* L.f.); Mimosaceae.)

Na : Eng.: Black siris, Fragrant albizia; Hindi: Kalasiris.

Hab : pl. see Gums.

Bark yields brown dye.

Alnus nitida – Endl.

11. *Alnus nitida* Endl.; Betulaceae.

Na : Eng.: Himalayan black cedar; Hindi: Kunis, Sarah; Kashmir: Chamh; Punjab: Sharol.

Hab : It is a large tree of the northwestern sub-tropical to temperate Himalayas.

The bark yields a dye.

Artocarpus heterophyllus Lamk.

12. *Arnebia hispidissima* (Lehm.) DC.; Boraginaceae.

Na : Hindi: Ratanjot.

Hab : It is a small herb.

Its roots yield a red dye, which is used for colouring hair oil.

13. *Artocarpus heterophyllus* Lamk.; (Syn.A. *integra* Thunb. Merr.; *A.integrifolia* L); Moraceae.

Na : Eng.: Jack tree, Jack fruit; Hindi: Kathal.

Hab : Pl. see Edible plants.

Wood, bark and roots yield yellow dye. It is stated that an extract of heart wood with alum is used by Buddhist priests for dying their robes yellow.

Bauhinia purpurea – L. flower

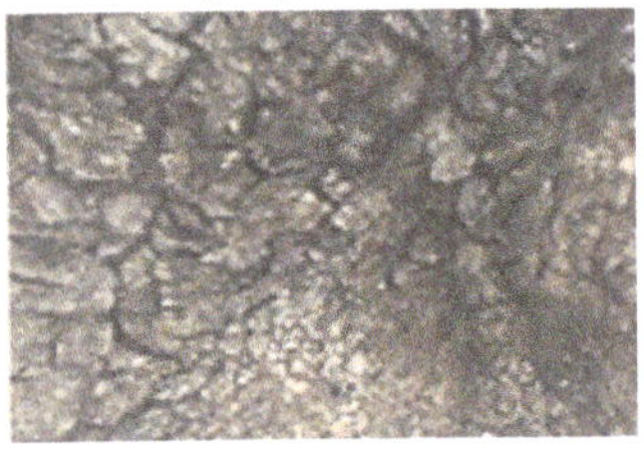

Bauhinia purpurea – L. Bark

14. *Bauhinia purpurea* L; Caesalpiniaceae.

Na : Eng.: Camel's foot tree, Pink Bauhinia; Hindi: Kaliar, Lalkachnar.

Hab : It is found in the Sub-Himalayan tracts (Pl. see gums).

The bark is the source of brown dye.

15. *B. variegata* L (Syn. *B. candida* Roxb.); Caesalpiniaceae.

Na : Eng.: Mountain ebony, Variegated Bauhinia; Hindi: Kachnar.

Hab : It is a tree with white flowers having red and yellow stripes. It is found in the Sub-Himalayan tracts, in drier parts in Central and South India. Pl. see Edible wild plants & Fatty oils.

The bark yields various fast shades of brown.

B. variegata L. – tree

16. *Berberis aristata* DC.; Berberidaceae.

Na : Eng.: Indian barberry; Hindi: Darhaldi, Rasaunt.

Hab : It is a spiny shrub found in the north-west temperate Himalayas. PI. see Medicinal plants.

The roots and stems yield yellow dye which is locally used for dyeing wool and leather.

B. variegata L. Branch

17. *B. lycium* Royle; Berberidaceae.

Na : Hindi: Darhaldi.

Hab : It is found in north-western temperate Himalayas.

Pl. see Medicinal plants.

Roots and stem yield yellow dye.

Berberis aristata DC - branch

18. *B. vulgaris* L; Berberidaceae.

Na : Eng.: Common barberry; Hindi; Chatrea, Kashmal.

Hab : It is found in north-western Himalayas.

Wood and bark yield yellow dye, which is used locally for dyeing wool and leather.

B. vulgaris L. - branch

19. *Bischofia javanica* Blume; Euphorbiaceae.

Na : Eng.: Bishop wood, Java cedar; Hindi: Bhillar.

Hab : Pl. see Tanning Materials.

Stem bark yields a red dye, which is used locally to stain cane baskets.

20. *Bixa orellana* L; Bixaceae.

Na : Eng.: Annatto, Arnato, Urucu; Hindi: Sinduriya.

Hab : The plant is native of Central America and is being grown elsewhere. In India it is grown to a small extent in Karnatak and coastal areas in Vishakhapatnam district of Andhra Pradesh. It is also met with as an escape in certain districts in Kerala, Maharashtra, Bengal and Assam. It is also grown in gardens for its ornamental flowers and fruits. There are two varieties one with white flowers and green capsules and the other with pink flowers and red capsules. Each capsule contains seeds and pulp.

Bischofia javanica Blume.

Bixa orellana –L.

Bixa orellana–L.fruit and seeds

The plant is propagated either from seeds or from stem cuttings. Plant cannot withstand frost. It produces fruits from the 3rd year after planting and about 0.5 kg of seed is collected. When the fruits are ripe the seeds are collected and mixed with ash and sown in the nursery with onset of monsoon. When the seedlings are 30 cm they are transplanted at an espacement of 2 × 2 m. In case of early ripening of fruits, depending on the locality factors, the seeds may be sown soon after collection and transplanted with the commencement of monsoon. Plants are kept in a bushy form by pruning the branches after collection of fruits to facilitate easy collection. The dye is obtained by macerating the pulp and bruising the seeds in hot water till it forms a suspension of dye. The seeds are separated. The dye known as annatto settles and it is dried and cut into cakes. The yield is 4.8 to 6% by weight of seeds. There is a colouring matter Bixin in the seeds, which is accompanied by small amounts of orellin a water soluble yellow substance, fat, resin and a bitter stuff. The seed either in the powder form or an extract of it viz., Bixin is used as a colouring substance. Both Bixin and annatto are soluble in essential oils and in fixed oils. They are also soluble in alkalis and are precipitated by acids. The seeds contain *only* a small amount of fatty oil. Annatto was once used for colouring silk and cotton but it is not a fast dye. On account of its non-toxic nature, it is now mainly used for colouring edible products like butter, ghee, margarine, cheese, chocolate and Vanaspathi. Small quantities of annatto seed are exported from India.

21. *Butea monosperma* (Lamk.) Taubert; (Syn. *B. frondosa*. Koenig ex Roxb.); Fabaceae (Papilionaceae.)

Butea monosperma. (Lamk) Taubert. branch with flowers

Na : Beng. & Mar.: Palas; Eng.: Flame of the forest, Bengal kino; Guj.: Khakra; Hindi: Dhak, Palas, Palasa; Kanerese: Muthule; Mal.: Palas in Samatha; Sans.: Palasha; Tam.: Parasa; Telugu: Modugu.

Hab : It is a small tree found practically throughout the tropical forests in the country (Pl. see Gums). The plant is raised by seed.

The flowers yield a brilliant but very fugitive yellow colouring matter. It is obtained in the form of a decoction or infusion from dried flowers. Addition of alum, lime or an alkali deepens the colour to orange and also makes it less, fugitive. The dye was being used for colouring textiles and fabrics. The colour is also sprinkled during Holy celebrations. The roots also yield red dye.

22. *B. superba* Roxb.; Fam: Fabaceae (Papilionaceae).

B. superba Roxb.

Na : Beng. & Sans.: Latapalasha; Guj.: Velkhakar; Hindi: Belpalas, Kinsuka; Kanerese: Ballimuthule; Mar.: Beliapalas; Tamil: Kodimarukkan; Telugu: Teegamothugu.

Hab : It is a woody climbing shrub found practically throughout tropical forests particularly in moister localities.

The flowers yield yellow or orange coloured dye.

23. ***Caesalpinia sappan*** **L; Caesalpiniaceae.**

Na : Eng.: Sappan wood; Hindi: Patang: Kan.: Pattanga; Mal.: Sappannam; Mar.:Patang; Sanskrit: Pattaranjaka; Tamil: Palungam; Telugu: Bakamu.

Hab : It is found is South India and also Bengal.

The heartwood yields a red dye. For extracting the colouring matter the *wood* is cut into chips or rasped into powder and extracted twice with water. The deep orange extract is allowed to ferment before use so that brazilin dye is converted into brazilifen. Sapon wood extract can be used for colouring cotton, woolen cloth and for making red ink.

The plant is raised by seed.

Caesalpinia sappan **L,branch with pods**

24. ***Careya arborea*** **Roxb.;Barringtoniaceae.**

Na : Eng.: Patana oak; Hindi: Kalikatbhi, Kumbhi.

Hab : Pl. see Edible plants.

The bark yields brown dye.

Cassia fistula **L.- tree**

25. ***Cassia fistula*** **L. Fam.: Caesalpiniaceae.**

Na : Eng.: Indian labernum, golden shower; Hindi: Amaltas.

Hab : Pl. see Medicinal plants and Tanning materials.

Bark yields red dye.

Cassia fistula **L. - flowers**

26. ***Casuarina equisitifolia*** **L (Syn.** ***C. muricata*** **Roxb.); Casuarinaceae.**

Na : Eng.: Casuarina, She-oak; Hindi: Junglisaru.

Hab : Pl. see Tanning materials.

The bark yields brown dye.

Casuarina equisitifolia **L.**

27. ***Ceriops tagal*** **(Perr.) C.B. Robins; (Syn. C.** ***candolleana*** **Arn.); Rhizophoraceae.**

Na : Hindi: Goran.

Hab : Pl. see Tanning materials.

Bark, which is mostly used for tanning purpose, yields red dye.

Casuarina equisitifolia **L.Bark**

28. ***Coscinium fenestratum*** **(Gaertn.) Colebr.; Menispermaceae.**

Na : Eng.: False calumba, Tree turmeric; Hindi: Jhad Haldi;

Kanerese: Marada arishna; Mal.: Maramanjal; Mar.: Jhade-haldi; Sannskrit: Daru Haridra; Tamil: Maramanjal; Telugu: Manupaspu.

Ceriops tagal (Perr.) C.B. Robins

Hab : It is a woody climbing shrub found in South India particularly in the Western Ghats, the Nilgiris in Tamilnadu and Kerala.

The plant is raised by seed. The stem yields a yellow dye for obtaining which the wood is broken into pieces, steeped in water, crushed in a rice-husking machine and the dye squeezed out.

29. *Cudrania javanensis* Trecul; Moraceae.

Na : Hindi: Manda.

Hab : Pl. see Edible plants.

Wood is the source of yellow dye.

30. *Cynometra ramiflora* L.; Caesalpiniaceae.

Na:-Beng.: Shingre; Hindi: shinge, Irapu; Kan.: Kanaga; Mal.: Irripa; Tamil: Irudbu, Nayppudukan; Uriya: Madhuka.

Cynometra ramiflora –L. plant

Hab : It is an evergreen tree found in the Sundarbans and the tidal forests in South India & Andamans.

It is often cultivated in gardens. Wood dyes purple colour.

31. *Datisca cannabina* L; Datiscaceae.

Na : Eng.: False hemp; Hindi: Akalber; Kashmir: Woftangal; Punjab:Bhangjala, Bujrabanga.

Hab : It is a herb found in the western Himalayas from Kashmir to Nepal.

The roots yield a yellow dye which has been used in Kashmir and other places in Himalayas for colouring wool, cotton and silk.

32. *Diospyros peregrina* (Gaertn.) Gurke (Syn. *D. embryopteris* Pers.); Ebenaceae.

Na : Na. Eng.: Indian persimon; Hindi: Gab.

Hab : Pl. see Edible plants.

The Fruits yield yellow dye.

Datisca cannabina –L. plant

33. *Diploknema butyraceae* (Roxb.) H.J. Lam. (Syn. *Bassia butyraceae* Roxb.; *Madhuca butyraceae* (Roxb.) Macb.; Sapotaceae.

Na : Eng.: Bassia fat, Indian butter tree; Hindi: Phalwara.

Hab : Pl. see Edible plants and also fatty oils,

Bark is the source of brown dye.

34. ***Erythrina variegata*** **L. var.** ***orientalis*** **(L.) Merr. (Syn.** ***E. indica*** **Lamk.); Fam.: Fabaceae (Papilionaceae).**

Na : Eng.: Coral tree; Bastard teak; Hindi: Pangra, Nasut, Ferrud.

Hab : Pl. see Edible plants.

The flowers, twigs and bark are the source of red dye.

Erythrina variegata **L.Plant (coral tree)**

Erythrina variegata **L.flowers**

35. ***Euonymus tingens*** **Wall.; Celastraceae;**

Na : Eng.: Spindle wood; Hindi: Badphali.

Hab : Pl. see Medicinal plants.

Bark yields a dye, which is reported to be used in Nepal for marking Tika on the forehead.

36. ***Ficus religiosa*** **L.: Moraceae.**

Na : Beng.: Ashathwa; Eng.: Peepul. Botree; Guj.: Jari, Pipro, Pipul. Hindi: Pipal, Aswatha; Kanerese: Arali, Ashwatha; Mal.: Arachu, Arayal, Ashwatham; Mar.: Ashvatha, Pimpala; Sans.: Ashwathha; Tamil: Arasu, Aswatham; Telugu: Ashvattham, Ragichettu.

Hab : It is found wild throughout India and also grown as an avenue tree. Bark, leaves and roots dye red.

Ficus religiosa **L.Tree**

Ficus religiosa **L. Twig**

37. ***Flemingia congesta*** **Roxb.; Fam.: Fabaceae (Papilionaceae).**

Na : Assam: Samnaskhat; Bengali: Bara-salpan, Bhalia; Bihar: Birbut, Otegarsul; English: Waras dye; Hindi: Bhalia, Barosalpan, Supta, Cusunt; Mal.: Kamantteri, Korkattachedi; Mar.: Dowdowla; Oriya: Bonokandulo.

Hab : It is an erect shrub common in the warmer parts throughout the country.

Gamet coloured hairs on the pod dye yellow.

38. ***F. grahamiana*** **Wight & Arn.; Fam.: Fabaceae (Papilionaceae).**

Na : Eng.: Warrus.

Hab : It is a small plant found in the hills in South India.

Fruits or pods are covered by garnet coloured hairs. The dried pods are rubbed over sieves. The coloured powder is known as warus. Warrus so prepared is a purplish red granular powder. This powder was being used for colouring wool and silk.

Flemingia congesta **–Roxb. shrub**

Galium verum **L. plant**

Garcinia atroviridis **Griff. tree**

G. xanthochymus Hook f. **tree & fruit**

Garuga pinnata **Roxb.**

39. *Fribraurea trotterii*

Na : Hindi (Manipur): Napoo.

Hab : It is a common plant in Manipur.

Roots dye yellow colour.

40. *Galium verum* L..; Rubiaceae.

Na : Eng.: Cheese rennet.

Hab : The plant occurs in the temperate regions of Western Himalayas.

Roots of the plant when cut and steeped in water yield a red dye which has been used for dyeing woolens.

41. *Garcinia atroviridis* Griff.; Guttiferae.

Hab : Pl. see Edible plants.

Fruits were used, with alum as a fixative, in dyeing silk.

42. *G. morella* Desr.; Guttiferae.

Na : Eng.: Ceylon gamboge; Hindi: Gota-ghanba.

Hab : Pl. see Edible plants.

Gamboge, a yellow emulsion obtained from the pith, leaves, flowers and fruits is used for preparing water colours and gold-coloured spirit varnishes for metals. It is used in Burma for dyeing silk robes of Buddhist priests.

43. *G. pedunculata* Roxb.; Guttiferae.

Na : Assam: Bor-thekera; Hindi and Bengali: Tikul, Tikur.

Hab : It is found in Sub-Himalayan tract in North East Bengal, Manipur; Goalpara.

Fruits dye yellow colour. They are used as a fixative or as a mordant for saffron dye.

44. *G. xanthochymus* Hook f.: (Syn. G *tinctoria* Dunn); Fam.: Guttiferae.

Na : Eng.: Egg tree; Hindi: Dampel, Tamala, Tumul.

Hab : Pl. see Edible plants.

Exudation from the under-ripe fruit, which is yellow in colour and the bark, are used in Assam for dyeing.

45. *Garuga pinnata* Roxb.: Burseraceae.

Na : Hindi: Ghogar, Toom.

Hab : Pl. see Tanning materials.

Galls, leaves and bark dye red.

46. *Geranium nepalense* Sweet; Fam.: Geraniaceae.

Na : Eng.: Nepal geranium, Nepalese cranes bill; Hindi: Bhanda; Kashmir: Roel; Punjab: Bhand.

Hab : It is a perennial herb found in the temperate regions in Kashmir, the Khasia Hills and the Nilgiris.

Roots yield red dye, which is used locally for colouring medicinal oils.

***Geranium nepalense* Sweet. Flowers**

47. *Gorodnia excelsa* Blume; Tanstroemiaceae.

Na : Lepcha: Chau-kung.

Hab : It is a tree of the Khasia Hills.

Its bark is the source of black dye, which is used for staining baskets.

***Indigofera tinctoria* . L. plant**

48. *Holigarna caustica* (Dennst.) Oken; (Syn. *H. longifolia* Buch. Ham. ex Roxb.); Anacardiaceae.

Na : Cachar: Bonsu-buphang; Hindi: Hulugiri; Lushai Hills: Katebel; Manipur: Kherai.

Hab : It is a tree found mainly in Bengal.

The juice obtained from the bark and rind of the fruit is used as a black varnish and for lacquer work.

***Juglans regia* – L. tree**

49. *Indigofera tinctoria* L; Fabaceae (Papilionaceae).

Na : Eng.: Indigo; Beng. & Hindi: Neel; Guj.: Gali, Gari, Nil; Mal.: Nilam; Kan., Mar., & Tamil: Nili; Sans.: Nilla, Nili, Nilika, Rangapatri; Telugu: Avivi Nili.

Hab : It is a common small shrub found in wastelands.

It was being cultivated in India, China and other countries of the East. The plant is propagated by seed normally by broadcast sowing or nursery raised seedlings. The spacing between the seedlings may be about 30 cm. Addition of farmyard manure in the field will boost the growth. Periodic irrigation will also help in boosting the growth. First harvest is done after 90 days and the subsequent harvests at an interval of 75 days. Plantation can be maintained with irrigation for 3-4 years. The plant is not browsed by cattle. Leaves and flowers yield Indigo deep-sea (blue) dye, which is commercially known as Indigo.

***Juglans regia* – L. Bark**

50. *Juglans regia* L.; Juglandaceae.

Na : Eng.: English walnut, Persian walnut; Hindi: Akhrot.

Hab : Pl. see Edible plants.

Bark yields a red dye. It is also used as a tooth stick to clean teeth.

***Kandelia candel* L.Druce**

51. *Kandelia candel* (L.) Druce (Syn. *K. rheedii* Wight & Arn.); Rhizophoraceae.

Na : Hindi: Goria.

Hab : It is a shrub or small tree found along the tidal swamps of Bengal, Andaman Islands and the Western parts of the country, Pl. see tanning materials.

Bark is the source of red dye.

52. *Lagerstroemia parviflora* Roxb.; Lythraceae.

Na : Assam: Sida; Bengal: Sida; Guj.: Kakria; Hindi: Sida, Bakli, Lendia; Kan.: Chennangi, Ventaku; Lepcha: Kunhilkung; Mal.: Cirnanii, Ventekku; Mar.: Lende, Bondga, Bondara; Punj.: Bakli, Dhaura; Tamil: Chennangi; Telugu: Chinnangi.

Hab : It is a large tree found in the dry deciduous forests almost throughout India.

The plant is raised by seed. The bark dyes black.

***Lannea coromandelica* (Houtt.) Merr.**

53. *Lannea coromandelica* (Houtt.) Merr. (Syn. *L. grandis* (Dennst.) Engle.: *Odina wodier* Roxb.); Anacardiaceae.

Na : Hindi: Jhingan.

Hab : Pl. see Edible plants.

Bark yields brown dye which is used as a preservative for fishing nets.

54. *Lawsonia inermis* L. (Syn. *L. alba* Lamk.); Lythraceae.

Na : Bengali: Mehedi. Mehendi; Eag.: Henna, Egyptian Privet; Guj.: Medi, Mendi; Hindi: Mehndi; Kan.: Mailancha, Gorante; Mal.: Mailanche, Ponttagi; Mar.: Mendhi; Sans.: Mendika, Rakta garbha, Ragangi; Tamil: Marithondi, Maruthani; Telugu: Goranti.

Hab : It is a shrub or small tree native of Arabia and Persia. In India it is found as an escape in wastelands and also cultivated as a hedge plant throughout the country. The plant is not festidious about soil. It is raised by cutting of 15-20 cm in rainy season. The plant may strike roots in about a month and they can be planted as hedge at a spacing of 30 × 15 cm. or 30 × 30 cm. If planted in the field the spacing may be 75 × 75 cm. Adding farmyard manure will boost yield. Watering will also help the growth. First harvest can be had after about a year. Subsequent harvests after every 6 months. The yield is about 200 kg for first harvest, and thereafter 500 kg with subsequent harvests. Harvested branches are dried and should be shaken to collect leave.

***Lawsonia inermis* –L. leaves**

Leaves are the source of a red dye, Lawsone, which is extensively used in Middle-Eastern countries for dying nails etc. For use as a colouring material powder of dried leaves is made into a paste with water and applied to the part to be dyed.

***Leonurus cardiaca* L.Herb**

55. *Leonurus cardiaca* L. (Syn. *Cardiaca vulgaria* Moench); Labiatae.

Na : Eng.: Common mother wort.

Hab : It is a perennial herb found in Kashmir, Kumaon. Pl. see Medicinal plants.

The leaves and flowers produce a dark olive green dye.

***Ligustrum vulgare* L. - Plant**

56. *Ligustrum vulgare* L; Oleaceae.

Na : Eng.: European privet.

Hab : It is a shrub or small tree found in warm and humid places.

Bark and fruits are the source of yellow-green and black dyes which are employed locally for colouring wool.

***Ligustrum vulgare* L. Plant with flowers & fruits**

57. *Mahonia nepalensis* DC.; (Syn. *Berberis Nepalensis* Spreng.); Berberidaceae.

Na : Nepal: Chatri.

Hab : It is found wild in the temperate regions in Bengal. Bark is the source of a yellow dye.

***Mahonia nepalensis* DC.**

58. *Mallotus philippiensis* Muell. Arg.; Euphorbiaceae.

Na : Eng.: Kamala; Hindi: Kamala; Rauni.

Hab : It is a small tree found commonly along the foothills of Himalayas from Bengal westwards, Orissa, Madhya Pradesh and Maharashtra States. Pl. see Medicinal plants and Fatty oils.

Kamala tree has long been valued as the source of dyeing material known in the trade as Kamala, Kamala powder, or Kamala dye and was used in dyeing of silk and wool. The colouring matter is present in the red glandular pubescence covering the ripe capsule and is usually collected when the fruits ripen. The dye is obtained either by agitating the fruits repeatedly or by stirring the fruits vigorously in water when the dye settles down as sediment, it is separated and dried. Principal colouring principles are the salmon coloured rottlerin and its yellow isomer isorottlerin. It produces a bright orange or flame colour which is fast to soap and alkalies and acids but fades in sunlight. With the advent of synthetic dyestuffs the use of this material has practically ceased.

***Mallotus philippiensis* Muell. Arg.**

Marsdenia tinctoria R.Br.

59. *Marsdenia tinctoria* R. Br.; Aslclepiadaceae.

Na : Assam: Meini-buli-likur, Manri; Bengal: Riyong; Hindi; Rion; Lepcha: Ryom; Telugu: Nitta Tuniki.

Hab : The species is found in Assam & Khasi Hills. Plants can be raised by seed. Leaves yield a dye which is similar to Indigo.

Melastoma malabathricum L.

60. *Melastoma malabathricum* L; Melastomaceae.

Na : Assam: Phatuka; Eng.: Indian Rhododendron; Hindi: Phulki; Kan.: Ankerki: Lepcha: Tungbram; Mal.: Kalampatti; Mar.: Palore; Tamil: Nakkukaruppan; Telugu: Pattuda.

Hab : It is a large ornamental shrub found in moist places in the humid parts of India.

Plants can be raised by seed. Fruits yield black or purple dye. It is stated that an ink can be prepared from it. Leaves and roots yield pink dye.

Memecylon edule Roxb. Tree

61. *Memecylon edule* Roxb.; (Syn. M. *tinctorium* Koenig), Melastomaceae.

Na : Eng.: Iron wood tree: Hindi: Anjan.

Hab : It is a small tree found in Orissa, Assam and the South Western parts of the country. Pl. see Edible plants.

The leaves yield yellow dye, which gives beautiful light shades; in combination with myrobalans and sappan wood it produces a bright red colour being much used for dyeing mats.

Mesua ferrea L.

62. *Mesua ferrea* L.; Fam.: Guttiferae.

Na : Eng.: Iron wood; Hindi: Naga Kesar, Nagchampa, Nageshwar champa.

Hab : It is a tree found in the Eastern Himalayas, Assam, West Bengal, Western Ghats and the Andaman Islands. Pl. see Edible plants & Fatty oils.

Flowers are the source of Yellow dye.

Mimusops elengi L

63. *Mimusops elengi* L; Sapotaceae.

Na : Eng.: Spanish cherry; Hindi: Maulsari.

Hab : Pl. see Edible plants.

Bark yields brown dye.

64. *Mirabilis jalapa* L; Nyctaginaceae.

Na : Eng.: Four o'clock plant; Hindi: Gulabbas.

Hab : Pl. see Essential oils.

Flowers are the source of a crimson dye. Bark and stem yield yellow dye.

Mirabilis jalapa L.

65. *Morinda angustifolia* Roxb.; Rubiaceae.

Na : Assam: Asu-goch; Beng.: Asho, Darharidra; Garo: Chennong; Hindi: Daru haridra, Banhardi; Khasi: Dieng-nong. Dieng-seroi; Lepcha: Haldikung; Lushai: Kawnpel.

Hab : It is a climbing shrub or tree found in Assam, the Khasia Hills and Sikkim.

The plants are raised by seed. Roots are the source of a yellow dye, which was formerly used for dyeing cotton yarn and cloth. Red shades are obtained on mordent yarn by using cold water extracts of the root.

Morinda angustifolia Roxb. Shrub

66. *M. bracteata* Roxb.; Rubiaceae.

Na : Eng.: Indian mulberry.

Hab : It is a shrub or small tree found in Orissa, Bengal, Rajasthan and the Andaman Islands.

Roots yield a red colour, which was formerly used to dye linen and basket work.

M. citrifolia L

67. *M. citrifolia* L Rubiaceae.

Na : Eng.: Indian mulberry: Beng., Guj., Mar., & Hindi: Bartundi, AI, Ach, Surangi; Kan.: Ainshi. Tagase Maddi; Mal.: Kattapitalavam, Mannanatti; Oriya: Achu, Pindre; Tamil: Nuna, Togaru; Telugu: Togaru, Togaru maddi.

Hab : It is found in Andaman Islands and the foothills of Eastern Himalayas, Bihar and Orissa.

The plant can be raised by seed. Root, bark and also the woody portion yield a Red dye. The root dye is known as Al dye. Formerly it was employed for colouring cotton, woolen and silk. The use of Al dye has been abandoned since the advent of synthetic dye stuffs.

M.tinctoria Roxb

68. *M. tinctoria* Roxb.; Rubiaceae.

Na : Eng.: Tagar wood; Hindi: Al.

Hab : The plant is found in Bihar and in central and south India. Pl. see Edible plants.

Roots are the source of a red dye, which was formerly used in dyeing cotton.

M. umbellata L.

69. *M. umbellata* L.; Rubiaceae.

Na : Eng.: Al.

Hab : It is a shrub of the hilly regions of Bengal, Khasia Hills and South-western India.

The plant can be raised by seed. The root bark contains the colouring principles, morindone and morindin. The red dye obtained from the bark is known as Al and was formerly used in dyeing textiles.

***Moringa oleifera* Lamk**

70. *Moringa oleifera* Lamk.: (Syn. *M. pterygosperma* Gaertn.; *Guilandina moringa* L.); Moringaceae.

Na : Eng.: Drum-stick tree, Horse radish tree; Hindi: Sahinjn, Soanjna.

Hab : Pl. see Medicinal & Edible plants.

Wood yields Indigo dye.

71. *Myrica nagi* Thunb.; (Syn. *M. rubra* Sieb. & Zucc.); Myricaceae.

Na : Eng.: Bay-berry, Box myrtle; Hindi: Kaiphal.

Hab : Pl. see Edible plants and Tanning materials.

Stem bark yields yellow dye.

***Myrica nagi* Thunb.**

72. *Nyctanthes arbortristis* L.; Oleaceae.

Na : Beng.: Sephalika, Seoli; Eng.: Tree of sorrow; Night flowering Jasmine; Guj.: Jayaparvati; Kanerese: Parijata; Hindi: Harsingar, Sephalika; Mal.: Pavizhammalli. Parijatakam; Mar.: Khurasli, Parijatak; Oriya: Godokodiko, Gunjoseyoli, Singaroharo; Tamil: Manjhapu, Pavazhamalligai; Telugu: Kapilanagadustu, Pagadamalle, Parijatamu

Hab : It is a small tree found in drier areas in Uttar Pradesh, Madhya Pradesh, Maharashtra and Andhra Pradesh and is also grown for its fragrant flowers.

An orange coloured dye is obtained from the bright orange corolla tubes of the flowers. The colouring matter is nyctanthin which is identical with a crocotin from saffron. The corolla tubes were formerly used for dyeing silk sometimes in combination with safflower, turmeric, indigo etc. Addition of lime or alum to the dye bath is reported to render the colour more permanent.

***Nyctanthes arbortristis* L.**

73. *Oldenlandia umbellata* L.; Rubiaceae.

Na : Eng.: Indian madder; Beng.: & Oriya: Surbuli; Hindi: Saya, Chirval; Mal.: Chayaver; Tamil: Chiruver, Imbuvaner, Sayaner, Telugu: Cheriveru.

Hab : This herb is found in Orissa, Bengal and South India. Root bark is the source of chay-root dye, once employed with mordents for imparting red colour to calico, wool and silk fabrics.

***Onosma echioides* L.**

74. *Onosma echioides* L.; Boraginaceae.

Na : Hindi: Ratanjot.

Hab : It is a perennial herb found in the temperate regions, in Kashmir and Kumaon Hills.

Roots yield a red dye known as Ratanjot which was used formerly for dyeing woolen cloth, oil and fat. The dye does not appear to be suitable for colouring edible fatty oils.

Oroxylum indic\um **Vent.**

75. *Oroxylum indicum* Vent. (Syn. *Calosanthes indica* Blume); Bignoniaceae.

Na : Hindi: Sonapatha, Sora, Arlu.

Hab : Pl. see Edible plants and Tanning materials.

The stem and root barks contain three flavone colouring matter.

Phyllanthus emblica **L. plant**

76. *Phyllanthus emblica* L.Syn. *Emblica officinalis* Gaertn.; Euphorbiaceae.

Na : Eng.: Emblic myrobalan; Hindi: Amla, Aonla.

Hab : Pl. see Medicinal and Edible plants.

The *pulp* of the fruit yields dark brown dye.

Pylllanthus reticulatus **Poir.**

77. *Phyllanthus reticulatus* Poir.; Fam.: Euphorbiaceae.

Na : Assam: Amluki; Beng.: Panjuli; Guj.: Datwan; Hindi: Panjoli Makhi, Buinowda, Neelbari; Kan.: Anamsule, Chippulinellu, Huliballe, Karesuli, Sanna hoge soppo; Mal.: Kottuniruri, Kilanelli, Nirnelli; Mar.: Pavana; Oriya: Jandaki; Punjab: Panjuli; Rajasthan: Kabonan; Sanskrit: Krishnakamboji; Tamil: Abiranji, Karunelli, Karuppupilanji, Nirppul, Kattukilanelli; Telugu: Nallapuli, Nallapurugudu, Pulisar.

Hab : It is a climbing shrub commonly found in Northern India ascending upto 1700 to 2000 m in the Himalayas in Khasi and Jaintia hills and in Andaman Islands.

Roots are the source of red dye.

Pinus roxburghii **Sar. Tree**

78. *Pinus roxburghii* Sar. (Syn. *P. longifolia* Roxb.); Pinaceae.

Na : Eng.: Long leaved pine, Three leaved pine, chir pine; Hindi: Chir, Salla.

Hab : Pl. see Essential oils and Resins.

Bark yields yellow dye.

P. wallichiana **AB Jackson.Tree**

79. *P. wallichiana* A. B. Jackson; (Syn. P. *excelsa* Wall.; *P. griffithii* M. clell); Pinaceae.

Na : Eng.: Indian blue pine, Five-leaved pine; Hindi: Kail.

Hab : Pl. see Resins.

Bark yields yellow dye, which was being used for colouring silk and

Pterocarpus marsupium Roxb.

P. santalinus Lf.

Rhus succedanea –L. tree

Rhus succedanea –L. Branch

wool.

80. *Psoralea plicata* Delile; Fabaceae (Papilionaceae.)

Na : Punjab: Bakhtmal.

Hab : It is a bushy under-shrub found in Punjab, Haryana and Delhi.

Its young pods yield a yellow dye.

81. *Pterocarpus marsupium* Roxb.; Fabaceae (Papilionaceae.)

Na : Eng.: Indian kino tree, Malabar kino; Hindi: Bijasal, Pitasara.

Hab : Pl. see Medicinal plants.

The gum and bark yield red dye.

82. *P. santalinus* Lf.; Fabaceae (Papilionaceae.)

Na : Eng.: Red sander; Guj.: Ratanjab; Beng. & Hindi: Lalchandan, Rakta chandan; Kan.: Agaru, Honne, Kempugandhachakke; Mal.: Patrangam. Tilaparnni; Mar.: Tambada, Chandana; Oriya:Raktachandan; Tamil: Atti, Sivappu, Chandanam; Telugu: Agarugandhamu, Raktagandhamu, Yerrachandanamu.

Hab : It is found in the districts of Cuddapah, Kurnool, Chittoor, Ananthapur and Nellore in Andhra Pradesh and to some extent in the adjoining districts in Tamilnadu.

Plantation can be raised by planting seedlings of seed origin or cuttings. Wood contains 16 per cent of a red colouring matter santalin. Santalin yields a blood-red solution with alcohol, yellow with ether and violet with ammonia and caustic alkalies, but is insoluble in water. The ground wood was chiefly used for dyeing wool, cotton and leather and for staining other woods.

83. *Rhus succedanea* L; Anacardiaceae.

Na : Eng.: Japanese wax tree; Hindi: Kakrasingi.

Hab : Pl. see Medicinal plants and Tanning materials.

A yellow liquid obtained from the mesocarp of the fruit is used in varnishes and furniture polish.

84. *Rubia cordifolia* L; Rubiaceae.

Na : Assam: Majathi; Bengal: Manjishtha; Bhutan: Soth; Eng.: Indian madder; Hindi: Manjit, Majith; Kan.: Siomalate; Siragathi, Manjushtha; Kashmir:

Dandu; Khasi: Ryhoic, Soh-misem; Lepcha: Vhyem; Mal.: Poont, Manjethi; Manipur: Moyum; Mar.: Manjeshta; Punjab: Kukharphali, Tiuru, Manjit, Sheni, Milu, Runang; Sans.: Manjistha, Kalameshika; Tam.: Shevelli, Manjiti; Telugu: Taamaravalli, Manjesta teega.

Hab : It is a prickly climbing herb common throughout tropical parts in the country and in the sub-tropical and temperate regions ascending upto 3750 m in the Himalayas.

***Rubia cordifolia* L. Tree**

The Indian madder of commerce consists of root-stocks and roots. But stem is also used for extraction of dye. The madder has been used for dyeing coarse cotton fabrics, blankets and carpets. For colouring, the cloth is steeped in an infusion of stem or root and mordented with alum. The colouring matter found in the roots is a mixture of purpurin and munjustin. Purpurin is the main colouring principle while munjustin an orange dye, occurs in the form of glycoside.

85. *R. khasiana* Kurz.; Rubiaceae.

Na : Eng.: Khasia madder.

Hab : It is commonly found in the eastern Himalayas.

Roots yield a red dye. This species is much richer in colouring matter than R. cordifolia L. Method of colouring is the same as described for *R. cordifolia L.*

***Rubia cordifolia* L. Root Stocks & Roots**

86. *R. sikkimensis* Kurz.; Rubiaceae.

Na : Eng. & Manipur: Moyum.

Hab : It is found from Nepal eastwards to Assam, Nagaland and Manipur ascending to an altitude of 1500 m.

Roots as well as the stem are the source of a brilliant red dye which is used by the hill tribes for dyeing cotton, woolens and hair. The dye is also used for decorating spears, ornaments, cane and bamboo articles. Roots contain purpurin and munjustin as the colouring matter.

87. *R. tinctorum* L; Rubiaceae.

Na : Eng.: Alizari, European madder; Hindi & Punjabi: Bacho.

Hab : It is a prickly climbing herb found in Kashmir.

Roots yield red dye. Until its displacement by synthetic dyes, the European madder had been one of the most important dye yielding plants, giving brilliant shades of red known for their beauty and fastness of tint to both light and water. Method of dyeing with the roots is the same as that for Indian madder. R. cordifolia L. Roots contain purpurin and alizarin as dyeing constituents.

***R. tinctorum* L**

***Sambucus ebulus* L.**

***Semecarpus anacardium* L.f. - tree**

***Semecarpus anacardium* L.f. - Fruits**

***Shorea robusta* Gaertn.f. Tree**

88. *Rumex dentatus* L; Polygonaceae.

Na : Hindi: Janglipalak, Lalbibi, Khatpalak, Ambavah, Amrula.

Hab : It is found in N.W. India, Kumaon, and the Western Ghats.

The roots yield a red dye.

89. *Sambucus ebulus* L; Caprifoliaceae.

Na : Eng.: Dwarf elder; Hindi: Mushkiara; Punjab: Gandal, Ganhula, Mushkiara, Richhkar, Siske, Tasar.

Hab : It is a small shrub found mainly in Kashmir and Himachal Pradesh in sub-tropical to temperate regions.

Fruits yield a blue dye which is used locally for colouring leather and yarn.

90. *Sarcochlamys pulcherrima* Goud.:

Na : Assam: Meraki; Cachar: Misagi-phang; Hindi & Garo: Dogal; Xhasi: Dieng-Iangshir, Dieng-sohkhwa-lali; Manipur: Sanmarti.

Hab : It is found in Assam and Khasia hills upto 300m altitude.

Leaves and twigs dye brown.

91. *Semecarpus anacardium* L.f. (Syn. *Anacardium orientale* L); Anacardiaceae.

Na : Eng.: Marking nut tree; Hindi: Bhilawa.

Hab : Pl. see Edible plants and Tanning materials.

Pericarp of the fruit yields black dye which is used for painting basket and wicker work. The extract of the pericarp is also used for the manufacture of varnishes, lacquers, enamels, moulding compositions and water-proofing and insulating materials.

92. *Shorea robusta* Gaertn. f.; Dipterocarpaceae.

Na : Eng.: Sal; Hindi: Sakriya, Sal, Karshya.

Hab : Pl. see Edible plants & Fatty oils.

Bark yields red dye.

93. *Symplocos crataegoides* Buch.-Ham. Ex D. Don; (Syn. S. *paniculata* Wall. ex D. Don; *Lodhra crataegoides* Dcne.); Symplocaceae

Na : Bombay & Kumaon: Lodh; Eng.: Sweet leaf, Sapphire berry; Hindi: Lodh; Jaunsar: Lodra; Khasi: Dieng-iong; Punjab: Lodar, Lodh; Sanskrit: Lodra.

Hab : It is a large shrub or small tree found in the sub-tropical to temperate regions in the Himalayas from Kashmir to Assam.

Leaves and bark yield a yellow dye.

94. *S. racemosa* Roxb.; Symplocaceae.

S. racemosa Roxb. Plant

Na : Assam: Kavirang, Bhomrati; Beng.: Lodh; Darjeeling: Khodai, Singen; Guj.: Lodar; Hindi: Lodh, Lodhra, Hara; Kan.: Balalodduginamara, Pachellu; Khasi: Lapong-dong; Mal.: Pachotti; Mar.: Lodh. Lodhra; Sans.: Lodhra, Marjana, Tilaka; Tamil: Vellilethi; Telugu: Lodduga, Erralodduga.

Hab : It is found in north-eastern India in Kumaon, Bihar and Assam up to 1400 m altitude.

The bark and leaves yield a yellow dye.

The dye was used for colouring textiles. It is one of the ingredients of Abir, a red powder used during the festival of Holi.

95. *S. sumuntia* Buch.-Ham; Symplocaceae.

Syzygium cumini (L). Skeels

Na : Eng.: Lodh tree; Hindi: Lodh, Pithani lodh; Lepcha: Singam.

Hab : The species is found in Assam, the Khasia hills, Manipur and Sikkim in the sub-tropical to temperate regions.

The bark and leaves yield a yellow dye with hot water, whereas with crude carbonate of soda a red dye useful for printing red on white is obtained.

96. *Syzygium cumini* (L) Skeels (Syn. *Myrtus cumini* L; *Eugenia jambo/ana* Lamk.); Myrtaceae.

Na : Eng.: Java plum, Jambolana; Hindi: Jamun, Jambu, Jambava.

Hab : Pl. see Edible plants.

Bark yields red dye which is used for colouring fishing nets.

97. *Tectona grandis* Lf.; Verbenaceae.

Tectona grandis Lf.;

Na : Assam:Chingjagu; Beng.: Segun; Eng. or Common name: Teak; Guj.: Saga. Sagach; Hindi: Teak, Sagwan; Lepcha: Ripnyok; Oriya: Singuru.

Hab : It is the most important timber species in the country found in the warmer parts particularly in south & central India and Bihar.

The bark is the source of yellow dye, which was used for staining baskets. Leaves also yield yellow red dye.

98. *Tephrosia purpurea* Pers.; Fabaceae. (Papilionaceae).

Tephrosia purpurea Pers. Herb

Na : Bengali: Ban-nil-gachh; Eng.: Purple Tephrosia; Guj.: Godakan, Jhila, Sarphankho; Hindi: Dhamasia, Sarphonka, Bannil, Jhojru, Pawad. Janglimothar; Kanerese: Empali, Vajaraneeli, Koggili; Mal.: Kozhenjil, Kaatamiri; Mar.: Sirapakha, Udhadi, Unbali; Munderi: Birca konda, Lilcakonda; Oriya: Kolothiyapokha, Mohisiako lothiga, Pokha, Soropokha; Sans.: Sharapunkha; Tamil: Kolungi, Paavali, Katkolingi, Kollukkay-velai; Telugu: Vempali, Neelavempali, Bontavempali.

Hab : It is a common herb.

Leaves dye blue.

Terminalia arjuna Roxb.ex DC Wight & Arn

99. *Terminalia arjuna* (Roxb. ex DC.) Wight & Arn.: (Syn. *T. glabra* Wight & Arn.; *Pentaptera arjuna* Roxb.; *P. glabra* Roxb.); Combretaceae.

Na : Eng.: Arjun; Hindi: Arjun, Maruthu, Kahu.

Hab : It is a large tree growing in south, central and northern India and also in the Punjab. Pl. see Tanning materials.

The bark yields a dye along with tannin, which is used in processing of leather.

Terminalia arjuna Branch

100. *T. bellirica* (Gaertn.) Roxb.; (Syn. *Myrobalanus bellerica* Gaertn.); Combretaceae.

Na : Eng.: Belleric myrobalan, Bastard myrobalan; Hindi: Bahera, Vibhitaka.

Hab : Pl. see Medicinal plants, Fatty oils and Tanning materials.

The fruit pulp yields a yellow dye. Bark yields brown dye. They are used locally for tanning and colouring hides.

T. bellirica (Gaertn.) Roxb.

101. *T. chebula* Retz.; Combretaceae.

Na : Eng.: Yellow myrobalan, Chebulic myrobalan; Hindi: Harir, Harda.

Hab : Pl. see medicinal plants and tanning materials.

Fruit pulp yields yellow dye alone with tannin which is used in leather industry.

T. chebula Retz. - branch

102. *T. tomentosa* Wight & Arn. Syn. *T. alata* Heyne ex Roth; Combretaceae.

Na : Bengal: Asan; Eng.: Indian laurel; Guj.: Sadar; Hindi: Asan. Sain, Saj; Kan.: Sadada; Oriya: Sahaju; Mar.: Ain; Tamil: Karramardi; Telugu: Nallamaddi.

Hab : The tree is found all over the tropical regions in the country. Pl. see Tanning material.

The bark dyes brown. It is used for tanning and colouring hides.

T. chebula Retz. - Fruits

103. *Toddalia asiatica* (L.) Lamk. (Syn. *T. aculeata* Pers.); Rutaceae.

Na : Eng.: Wild orange tree; Hindi: Kanj, Tindupara.

Hab : Pl. see Edible plants.

The roots yield a yellow dye.

Toddalia asiatica **(L) Lamk. Plant**

104. *Toona ciliata* Roem.; (Syn. Cedrela toona Roxb. ex Rottl. &Willd.); Meliaceae.

Na : Eng.: Cedrela tree, Red cedar; Hindi: Toon.

Hab : Pl. see Gums and Tanning materials.

The flowers yield a yellowish-red dye. The seeds also yield a red dye.

Toona ciliata **Roem. Trunk**

105. *Trifolium pratense* L; Fabaceae (Papilionaceae).

Na : Eng.: Purple clover, Broad leaved clover, Red clover; Hindi: Chit-batto, Trepatra.

Hab : It is a common wild herb.

The flowers yield yellow dye.

Toona ciliata **Roem. Branch**

106. *Ventilago Madraspatana* Gaertn.; Rhamnaceae.

Na : Hindi: Bika Kalibel, Pitti.

Hab : Pl. see Edible plants.

The root bark is the source of a red dye. 'Ventilagi' which is used locally for colouring mordented cotton, wool etc. By using various techniques the dye can be used to produce several fast colours, for which cochineal or madder is generally used. The bark is marketed in the form of dark red or brown chips and sold under the names, Raktapita, Pappilichakka, Suralpattai, Lokandi and Konwait.

Trifolium pratense **L. Herb**

107. *Woodfordia fruiticosa* (L) Kurz; (Syn. W. floribunda Salisb.; *Lythrum fruticasum* L); Lythraceae.

Na : Hindi: Dhauta, Dhawi, Jaju.

Hab : Pl. see Gums and Tanning materials.

The flowers yield yellow-red dye.

Woodfordia fruiticosa **(L) Kurz, Shrub**

108. *Wrightia tinctoria* R. Br.; (Syn. *W. rothii* G. Don: *Nerium tinctorum* Roxb.); Apocynaceae.

Na : Beng.: Indrajan; Eng: Sweet Indrajao; Guj.: Indrajan, Runchalledudhlo, Dudhlo; Hindi: Indrajau, Meetha Indrajao, Dudhi; Kan.: Kodamurki, Bepalle, Kodesige; Mal.: Kotakappalla, Aiyapala; Mar.: Kalakuda, Indrajau; Oriya: Pitakaruan, Dudhokriya, Krya; Sans.: Hyamaraka; Tamil: Veypale, Irumpalai, Thonthapalai.

Woodfordia fruiticosa **(L) Kurz. Dried flowers**

***Wrightia tinctoria* R. Br.**

***W. tomentosa* Roem. & Schult**

***Ziziphus mauritiana* Lamk. Branch**

Hab : It is a tree found in Rajasthan, Madhya Pradesh, Maharashtra, Andhra Pradesh and Tamilnadu States.

The wood is white pale yellow and light in weight, used for toys. Fruits ripen by January to March. Seeds are thin and papery and are easily blown away by wind. After collection the seed is soaked in water and sown. Cuttings of 8-10 cm after treating with IBA 4000 PPM for 15 seconds can also be planted in poly bags. After 40 days rooting is noticed. Thereafter they can be planted in the field. The flowers are the source of a yellow dye. Leaves are the source of blue dye Indigo.

109. *W. tomentosa* Roem. & Schult. (Syn. *W. mollissima* Wall.: *Nerium tomentosum* Roxb.): Apocynaceae.

Na : Hindi: Dharauti, Dudhi.

Hab : Pl. see Edible plants.

The seeds and roots yield a yellow dye.

110. *Ziziphus mauritiana* Lamk; (Syn. Z. *jujuba* Lam. non Mill.); Rhamnaceae.

Na : Eng.: Chinese date, Jujube; Hindi: Ber, Bedara, Kar Kandhu, Kuvala, Pemdimber.

Hab : Pl. see Edible plants.

Bark yields red dye.

111. *Z. xylopyrus* Willd: Rhamnaceae.

Na : Hindi: Katber, Gothber.

Hab : Pl. see Tanning materials.

Fruits are the source of red dye.

12

Sources of Katha and Cutch

Katha and cutch are produced from the heartwood of *Acacia catechu* Lf. willd. and A. *chundra*. Katha is mainly used in betel leaf (pan) as masticatory and valued for its catechin content. Cutch, the bye product of Katha industry is largely used for – dyeing, colouring pulp in paper industry, water softening and in deep oil drilling operations.

1. *Acacia catechu* L.f. willd.; Mimosaceae.

Na : Eng.: Black Cutch, Catechu; Hindi: Khair, Katha.

Hab : Pl. see Gums.

2. *A. chundra;* Mimosaceae.

Na : Hindi: Sundra; Telugu: Sundra.

Hab : It is found mainly in southern states of Andhra Pradesh, Karnataka, Tamilnadu and Maharashtra.

Acacia chundra

Process: Forest Research Institute (Chemistry of forest products division), Dehradun has developed a cottage-scale process for the manufacture of good quality katha and cutch. This is an improvement on the traditionally followed method of katha manufacture in the forest. The traditional method has certain disadvantages as large number of earthen pots are purchased year after year and transported over long distances, quite a few of them break in transit and also in the process of manufacture of katha. When they break during the process of katha manufacture there is loss of valuable extract/concentrate. The earthen pots being narrow mouthed, the extract takes longer time to concentrate and some katha which otherwise would have settled down gets converted to cutch and other decomposition products.

The improved method has the following advantages:

(i) Higher yield with better quality of katha.

(ii) Shorter period for processing.

(iii) Recovery of cutch (which is wasted in the indigenous process).

(iv) Elimination of direct contact of the chips with the heated surface of the vessel during extractions.

(v) The finished product takes a short period for drying.

Boiling of khair chips in earthen pots

The equipment required for the above process on a cottage scale are extractor, concentration pan (*Karahi*), basket (wire-net cage) with handle-buckets, wooden frames provided with perforated wooden plates, cutter (made of aluminium wire or knife), filter press (hand operated with 12 round plates), screw press, hydrometer, muslin cloth, axes. Instead of iron, aluminium utensils may be used.

***Extraction and crystallization*:** The heartwood of Khair is converted into thin fine chips by an axe. Eight to nine kilograms of heartwood chips are taken in a copper wire-net cage to avoid direct contact of the chips with heated surface of the extractor.

These cages are placed in extractors handle-buckets along with 25 to 27 litres of water (wood, water ratio = 1:3) for each extraction. Two extractions of 3 hours each are made by boiling the chips with water. The second extract can be used for the first extraction of the fresh chips. The extracts are mixed after filtration through muslin cloth and concentrated in an open (Karahi) on fire to a density of

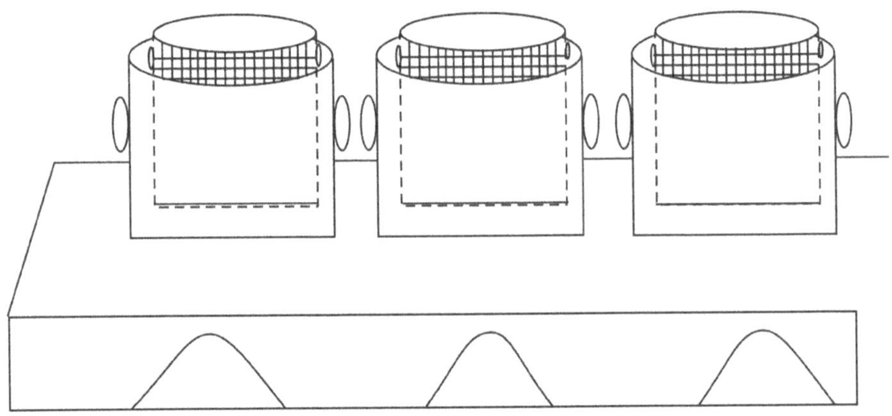

Extractors with cages

1.07 to 1.08 as read by hydrometer made of special material .This much of concentration is enough in the cold winter climate for katha to crystallize out. But in hotter places where winter cold is mild it may be necessary to concentrate the liquor a little more for seperating katha on cooling. Similarly the extract has to be concentrated to a density of 1.11 or even slightly higher for proper crystallization of katha if Khair wood happens to be of poor quality and of low yield. It is always advantageous to determine by an initial experiment the density of concentrate required for proper crystallization of katha. Unnecessary over concentration has to be avoided as otherwise cutch becomes gummy and retards crystallization of katha. Katha from over concentrated extract turns red. It is allowed to stand aside in shade for crystallization for a couple of days or more.

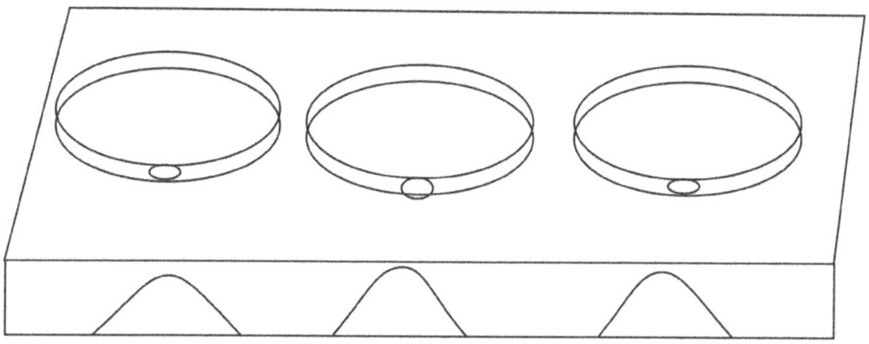

Concentration pans

Crystallisation is allowed to take place in the pans themselves and the solid material (Katha), gradually settles down to the bottom of the pans. Crystallization may be induced by adding a small quantity of catechin crystals.

***Filtration*:** After complete crystallization, the mass is passed through a filter press (Frame and plate type) using a thick filter cloth on the plates. Filter press is operated by hand. Frame and plate are made of stainless steel. A wash with fresh cold water improves the quality of katha. The cold aqueous solution containing cutch is collected in buckets. The katha is then taken out from the press using wooden or nickel spatula and is pressed again using a Screw press (book-binder type) in a wooden frame provided with canvas cloth, the last traces of cutch in solution are thus expelled out. Katha is then cut into uniform tablets of 5 cm × 5cm by means of a copper wire cutter or a knife and dried in sheds. The average overall yield of katha is about 3 to 4%. The yield is maximum when shavings of the wood are employed. Details of the new process for manufacture of katha (catechin) and cutch are covered by Indian Patent No. 91662.

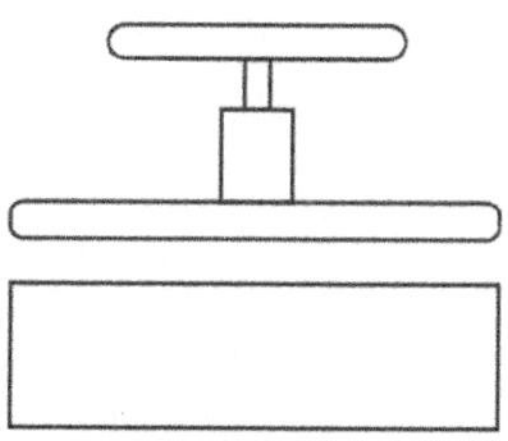

Filter press

***Cutch*:** The mother liquor after the removal of katha is further concentrated in an open pan till it becomes viscous. This viscous mass is poured into wooden frame of suitable size and allowed to cool. Cutch takes comparatively longer time to dry in shade (Yield, 6-8%).

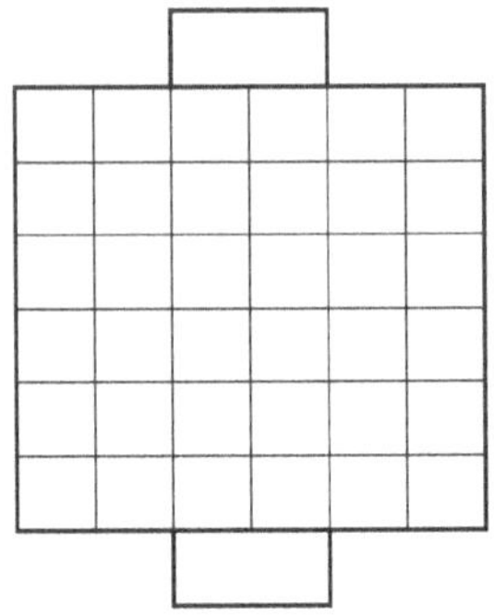

Wire Cutter

List of Botanical and Zoological Names

B

D

I

J

K

L

M

N

O

Q

R

S

T

U

V

W

X

Z

List of Local, English and Trade Names

B

E

I

J

Jangliaushbah 70

K

O

P

Q

R

S

Gedeon J., 1954 Saponins from Indian Soapnuts, Journal of Science and Industrial Research Vol. 13 B, p. 427-8.

Gedeon J., 1955, Saponins from Indian Soapnuts, Chemical abstracts, Vol. 49, No. 1, January p. 646.

Guenther F-1952, The Essential Oils, D.Van Nostrand C., Inc., New York.

Gupta M.P. Dr., Jain M.S., 1980, Experience in breeding of Musk deer for the Production of Musk, Indian Forester, Vol.106, No.5, pp. 357-362

Hardy Eric 1942, Herbs that produce saponin - plant substitutes in the Soap shortage, Perfumery & Essential Record, Vol. 33, No. 7, July, p.203.

Howes F.N., 1949, Vegetable gums and resins Published by the Chronica Botanica Company Wallham Mass, U.S.A.

Howes F.N., 1955, Vegetable Tanning materials, Butterworths Publications Ltd., 88 Kingsway, London, W.C. 2.

Karnik M.G., Sharma O.P., Inder Dev. 1971 Studies in the Chemical composition and possible utilities of Soapnuts *Sapindus mukorossi* Gartn. Indian Forester Vol. 97, No. 8, pp. 462-465.

Kirtikar K.R., Basu B.D., revised by Blatter E., Caices J.F., and Mhaskar K.S., 1935, Indian Medicine plants (4 vols.) Published by Lalit Mohan Basu, Allahabad.

Krishna S., Puntambekar S.V., and Raizada M.B., 1936, Minor Forest Products of Chakrata, Dehra Dun, Saharanpur, etc., and neighbouring forest divisions, Part 1, the oil bearing seeds (Indian Forest Records), New Series, Vol. 1, No. 1, Manager of Publications, New Delhi.

Krishnamurthy T., 1964, A note on Rudraksha, *Elaeocarpus sphaericus* (Gaertn.), K.Schum., Indian Forester Vol.90, No.11, pp. 774-776.

Krishnamurthy T., and Karira G.V., *Rauvolfia serpentina* Benth, (Rauvolfia, Sarpagandha), published by Indian Council of Agricultural Research, Krishi Bhavan, New Delhi.

Krishnamurthy T., Karira G.V., Sharma B.K. and Kuldip Bhatia, 1965, Cultivation and exploitation of *Podophyllum hexandrum* Royle (Syn. *P. emodi* Wall. Ex Hook. f. Thomas) Indian Forester Vol. 91 No. 7, pp. 470-475.

Krishnamurthy T., Karira G.V., and Sood V.K., 1965 Note on *Eucalyptus fruticetorum* F. Muel. (Syn. *E. polybractea* R.T. Baker, Indian Forester, Vol. 91, No. 11, pp. 818-820.

Krishnamurthy T., and Sethi H., Unpublished notes on *Cinnamomum zyylanicum* Broyn.

Krishnamurthy T., and Sharma B.K., Unpublished notes on cultivation and exploitation of *Carum carvi* L. in Chakrata hills and its essential oil.

Krishnamurthy T., Sharma B.K., and Beri R.M., 1965 A note on cultivation and utilization of *Ephedra gerardiana* Wall. Indian Forester, Vol. 91, No. 3, pp. 196-198.

Krishnamurthy T., 1974 Minor Forest Products of India their availability and Demand, Utilization and exploitation and suggestions for the development. Proceedings of the Commonwealth forestry Conference held at Oxford.

Krishnamurthy T., and Srikanta Rao P.V., 1965, Utilization of Kusum (*Schleichera oleosa* (Lour.) Merr.; syn. *S. trijuga* willd.); Indian Oil seeds Journal, Vol, 9, No. 3, pp. 220.

Krishnamurthy T., & Juyal S.P., Fo rest Research Institute & Colleges, Dehra Dun, Sal (*Shorea robusta* Gaertn.) Seed- A. potential source of vegetable fat (Unpublished).

Kushalappa K.A., Silviculturist, Bangalore 1978, Indian Lavender oil plantation in Karnataka, Myforest, September 1978, pp. 153-160.

Lock, G.V., 1961, Sisal, Longmans Green & Co., Ltd., 48 Grosvenor Street, London W I.

Maheswari P., and Umrao Singh, 1965, Dictionary of Economic plants in India, Indian Council of Agricultural Research, New Delhi.

Mathur S.C., & Saxena V.S., Studies on Lead tree. *Leucaena Leucocephala* Benth. 1978, Technical Bulletin, No. 2, Forest Dept. Govt. of Rajasthan, Jaipur.

Nadkarni H.K., Indian Materia Medica 2 volumes, published by Ramdas & Bhatkal for popular prakashan private limited, 35 C Tardeo Road, Popular Press Building, Bombay.

Paranjpe D.R., and Ayyar P.R., 1929, Oil from seeds of *Sapindus trifoliatus* L. (Soapnut.tree), Journal of Indian Institute of Science No. 12A pp.179-184.

Perkin A.G., and Everest A.E. 1918, The Natural Organic colouring matters, Longmans Green and Co., 39 Paternoster Row, London.

Pharmacopoeia of India 1955, Published by the Manager of Pubcation, Delhi.

Prasad K., Assistant Director, Wildlife preservation A2 E, Curzon Road Barraks, New Delhi. African Ivory Trade in India. 1980, paper submitted to the Seminar on Wildlife resources for rural development held at Hyderabad July 7-11, 1980.

Ramaswamy M.N., 1945, Minor Forest Products in Mysore (A survey), Superintendent, Government Press, Bangalore.

Rao P.S., Chief Conservator of Forests, Andhra Pradesh Forest Department, Hyderabad, 1976, A position paper on Cashew in the Reserved forests in Andhra Pradesh.

Rao. P.S., Chief Conservator of Forests, Andhra Pradesh Forest Department, 1979, The place of Cashew in Forest Land, Indian farming March, 1979

Ridley H. N., 1912, Spices, Macmillan and Co., Ltd., St. Martin's Street, London.

Royle J.F., 1855, The fibrous plants of India, Smith, Elder and Co., 65, Cornhill, London.

Said M., 1963, Minor Forest Products of West Pakistan, Pakistan Journal of Forestry, No. 3, pp. 276-278.

Saunders C.F., 1920, Useful wild plants of the United States and Canada, Robert M., McBridge & Co., New York

Seshagiri Rao A.R., & Hemadri K., 1979 Andhra Pradesh Mandumokkalu (Telugu), Medicinal Plants of Andhra Pradesh.

Singer R., 1961, Mushrooms and Truffles, Leonar Hill (Books) Ltd., 9 Eden Street, London N. WI.

Singh G.D., 1957, Departmental collection of honey and wax, Bhinga range, Baharaich Forest Division, Uttar Pradesh, Indian Forester, February, pp. 113-117.

Singh J.A., 1935, Note on Sigekai, Working plan for Satara forest division, Maharashtra.

Srinivasan M.M., 1956, Lac - its cultivation and manufacture in India; Indian Forester No. 2, pp. 93 -96

Watson S.J., 1950, Grassland and Grassland products, Edward Arnold & Co., London.

Watt G., 1889 - 1893 A dictionary of the Economic Products of India, Government of India Press, Calcutta, Vols. I to VI.

Wealth of India, 1976, vols, II published by the Publication & Information Directorate, New Delhi.

9 789352 300587

Printed at Repro India Ltd.